Benchmark Papers in Geology Series

Editor: Rhodes W. Fairbridge, Columbia University

A selection from the published volumes in this series

GLACIAL ISOSTASY / *John T. Andrews*
TETHYS: The Ancestral Mediterranean / *Peter Sonnenfeld*
MAGNETIC STRATIGRAPHY OF SEDIMENTS / *James P. Kennett*
SEAFLOOR SPREADING CENTERS: Hydrothermal Systems / *Peter A. Rona and Robert P. Lowell*
MEGACYCLES: Long-Term Episodicity in Earth and Planetary History / *G. E. Williams*
RIFT VALLEYS: Afro-Arabian / *A. M. Quennell*
OROGENY / *John G. Dennis*
GEOSYNCLINES: Concept and Place Within Plate Tectonics / *F. L. Schwab*
OPHIOLITIC AND RELATED MELANGES / *G. J. H. McCall*
MANTLE FLOW AND PLATE THEORY / *Zvi Garfunkel*
EARTH TIDES / *J. C. Harrison*
CONTINENTAL DRIFT / *James H. Shea*
PLATE TECTONICS / *James H. Shea*
CONTINENTAL RIFTS / *A. M. Quennell*

A complete listing of volumes published in this series begins on page 343.

CONTINENTAL DRIFT

Edited by

JAMES H. SHEA

University of Wisconsin-Parkside

A Hutchinson Ross Publication

VNR VAN NOSTRAND REINHOLD COMPANY

New York

Benchmark Papers in Geology, Volume 88
Library of Congress Catalog Card Number: 85-3242
ISBN: 0-442-28240-0

Manufactured in the United States of America.

Published by Van Nostrand Reinhold Company Inc.
135 West 50th Street
New York, New York 10020

Van Nostrand Reinhold Company Limited
Molly Millars Lane
Wokingham, Berkshire RG11 2PY, England

Van Nostrand Reinhold
480 Latrobe Street
Melbourne, Victoria 3000, Australia

Macmillan of Canada
Division of Gage Publishing Limited
164 Commander Boulevard
Agincourt, Ontario MIS 3C7, Canada

15 14 13 12 11 10 9 8 7 6 5 4 3 2 1

Library of Congress Cataloging in Publication Data
Main entry under title:
Continental drift.
(Benchmark papers in geology; v. 88)
"A Hutchinson Ross publication."
Includes indexes.
1. Continental drift—Addresses, essays, lectures. I. Shea, James H.
II. Series: Benchmark papers in geology; 88.
QE511.5.C64 1985 551.1'36 85-3242
ISBN 0-442-28240-0

CONTENTS

SERIES EDITOR'S FOREWORD

The philosophy behind the Benchmark Papers in Geology is one of collection, sifting, and rediffusion. Scientific literature today is so vast, so dispersed, and, in the case of old papers, so inaccessible for readers not in the immediate neighborhood of major libraries that much valuable information has been ignored by default. It has become just so difficult, or so time consuming, to search out the key papers in any basic area of research that one can hardly blame a busy person for skimping on some of his or her "homework."

This series of volumes has been devised, therefore, as a practical solution to this critical problem. The geologist, perhaps even more than any other scientist, often suffers from twin difficulties—isolation from central library resources and immensely diffused sources of material. New colleges and industrial libraries simply cannot afford to purchase complete runs of all the world's earth science literature. Specialists simply cannot locate reprints or copies of all their principal reference materials. So it is that we are now making a concerted effort to gather into single volumes the critical materials needed to reconstruct the background of any and every major topic of our discipline.

We are interpreting "geology" in its broadest sense: the fundamental science of the planet Earth, its materials, its history, and its dynamics. Because of training in "earthy" materials, we also take in astrogeology, the corresponding aspect of the planetary sciences. Besides the classical core disciplines such as mineralogy, petrology, structure, geomorphology, paleontology, and stratigraphy, we embrace the newer fields of geophysics and geochemistry, applied also to oceanography, geochronology, and paleoecology. We recognize the work of the mining geologists, the petroleum geologists, the hydrologists, and the engineering and environmental geologists. Each specialist needs a working library. We are endeavoring to make the task of compiling such a library a little easier.

Each volume in the series contains an introduction prepared by a specialist (the volume editor)—a "state of the art" opening or a summary of the object and content of the volume. The articles, usually some twenty to fifty reproduced either in their entirety or in significant extracts, are selected in an attempt to cover the field, from the key papers of the last century to fairly recent work. Where the original works are in foreign languages, we have endeavored to locate or commission translations. Geologists, because of their global subject, are often acutely aware of the oneness of our world. The selections cannot therefore be restricted to any one country, and whenever possible an attempt is made to scan the world literature.

To each article, or group of kindred articles, some sort of "highlight commentary" is usually supplied by the volume editor. This commentary should serve to bring that article into historical perspective and to emphasize its particular role in the growth of the field. References, or citations, wherever possible, will be reproduced in their entirety—for by this means the observant reader can assess the background material available to that particular author, or, if desired, he or she too can double check the earlier sources.

A "benchmark," in surveyor's terminology, is an established point on the ground that is recorded on our maps. It is usually anything that is a vantage point, from a modest hill to a mountain peak. From the historical viewpoint, these benchmarks are the bricks of our scientific edifice.

RHODES W. FAIRBRIDGE

PREFACE

The concept of continental drift has always been outrageous, even to geologists who prided themselves on their ability to conceive of the solid earth as consisting of a mass of material where things are constantly moving: Enormous glaciers advance hundreds of miles and then retreat; colossal mountain ranges are folded and uplifted tens of thousands of meters and then eroded down to their roots; the seas advance to cover half a continent and then retreat; and giant earthquakes cause the earth to ring like a bell. Despite the knowledge of these occurrences, most geologists drew the line at drift and refused to countenance the idea that the largest positive relief features of the globe actually move with respect to one another through distances of thousands of kilometers. Nevertheless, the evidence was there, and eventually it had to be faced.

The theory of continental drift has had at least three distinct beginnings. Its first was in the middle to late nineteenth century when the idea was associated with catastrophic events such as the biblical deluge and the fission of the moon from the earth. The second was engineered by a German meteorologist and an American geologist early in this century. In this instance the theory was more solidly supported with evidence gathered from a wide variety of geological disciplines and geographic areas. It built up a certain momentum, but by 1950 most geologists once again had rejected it as untenable. Ironically, just as the theory was pronounced irretrievably dead by being ignored at a major international symposium on the crust of the earth, new observations were being made that would resurrect the theory, lead to its total reconstruction into a revolutionary theory called *plate tectonics,* and then, in a matter of a few years, convince virtually the entire scientific community of the validity of the new theory.

The purpose of this Benchmark volume is to provide easy access to a number of important papers that can serve as benchmarks to the historical growth of continental drift theory and evidence relevant to it up to about 1950. The volume is divided into five parts, each of which deals with a particular line of investigation. Part I covers Gondwana, its former existence and extent and its late-Paleozoic glaciation. Part II discusses geophysics, the nature and significance of paleomagnetic and seismic evidence relevant to continental drift. Part III investigates the importance of oceanic ridges and faults to submarine topography and structure. Part IV introduces three particular theories and their role in stimulating thought and observation. Finally, Part V discusses regional geologic investigation by conventional geologic methods of three types of crustal movement.

Several other Benchmark volumes that deal with specific topics relevant to continental drift include reprints of papers that might have been reprinted in this volume but that were omitted for that reason. Following is a list of those volumes: W. G. Ernst, *Metamorphism and Plate Tectonic Regimes* and *Subduction Zone Metamorphism;* W. Walker, *Metallogeny and Global Tectonics;* C. A. Ross, *Paleobiogeography;* B. Voight, *Mechanics of Thrust Faults and Decollement;* P. H. Mattson, *West Indies Island Arcs;* M. B. Deacon, *Oceanography: Concepts and History;* J. B. Wright, *Mineral Deposits, Continental Drift, and Plate Tectonics;* P. A. Rona and R. P. Lowell, *Seafloor Spreading Centers;* G. E. Williams, *Megacycles: Long-Term Episodicity in Earth and Planetary History;* A. M. Quennell, *Rift Valleys: Afro-Arabian;* J. G. Dennis, *Orogeny;* F. L. Schwab, *Geosynclines;* G. J. H. McCall, *Ophiolites and Related Mélanges;* and Z. Garfunkel, *Mantle Flow and Plate Theory.*

This Benchmark volume is unique because, in addition to providing reprints of 22 important papers related to continental drift, including new translations of 5 papers, it includes an original work of historical research, *A Documented Chronology of Plate Tectonics.* The *Chronology* consists of a chronological listing of 509 significant events (benchmarks, if you will) in the history of plate tectonics from 1750 to 1975, each carefully documented with references. It is anticipated that the *Chronology* will prove useful and convenient to historians of science and to workers in various specialties by providing ready citations to important papers in their fields and by placing the events of the history of plate tectonics in clear temporal relationship to one another.

JAMES H. SHEA

CONTENTS BY AUTHOR

CONTINENTAL DRIFT

INTRODUCTION

The idea that Earth's continents are not fixed in place with respect to each other but that they move about over the surface of the globe was first proposed in 1858 by Antonio Snider. Snider suggested that the Atlantic continents formed a single land mass during the Carboniferous period, that they split apart during the biblical deluge, and that in one catastrophic episode, the Atlantic Ocean was formed. Similar catastrophic theories of drift were later proposed by Osmond Fisher (Paper 17), Pickering (1924), and Baker (1911), all of whom invoked the separation of the moon from the earth as the immediate cause of drift.

During the last half of the nineteenth century, two important lines of evidence emerged that were later seen as strongly supporting the idea of continental drift. First, geologists recognized that during the late Paleozoic four Gondwana continents (Australia, India, Africa, and South America) supported their own unique, endemic flora, the *Glossopteris* flora, which did not occur in either Eurasia (outside of India) or North America (see H. F. Blanford, 1875; W. T. Blanford, 1886; Rubidge, 1859; Tate, 1867; Zeiller, 1895). Antarctic *Glossopterids* were not discovered until 1928, by W. N. Edwards (Paper 3). Similarly, the distribution of many fossil animals showed the same puzzling pattern. Eventually, two basic hypotheses were developed to explain these instances of disjunct endemism. First was the idea of an enormous supercontinent, parts of which were said to have subsequently sunk beneath the deep waters of the Atlantic, Pacific, and Indian Oceans (see Suess, 1885-1909; Gregory, 1929, 1930; H. F. Blanford, 1875; W. T. Blanford 1884, 1890).

The second hypothesis envisioned a series of land bridges that once connected the Gondwana continents and allowed free migration of plants and animals across them (see Arldt, 1917-1922; Ihering, 1931; Schuchert, 1932). It was soon realized by a number of workers, however, that the existence of suboceanic continental crust constitutes a rather severe gravitational (isostatic) problem. What was holding the low-density crust down, and why was there no indication of the situation in the strength of the gravitational field above the areas of sunken continental crust? Recognizing these problems, Bailey Willis (1932) proposed that the land bridges consisted of relatively narrow "isthmian links," and other workers reduced them to chains of "island stepping stones." Curiously, the idea of land bridges persisted to as late as 1952 (see Mayr, 1952).

The second major line of evidence that was extensively developed during the late nineteenth century was the extensive Carboniferous glaciation of the Gondwana continents as discovered by Blanford, Blanford, and Theobald (1859) in India; Selwyn (1859) in Australia; Sutherland (1868) in Africa; and Derby (1888) in South America. The problem posed by this widespread evidence of glaciation was that at the same time virtually the entire southern hemisphere and India were experiencing a glacial climate, large areas of the United States, western Europe, and even Spitzbergen were the site of tropical climates, as indicated by the enormous deposits of Carboniferous coal (from whence the name *Carboniferous* originated). In hindsight, it is relatively easy to see that, if the continents are kept in their present relative locations, there is simply no way to explain the well-established pattern of Carboniferous climates.

Thus, when Alfred Wegener (1912) conceived his theory of continental drift and set about looking for evidence to support the rather outrageous idea, he quickly became aware of these two geological puzzles and recognized that drift provides a much more satisfactory explanatory mechanism than anything thus far proposed. He also extended his original theory, which was limited to the Atlantic continents, to include India, Australia, and Antarctica. Eventually, Wegener expanded his original short papers into a book that went through four editions and was translated into several languages, including English, French, and Spanish (see Wegener 1915, 1924).

Because of the detailed nature of his theory of drift, his extensive writings on the subject, and his comprehensive treatment of the relevant evidence, Wegener is often credited with being the founder of continental drift theory, and other contributors are sometimes ignored. In addition to the precursors already mentioned, however, Wegener was anticipated by F. B. Taylor (1910). Taylor had a number of interesting ideas, including the idea that the Mid-Atlantic Ridge is the line from which Europe-Africa and the Americas separated. As indicated by the title of his major work on drift (*Bearing of the Tertiary Mountain Belt on the Origin of the Earth's Plan*), Taylor based his concept almost entirely on the geographic distribution and shape of Tertiary mountain belts and their relationship to the continents of which they are a part. Taylor's basic idea was that the continents have moved away from the earth's rotational poles and toward the equator. In marked contrast to Wegener, he ignored most other lines of evidence.

During the period from 1920 to 1950, the debate within the geological community about continental drift was intense but sporadic. Wegener acquired a number of followers (mostly, but not exclusively, from the Gondwana continents) including A. L. Du Toit (1937), who wrote an influential book entitled *Our Wandering Continents,* but most northern hemisphere geologists opposed the idea, many vehemently. Three major symposia held during this period dealt with continental drift. The first, convened in New York in 1926, was organized by the American Association of Petroleum Geologists and included both Wegener and Taylor. Most of the participants strongly opposed the drift idea, and only a strong defense by W. A. J. M. van Waterschoot van der Gracht saved the day for the proponents (see van der Gracht et al., 1928). Thirteen years later, in 1939, a similar debate was staged in Frankfurt, Germany (see Anonymous, 1939), and again the opponents of drift strongly outnumbered the proponents, among whom were A. L. Du Toit and Kurt Wegener (Alfred's brother; Alfred died in 1930 while on a Greenland expedition). So low had the fortunes of drift fallen by 1954 that a major international symposium organized in New York by Columbia University included no supporters and resulted in the publication of a book entitled *The Crust of the Earth* (Poldervaart, 1955) that does not even mention the idea of drift.

Although a simplistic view might suggest that continental drift rapidly fell almost completely from favor from 1920 to 1950, in fact many new drift theories were proposed during that period, some by eminent scientists such as Beno Gutenberg (1927), Arthur Holmes (1933), and Reginald A. Daly (1923), and important investigations related to drift were undertaken. For example, the great Alpine geologists continued their efforts to decipher the enormously complex structure of that range, and nearly all of them came to feel that the Alps were formed when the continent of Africa drifted against and thrust over the southern margin of the European continent. They accepted drift primarily because it was the only theory of tectonics that could satisfactorily account for the structures and stratigraphy they were mapping and analyzing (see part V).

Similarly, investigations into the topography and structure of the sea floor increased in number and sophistication during this period. The German research vessel *Meteor* carried out its extensive investigations of the Atlantic in the mid-1920s (see Mauer and Stocks, 1933) and, among other things, mapped the Mid-Atlantic Ridge with its newly developed echo sounder. This work was followed up after World War II by workers from the Lamont Geological Observatory (see Paper 16). In the Indian Ocean the Danes (Schmidt, 1932) and the English discovered and mapped the Indian Ocean Ridge (Paper 14), and in the Pacific the nonmagnetic research ship *Carnegie* mapped the East Pacific Rise (see Carnegie Institution of Washington, 1942–1944).

Also developing toward the end of this period of decline of drift theory were seismic refraction methods that would eventually allow the determination of the structure of the oceanic crust, the thickness of deep-sea sediments, and the nature of the continent/ocean transition, all of which would prove crucial to the acceptance of continental drift as a part of sea-floor spreading and plate tectonics (see Papers 11 and 12; Ewing, Crary, and Rutherford, 1937). Another line of investigation whose results were not understood at this time was the work on deep-focus earthquakes. The extreme depth of these quakes was discoverd by Turner (1922), and the geometry of their occurrence was first worked out by Wadati (Paper 10) and later extended by, among others, Leith and Sharpe (1936), Visser (1936), and Gutenberg and Richter (1938, 1939). Related to the work on deep-focus earthquakes was F. A. Vening Meinesz's discovery (1931) of the enormous linear negative gravity anomalies that stretch continuously for thousands of kilometers along the Java Trench and the western boundary of the Pacific Ocean, which perplexed theorists for many years.

In summary, it can be seen that the decline of continental drift up to about 1950 was, in many respects, an illusion. During this period, many of the crucial lines of investigation (notably, sea-floor topography and structure, structure of continental margins and fold-mountain belts, marine gravity anomaly patterns, paleomagnetism, and deep-focus earthquakes) that eventually led to the theories of sea-floor spreading and plate tectonics were being actively pursued, and crucial observations were being made. However, it is also clear that, until those investigations were thoroughly developed, the observational basis on which the theory would eventually advance was not available, and such a basis was not found until after 1950, the point at which the coverage of this volume of the Benchmark series ends.

REFERENCES

Anonymous, 1939, Atlantis 1939, *Geol. Rundschau* **30:**1–400.

Arldt, T., 1917–1922, *Handbuch der Palaeogeographie,* 2 vols., Borntraeger, Leipzig, 1647p.

Baker, H. B., 1911, Origin of the Moon; Interesting Theory Advanced by Dr. Howard B. Baker, *Detroit Free Press,* 23 April, Feature Magazine section, p. 7.

Blanford, H. F., 1875, On the Age and Correlations of the Plant-Bearing Series of India, and the Former Existence of an Indo-Oceanic Continent, *Geol. Soc. London Quart. Jour.* **31:**519-542.

Blanford, W. T., 1884, Presidential Address to Section C (Geology), *British Assoc. Adv. Sci. Rept.*, 54th Ann. Mtg., pp. 691-711.

Blanford, W. T., 1886, On Additional Evidence of the Occurrence of Glacial Conditions in the Palaeozoic Era, and on the Geological Age of the Beds Containing Plants of Mesozoic Type in India and Australia, *Geol. Soc. London Quart. Jour.* **42:**249-260.

Blanford, W. T., 1890, The Permanence of Ocean Basins (Anniversary Address of the President), *Geol. Soc. London Quart. Jour.* **46:**59-109.

Blanford, W. T., H. F. Blanford, and W. Theobald, 1859, The Geological Structure and Relations of the Talchir Coal-Field in the District of Cuttach, *Geol. Survey India Mem.* **1:**33-88.

Carnegie Institution of Washington, 1942-1944, *Scientific Results of Cruise VII of the Carnegie during 1928-1929* (4 vols.), Carnegie Institution of Washington Pub. 544.

Daly, R. A., 1923, The Earth's Crust and its Stability, *Am. Jour. Sci.* **205:**349-371.

Derby, O. A., 1888, Uber Spuren einer carbonen Eizeit in Sudamerica, *Neues Jahrb. Mineralogie Geologie Paläontologie* **2:**172-176.

Du Toit, A. L., 1937, *Our Wandering Continents,* Hafner Pub. Co., New York, 366p.

Ewing, W. M., A. P. Crary, and H. M. Rutherford, 1937, Geophysical Investigations in the Emerged and Submerged Atlantic Coastal Plain, Part I—Methods and Results, *Geol. Soc. America Bull* **48:**753-802.

Gregory, J. W., 1929, The Geological History of the Atlantic Ocean, *Geol. Soc. London Quart. Jour.* **85:**68-122.

Gregory, J. W., 1930, The Geological History of the Pacific Ocean, *Geol. Soc. London Quart. Jour.* **86:**72-136.

Gutenberg, B., 1927, Die Veranderung der Erdkrusts durch Fliessbewegungen der Kontinentalscholle, *Gerlands Beitr. Geophysik* **16:**239-247,**18:**281-291.

Gutenberg, B., and C. F. Richter, 1938, Depth and Geographical Distribution of Deep-Focus Earthquakes, *Geol. Soc. America Bull.* **49:**249-288.

Gutenberg, B., and C. F. Richter, 1939, Depth and Geographical Distribution of Deep-Focus Earthquakes, *Geol. Soc. America Bull.* **50:**1511-1528.

Holmes, A., 1933, The Thermal History of the Earth, *Washington Acad. Sci. Jour.* **23:**169-195.

Ihering, H. von, 1931, Land Bridges across the Atlantic and Pacific Oceans during the Kainozoic Era, *Geol. Soc. London Quart. Jour.* **87:**376-391.

Leith, A., and J. A. Sharpe, 1936, Deep-Focus Earthquakes and Their Geological Significance, *Jour. Geology* **44:**877-917.

Mauer, H., and Stocks, T., 1933, *Die Echolotungen des Meteor: Wissenschaftliche Ergebnisse der Deutschen Atlantischen Expedition auf dem Forschungs- und Vermessungsschiff Meteor 1925-1927,* Band II, Part 1 (text), Part 2 (Echolotprofile), W. de Gruyter & Co., Berlin and Leipzig.

Mayr, E., ed., 1952, The Problem of Land Connections across the South Atlantic Ocean with Special Reference to the Mesozoic, *Am. Mus. Nat. History Bull.* **99**(Art. 3):79-258.

Pickering, W. H., 1924,The Separation of the Continents by Fission, *Geol. Magazine* **61:**31-35.

Poldervaart, A., ed., 1955, Crust of the Earth, *Geol. Soc. America Special Paper 62,* 762p.

Rubidge, R. N., 1859, On Some Points in the Geology of South Africa, *Geol. Soc. London Quart. Jour.* **15:**195-198.
Schmidt, J., 1932, *Dana's togt omkring jorden 1928-1930,* Gyldendal, Copenhagen, 255p.
Schuchert, C., 1932, Gondwana Land Bridges, *Geol. Soc. America Bull.* **43:**875-915.
Selwyn, A. R. C., 1859, Geological Notes of a Journey in South Australia from Cape Jervis to Mount Serle, *Adelaide Parliamentary Papers, South Australia 20,* 4p.
Snider, A., 1858, *La Creation et ses mysteres devoiles,* A. Franck and E. Dentu, Paris, 487p.
Suess, E., 1885-1909, *Das Antlitz der Erde,* F. Tempky, Prague and Vienna, 5 volumes totalling 2936p.
Sutherland, P. C., 1868, *On the Geology of Natal,* Pietermaritzburg.
Tate, R., 1867, On some Secondary Fossils from South Africa, *Geol. Soc. London Quart. Jour.* **23:**139-175.
Taylor, F. B., 1910, Bearing of the Tertiary Mountain Belt on the Origin of the Earth's Plan, *Geol. Soc. America Bull.* **21:**179-226.
Turner, H. H., 1922, On the Arrival of Earthquake Waves at the Antipodes, and the Measurement of the Focal Depth of an Earthquake, *Royal Astron. Soc. Monthly Notices, Geophys. Supp.* **1:**1-13.
van der Gracht, W. A. J. M. van Waterschoot, et al., 1928, *Theory of Continental Drift,* American Association of Petroleum Geologists, Tulsa, Okla., 240p.
Vening Meinesz, F. A., 1931, Gravity Anomalies in the East Indian Archipelago, *Geog. Jour.* **77:**323-337.
Visser, S. W., 1936, Some Remarks on the Deep-Focus Earthquakes in the International Seismological Summary, *Gerlands Beitr. Geophysik* **48:**254-267.
Wegener, A., 1912, Die Entstehung der Kontinente, *Petermanns Geog. Mitt.* **58:**185-195, 253-256, 305-309.
Wegener, A., 1912, Die Entstehung der Kontinente, *Geol. Rundschau* **3:**276-292.
Wegener, A., 1915, *Die Entstehung der Kontinente und Ozeane,* F. Vieweg und Sohns, Braunschweig, 94p.
Wegener, A., 1924, *The Origin of Continents and Oceans* (translated from the 3rd German edition by J. G. A. Skerl), E. P. Dutton, New York, 212p.
Willis, B., 1932, Isthmian Links, *Geol. Soc. America Bull.* **43:**917-952.
Zeiller, R., 1895, Sur quelques empreintes vegetales des gisements houillers du Bresil meridional, *Acad. Sci. Paris C. R.* **121:**961-964.

Part I

GONDWANA

Editor's Comments on Papers 1 Through 6

1 **TATE**
Excerpts from *On Some Secondary Fossils from South Africa*

2 **OLDHAM**
Memorandum on the Correlation of the Indian and Australian Coal-Bearing Beds

3 **EDWARDS**
The Occurrence of Glossopteris *in the Beacon Sandstone of Ferrar Glacier, South Victoria Land*

4 **STOW**
Appendix from *On Some Points in South-African Geology*

5 **SUTHERLAND**
Notes on an Ancient Boulder-Clay of Natal

6 **DU TOIT**
Excerpts from *The Carboniferous Glaciation of South Africa*

Although its full significance was not recognized at the time, a great deal of important evidence supporting continental drift was gathered during the nineteenth century, before continental drift was seriously considered as a hypothesis by most geologists. Much of the evidence was in the form of two distinct but related findings, both mostly in the southern hemisphere. The first consisted of occurrences of fossil plants belonging to the *Glossopteris* flora. *Glossopteris* itself was first described and named from Australia and India in 1828 by the French paleobotanist Adolphe Brongniart in his monumental *Histoire des Vegetaux Fossiles.* Brongniart's discoveries were supported and extended in Africa by R. N. Rubidge (1859) and R. Tate (Paper 1), in Australia by R. D. Oldham (Paper 2), in Antarctica by A. C. Seward (1914, from materials collected by members of the ill-fated R. F. Scott expedition) and W. N. Edwards (Paper 3), and in South America by A. Hettner (1891) (see also H. F. Blanford 1875; and W. T. Blanford 1876 and 1886).

For many years the meaning of this best known of all cases of disjunct endemism was the cause of heated controversy in the geological world. At first there were two main contending hypotheses, neither of which involved any relative horizontal movement of the continental blocks:

1. A supercontinent (named Gondwanaland by Eduard Suess (1885-1909) in his influential *Das Antlitz der Erde*) comprised of all the aforementioned areas had once existed. Parts of this continent later subsided to become the floor of portions of the deep, now intervening, oceans. (See also Stow (Paper 4) and Marcou, 1857-1860; Sclater, 1864; H. F. Blanford, 1875; W. T. Blanford, 1884, 1890; White, 1889; Frech, 1902; Arber, 1905; Haug, 1907-1927; Gregory, 1929, 1930)
2. Some sort of migratory route such as a land bridge, isthmian link, or chain of islands had once connected the separated land masses (Arldt, 1917-1922; Ihering, 1931; Schuchert, 1932; Willis, 1932; Du Toit, 1921; Mayr, 1952).

In 1908, F. B. Taylor (see Taylor, 1910) and, in 1912, A. Wegener (see Wegener, 1924) entered the field with a third hypothesis.

3. The areas where the *Glossopteris* flora occurs had been parts of the same land mass but these land masses later broke apart, drifted to their present positions, and in fact are still moving.

The second major line of evidence was, if anything, even more troublesome to conventional geology because it could not be accounted for by simply erecting and then dismantling land bridges or supercontinents at will. This evidence seemed impossible to reconcile with the earth's basic symmetrical, latitudinal climatic zoning. In 1859 evidence of a late-Paleozoic glacial age was discovered in India by W. T. Blanford, H. F. Blanford, and W. Theobald and in Australia by A. R. C. Selwyn (see also Paper 2). These startling discoveries were followed by similar findings in 1868 in Africa by P. C. Sutherland (Paper 5) and in 1888 by O. A. Derby in South America and, more than a hundred years after the original find, in 1962 by W. E. Long in Antarctica. In 1921, the great South African geologist Alexander Du Toit, Alfred Wegener's most prolific and respected supporter, demonstrated almost beyond refutation that the glaciation of South Africa can best be explained by continental drift (Paper 6).

The paleoclimatic and geologic questions raised by these discoveries were enormous:

How could the southern hemisphere of Earth have been glaciated while large areas in the northern hemisphere were experiencing tropical conditions?
How could the anomalous ice-flow directions (up out of ocean basins and onto continents in some cases and away from the equator in others) be explained?
Were the oceans themselves completely frozen in the southern hemisphere?

There seemed to be no other acceptable explanation of the evidence for those who were familiar with the evidence. Consequently, many geologists from the Gondwana continents (for example, A. L. Du Toit, 1937; B. Sahni, 1936; L. C. King, 1962) and others who worked in those areas (for example, H. A. Brouwer, 1921; A. C. Seward, 1931) became firm believers in continental drift despite hostile criticism from their Laurasian colleagues.

The articles reprinted herein were selected on the basis of their contribution to the overall picture of a unified Gondwana continent. Only two of the six (Du Toit and Edwards) refer to continental drift; the others were written before that hypothesis was taken seriously. Nevertheless, those papers and others like them provided much of the solid supporting evidence for continental drift, evidence that was all the more important because of the mechanical problems of the early continental drift model. To paraphrase one eminent geologist of the time, "If continental drift did happen, it could happen."

REFERENCES

Arber, E. A. N., 1905, *Catalogue of the Fossil plants of the* Glossopteris *Flora in the Department of Geology, British Museum (Natural History),* Brit. Mus. (Nat. Hist.), London, 255p.
Arldt, T., 1917-1922, *Handbuch der Palaeogeographie* (2 vols.), Borntraeger, Leipzig, 1647p.
Blanford, H. F., 1875, On the Age and Correlations of the Plant-Bearing Series of India, and the Former Existence of an Indo-Oceanic Continent, *Geol. Soc. London Quart. Jour.* **31:**519-542.
Blanford, W. T., 1876, Note on the Geological Age of Certain Groups Comprised in the Gondwana Series of India, etc., *Geol. Survey India Record* **9**(pt. 3):79-86.
Blanford, W. T., 1884, Presidential Address to Section C (Geology), *British Assoc. Adv. Sci. Rept.,* 54th Ann. Mtg., pp. 691-711.
Blanford, W. T., 1886, On Additional Evidence of the Occurrence of Glacial Conditions in the Palaeozoic Era, and on the Geological Age of the Beds Containing Plants of Mesozoic Type in India and Australia, *Geol. Soc. London Quart. Jour.* **42:**249-260.

Blanford, W. T., 1890, The Permanence of Ocean Basins (Anniversary Address of the President), *Geol. Soc. London Quart. Jour.* **46:**59-109.

Blanford, W. T., H. F. Blanford, and W. Theobald, 1859, The Geological Structure and Relations of the Talchir Coal-Field in the District of Cuttach, *Geol. Survey India Mem.* **1:**33-88.

Brongniart, A., 1828, *Histoire des vegetaux fossiles, ou recherches botaniques et geologiques sur les vegetaux renfermes dans les diverses couches du globe,* 2 vols., Dufour and D'Ocange Paris, (reprinted 1965, A. Asher, Amsterdam).

Brouwer, H. A., 1921, De alkaligesteenten van de Serra do Gericino ten Noordwesten van Rio de Janerio en de overeenkomst der eruptiefgesteenten van Brasilie en Zuid-Africa, *Koninkl. Nederlandse Akad. Wetensch. Proc.,* part 29:1005-1020.

Derby, O. A., 1888, Uber Spuren einer carbonen Eizeit in Sudamerica, *Neues Jahrb. Mineralogie Geologie Paläontologie* **2:**172-176.

Du Toit, A. L., 1921, Land Connections between the Other Continents and South Africa in the past, *South African Jour. Sci.* **18:**120-140.

Du Toit, A. L., 1937, *Our Wandering Continents, An Hypothesis of Continental Drifting,* Hafner Publishing Co., New York, 366p.

Frech, F. D., 1902, *Lethaea geognostica, Theil I, Lethaea palaeozoica,* Bd. 2, Lief 3-4: Die Dyas, Stuttgart, 354p.

Gregory, J. W., 1929, The Geological History of the Atlantic Ocean, *Geol. Soc. London Quart. Jour.* **85:**68-122.

Gregory, J. W., 1930, The Geological History of the Pacific Ocean, *Geol. Soc. London Quart. Jour.* **86:**72-136.

Haug, E., 1907-1927, *Traite de Geologie,* 3 vols., Colin, Paris, 1396p. (see vol. 2).

Hettner, A., 1891, Das sudlichste Brasilien (Rio Grande do Sul), *Zeitschrift Gesell. Erdkunde Berlin* **26:**85-144.

Ihering, H. von, 1931, land Bridges across the Atlantic and Pacific Oceans during the Kainozoic Era, *Geol. Soc. London Quart. Jour.* **87:**376-391.

King, L. C., 1962, *The Morphology of the Earth,* Hafner Publishing Co., New York, 699p.

Long, W. E., 1962, Sedimentary rocks of the Buckeye Range, Horlick Mountains, Antarctica, *Science* **136:**319-321.

Marcou, J., 1857-1860, *Lettres sur les roches du Jura et leur distribution geographique dan les deux hemispheres,* F. Klincksieck, Paris, 364p.

Mayr, E., ed., 1952, The Problem of Land Connections across the South Atlantic Ocean with Special Reference to the Mesozoic, *Am. Mus. Nat. History Bull.* **99**(Art. 3):79-258.

Sahni, B., 1936, Wegener's Theory of Continental Drift in the Light of Paleobotanical Evidence, *Indian Bot. Soc. Jour.* **15:**319-332.

Rubidge, R. N., 1859, On Some Points in the Geology of South Africa: *Geol. London Quart. Jour.* **15:**195-198.

Schuchert, C., 1932, Gondwana Land Bridges, *Geol. Soc. America Bull.* **43:**875-915.

Sclater, P. L., 1864, The Mammals of Madagascar, *Quart. Jour. Sci.* **1:**213-219.

Selwyn, A. R. C., 1859, Geological Notes of a Journey in South Australia from Cape Jervis to Mount Serle, *Adelaide Parliamentary Papers, South Australia 20,* 4p.

Seward, A. C., 1914, *Antarctic Fossil Plants, British Antarctic* Terra Nova *Expedition, Geology* **1:**1-49.
Seward, A. C., 1931, *Plant Life through the Ages,* Cambridge University Press, Cambridge, 601p.
Suess, E., 1885-1909, *Das Antlitz der Erde,* F. Tempky, Prague and Vienna, 5 volumes totalling 2936p.
Taylor, F. B., 1910, Bearing of the Tertiary Mountain Belt on the Origin of the Earth's Plan, *Geol. Soc. America Bull.* **21:**179-226.
Wegener, A., 1924, *The Origin of Continents and Oceans* (translated from the third German edition by J. G. A. Skerl), E. P. Dutton and Co., New York, 212p.
White, C. D., 1889, Carboniferous Glaciation in the Southern and Eastern Hemispheres—With Some Notes on the *Glossopteris* Flora, *American Geologist* **3:**299-330.
Willis, B., 1932, Isthmian Links, *Geol. Soc. America Bull.* **43:**917-952.

1

Reprinted from pages 139, 140, and 142 of *Geol. Soc. London Quart. Jour.* **23:**139–175 (1867), by permission of the publisher, Scottish Academic Press, Edinburgh.

ON SOME SECONDARY FOSSILS FROM SOUTH AFRICA

Ralph Tate

[Editor's Note: Only the introduction, the discussion of *Glossopteris browniana,* and the general remarks about the Karoo beds are reprinted here.]

I. Introduction.

As a large number of fossils from the Secondary rocks of South Africa remain undescribed in the Society's Museum, it seems desirable that an account of them should be submitted to the Society, as a sequel to the Monographs on Secondary Fossils from South Africa which have appeared in the Society's Transactions and Quarterly Journal.

This paper contains descriptions of the Plants of the Karoo beds, the Reptilian and Molluscan life of which has been made known by the labours of Professors Owen and Huxley, and the late Mr. Daniel Sharpe—of the vegetable remains from the phytiferous beds of Geelhoutboom, and of the fossils from the Jurassic strata of the Zwartkop and Sunday's River (the last part being a Supplement to Mr. Daniel Sharpe's "Descriptions of Fossils from the Secondary Rocks of South Africa"). The study of the organic remains of the beds enumerated has enabled me to offer, in addition, some generalizations on the affinities of each of the above-mentioned groups of Secondary strata developed in this region.

II. Karoo Beds.

a. *Fossil Flora of the Karoo Beds.*

1. Glossopteris Browniana, Pl. VI. figs. *5a*, *5b*, *7a*, *7b*.

The Society is in possession of several impressions of a *Glossopteris*, in a friable sandstone, obtained by Dr. W. G. Atherstone, F.G.S., from Heald Town, near Fort Beaufort, Eastern Province, South Africa,—and many others of the same form, in a dark shale, from Bloemkop *, collected by Mr. C. J. Powell, of Graaf Reinet, and presented by Dr. R. N. Rubidge, F.G.S. There are also in the Collection some well-preserved specimens from Natal, that were collected and sent to England by Dr. P. C. Sutherland. These all agree, in the main, with *Glossopteris Browniana*.

The following words employed by Sir C. Bunbury†, on comparing the Indian *Glossopteris* with the common Australian *G. Browniana*, are applicable when the *Glossopteris* from Natal is compared with the *G. Browniana*:—"I can find no satisfactory specific distinction; the venation is essentially the same. We may, indeed, find African specimens in which the meshes of the reticulation near the midrib are larger and broader than in the ordinary Australian plant, and

* Quart. Journ. Geol. Soc. vol. xv. p. 198; and vol. xvii. p. 329.
† *Ibid.* vol. xvii. p. 328.

others in which the veins are more strongly recurved; but in both the Indian and Australian plants I find so many shades of variation in these particulars that I cannot attach much importance to them."

The apex of the African plant is obtuse, like that of the Australian plant, and never acute as in the Indian. The frond is more attenuated below in the African fern; but there seems to be some degree of variation in this particular. It has a more firm and rigid appearance than the Indian and Australian forms.

b. *General Remarks on the Fossil Plants of the Karoo Beds.*

1. A marked feature in the ancient flora of the Karoo district is the paucity of specific forms (five in number), in proportion to the abundance of specimens (forty in the Collection).

2. This flora presents close analogy with that of the Coal-formation of Eastern Australia and the plant-bearing beds of Burdwan and Nagpur in India. The characteristic plant in each of these deposits and in the Karoo beds is a *Glossopteris*; and it seems that the Indian, Australian, and South-African plants are specifically identical.

The association of the genera *Glossopteris, Phyllotheca,* and *Dictyopteris* (?) affords some evidence of Mesozoic affinities in this fossil flora.

3. From the very characteristic Jurassic flora presented by the Geelhoutboom shales and limestones (which, from the structure of the country, may be confidently regarded as being higher in the geological scale), the flora of the Karoo beds (which has no affinity with the former) cannot be regarded as contemporaneous with the Jurassic flora of Europe; but, as it possesses a Mesozoic facies, it may be considered to be of Triassic age.

4. The nature of the flora conspires with the fauna to establish the lacustrine origin of the Karoo series. The following synopsis (by Prof. RUPERT JONES) of these beds and their vertebrate and other fossils (as far as hitherto described) embraces the characteristics of this remarkable formation.

2

Reprinted from *Geol. Survey India Record* **19**:39-47 (1886)

Memorandum on the Correlation *of the* Indian *and* Australian coal-bearing beds, *by* R. D. Oldham, A.R.S.M., *Deputy Superintendent Geological Survey of India.*

In the spring of 1885 I was deputed, in extension of privilege leave, to visit and examine the coal-measures of New South Wales and Victoria, with a view to elucidating, if possible, the vexed question of their relations to our Indian coal-measures. Accordingly I reached Melbourne on the 11th August and went on to Sydney, whence I proceeded to Newcastle to examine what I may call a classical section in Australian geology.

I had intended, besides, to examine in detail the section in the southern coal-fields and in the Blue Mountains as well as the beds of Bacchus Marsh in Victoria; but a relapse of malarious fever, originally contracted in India, interfered considerably with my movements, and I was compelled to sail from Melbourne on the 2nd September, having only been able to examine the section west of Newcastle and pay a flying visit to the Blue Mountains. With respect to the other localities mentioned above, I had to be content with such information as could be obtained from published reports, and in conversation with those who had examined the ground. I am consequently unable to give the connected account of the Australian coal-bearing series I should have wished to; but as regards the question of the relative age of our Indian coal-measures and those of New South Wales this is of little importance, seeing that the Australian geologists have not yet been able to connect the beds as exposed in the different coal-fields with each other, and with the typical section west of Newcastle to which they must all be ultimately referred. Such observations as I was able to make will be embodied in the following note.

Though it has practically no bearing on the question, it may not be out of place to note the resemblance in lithological characters between certain members of the Australian carbonaceous series and of the Gondwana series in India; thus, at Newcastle may be seen beds of coarse-grained soft whitish sandstone containing a considerable proportion of decomposed felspar and very like some of the Barakar sandstones; again, the boulder beds below the coal-measures are often identical with the similar boulder beds of the Talchirs in everything but the presence of marine fossils; but I attach no value to these lithological resemblances except in the latter case, and then only in so far as they indicate a condition of deposit which, considering the latitudes they are found in and the nature of the beds they are intercalated with, must have been of an exceptional and transient nature.

Introductory.

The relative ages of the Indian and Australian coal-measures have long been a question of great interest to geologists, and in spite of the amount that has been written on the subject and the ability of the writers there is probably no geological question more involved in doubt and difficulty.

The points in dispute are four—1st, the age of the members of the Indian coal-bearing formation, or Gondwana series, as compared with the coal-bearing series in Australia, and more especially of the coal-measures in either country; 2nd, the ages of the members of the Gondwana series as compared with the members of the geological sequence in Europe; 3rd, those of the Australian beds compared with the same sequence; and 4th, the relative ages of beds exposed in different parts of Australia.[1]

When first the plant fossils of the coal-measures of India (Damudas) and of Australia (Newcastle beds) became at all known, a similarity between the two floras was noticed which naturally led to their being considered as of contemporaneous origin. In both *Glossopteris* was abundant, and the species *G. browniana* (as well as other species then supposed to be identical) was found in both; both contained a species of *Vertebraria*, another of *Phyllotheca*, which were believed to be identical, and are certainly closely allied to each other, while the former genus was till recently unkown from any other formation but the two in question.

Conflict of opinion.

The contemporaneity of the Indian and Australian coal-measures being taken for granted, the second and third of the points in dispute became connected with each other. In Australia two schools arose,—one, headed by the late Rev. W. B. Clarke and those who had examined the beds in the field, reasoning from the close connection of the Newcastle coal-measures with beds containing a marine fauna closely allied to that of the carboniferous beds of Europe, argued that the Newcastle beds were of carboniferous or at any rate late palæozoic age; the other, headed by Professor McCoy and others who had never seen the beds *in situ*, reasoned on what they declared to be the characteristically mesozoic facies of the flora, and on this ground alone, not satisfied with declaring that the beds must be of jurassic age, impugned the accuracy of some of the Rev. W. B. Clarke's statements. In accordance with this divergence of opinion regarding the Australian beds there was a corresponding difference as to the age of our Indian coal-measures; some of the most prominent of the Europe palœophytologists looked upon them as of mesozoic age, while the members of the Indian Geological Survey who had examined the same beds in the field regarded them as of late palæozoic age on the ground of the resemblance of their flora to that of the Australian coal-measures, accepting with regard to these last the opinion of those who had examined them in the field. But as regards the beds above the coal-measures, and so intimately associated with them that they could not be looked upon otherwise than as subordinate members of one great series embracing the coal-measures also, it was acknowledged in both countries and by both parties that they should be regarded as of mesozoic age.

Such was the state of opinion on these questions when Dr. Feistmantel examined the floras of the sub-divisions of the Gondwana series and not only classed them all as of mesozoic age, but even attempted to correlate them to particular horizons of that period; and moreover after a detailed examination of the flora of the Australian coal-measures declared that it was not nearly so closely related to

[1] The Australian series as exposed in the districts between Sydney and Newcastle and the country west of that place is divided by Australian geologists as follows. The nomenclature of the subdivisions, especially in the lower part of the series, is not fixed, but the terms given below will be used in this paper. For the series as a whole I shall use the term Carbonaceous, suggested by Prof. McCoy.

1. Wianamatta shales.
2. Hawkesbury sandstones.
3. Newcastle beds,—or upper coal-measures.
4. Upper marine beds with carboniferous fauna.
5. Stony Creek beds,—or lower coal-measures.
6. Lower marine beds with carboniferous fauna.

that of the Indian beds as had been supposed, and that it could consequently not be appealed to as proving the palæozoic age of the Damudas in India.

Conflict as to facts of superposition in Australia.

But in Australia the dispute was not confined to the conclusion to be drawn from acknowledged facts, but, strange as it may seem, the very fundamental facts of the superposition of the beds was disputed. At the base of the sandstones and shales containing seams of coal and known as the Newcastle beds there is a great thickness of beds from which a marine fauna, closely related to that of the carboniferous period in Europe, has been obtained.[1] These marine beds are divided into an upper and a lower sub-division by a band of sandstones and fine conglomerates containing some seams of coal, and as these beds yielded species of *Glossopteris* and other similar forms which were declared to be characteristically mesozoic, it was assumed that the superposition of these beds by others containing a palæozoic fauna must be only apparent and in reality due to inversion or faulting.

Description of section.

Such a conclusion however could not be allowed by any one who had seen the ground where these beds are exposed. The section is fortunately easily accessible by the Great Northern Railway starting from Newcastle and the beds are well exposed in the frequent cuttings. There are two exposures of these lower coal-measures on opposite sides of an anticlinal, one at Stony Creek, 2 miles west of Branxton, and the other at Greta, 10 miles further west. At both places the dip is moderate and steady, to east-south-east at Stony Creek, to west north-west at Greta; at both places the section is practically continuous, and the marine beds may be traced dipping under the coal seams and a short way above them again reappearing. The re-appearance of the seam on the opposite side of the anticlinal, and the absence of any duplication of the seam are conclusive against any theory that the appearances are due to inversion or strike fualts, while if further proof were necessary it would be found in the fact that both at Stony Creek and Greta shafts have been sunk through the marine beds into the coal, and at the former place through it into more marine beds,[2] thus clearly showing that the coal measures are inter-bedded with the marine beds.

No Gondwana horizon directly referable to the New South Wales section,

It is unfortunate for our present purpose that none of the sub-divisions of the Gondwana series in India can be definitely and directly correlated with any of those of the carbonaceous series as exhibited in the Newcastle section where their relative position is clear and free from doubt. But in Victoria there are some beds containing *Gangamopteris*, known as the Bacchus Marsh beds, which seem to be the equivalents of the Talchirs. The flora of these Bacchus Marsh beds is poor, consisting, so far as is known, of only three species of *Gangamopteris*;

but Talchirs represented in Victoria.

[1] According to De Koninck (quoted in Rev. W. B. Clarke's "Remarks on the sedimentary formations of New South Wales," 4th Edition, Sydney, 1878, Appendix XVI "C," pp. 144-148) out of 249 species known, 81 are found in the carboniferous beds of Europe.

For a fuller account of the stratigraphy of these beds, and the history of this dispute, see Rev. W. B. Clarke's "Remarks on the sedimentary formation of New South Wales," 4th Edition, Sydney, 1878.

[2] I am indebted to Mr. C. S. Wilkinson for this information.

but, as Dr. Feistmantel has shown, of these three species one is identical with, another closely allied to, species found in the Talchir beds of India.

Palæontological

But their palæontology is not the only connection between the two, for, like the Talchirs, the Bacchus Marsh beds contain abundant evidence of the action of floating ice. According to the late Sir R. Daintree, there are "strata, mainly composed of fine mud, dotted throughout with various sized, generally rounded, pebbles, and those pebbles mostly unknown in the vicinity, and some not yet seen in place so far as the Geological Survey has extended a minute examination;" [1] further on he says that "blocks of granite," in some instances over a ton in weight, are found embedded in a matrix of soft mud;" [2] and in the last progress report by the Secretary for Mines in Victoria, Mr. Murray states, on the authority of the late Sir R. Daintree, that some of these granite boulders resemble no granite that occurs as a rock-mass nearer than Queensland.[3]

and physical resemblances.

It is impossible to account for the formation of such beds as these except by the agency of floating ice in large masses, and as both the Talchirs and the Bacchus Marsh beds show that when they were deposited the climate was much more severe than that now prevalent, we may take this as indicating that during their deposition there was a widespread glacial epoch corresponding to that which is known to have occurred in post-tertiary times. This consideration, whatever weight might be attached to it if stood by itself, may certainly be said to corroborate the fossil evidence, and we may consequently take it as certain that the Talchir and Bacchus Marsh beds are the representatives of each other.[4]

Evidence of glacial action.

I am not aware that any attempted correlation of these beds with a definite horizon in New South Wales was published before Dr. Feistmantel in 1880 gave it as his opinion that they were the equivalents of the Hawkesbury sandstones. This opinion, so far as I can glean from his published writings, was based on the so-called lower mesozoic facies of the Bacchus Marsh flora, and was supposed to be confirmed by Mr. C. S. Wilkinson's discovery of what he believed to be evidence of glacial action in the Hawkesbury sandstones. Mr. Wilkinson thus describes this evidence—"In the sections exposed in the quarries at Fort Macquarrie, Woolloomooloo, Flagstaff hill and other places, may be seen angular boulders of shale [5] of all sizes up to 20 feet in diameter, embedded in the sandstone in a most confused manner, some of them standing on end as regards their stratification, and others inclined at all angles. They contain the same fossil plants that are found in the beds of shale from which they have evidently been derived. These angular boulders occur nearly always immediately above the shale-beds,

Correlated with Hawkesbury beds by Dr. Feistmantel.

Glacial action in Hawkesbury sandstones.

[1] Geological Survey. Report on the Geology of the District of Ballan by Richard Daintree. Melbourne, 1866, p. 10.

[2] *Ibid.*, p. 10

[3] Geological Survey. Progress report by the Secretary for Mines, Melbourne, 1881, p. 80.

[4] See *Feistmantel*, Rec. G. S. I. XIII. 257 (1880).

[5] Which is interbedded with the sandstones.

and are mixed with very rounded pebbles of quartz; they are sometimes slightly curved as though they had been bent while in a semiplastic condition, and the shale-beds occasionally terminate abruptly as though broken off."[1]

Not comparable with that of Bacchus Marsh.

Although it is difficult, if not impossible, to account for these appearances, except by the action of ice in some form or other, yet it is evident that they are by no means comparable with the proofs of glacial action exhibited by the Bacchus Marsh beds. The angular form of the fragments of shale shows that in some manner or other they must have been indurated before disturbance, and it is impossible to account for this induration of what must then have been recently deposited mud except by the freezing of the interstitial water. This supposition would accord with the general nature of the evidence, which indicates the action of ground-ice such as is formed during the severe winters of North America rather than the presence of large masses of floating ice; and hence does not necessarily indicate so severe a climate as that afforded by the Bacchus Marsh beds of Victoria.

Glacial action in lower carboniferous marine beds of New South Wales.

But there are in New South Wales conglomeratic beds which are strictly comparable with those of Bacchus Marsh; these are the marine beds with carboniferous fauna, Nos. 4 and 6 of the scheme on page 40 (note). I am not aware of any previous published notice of the evidence of glacial action they afford, but as long ago as 1861, the lithological resemblance between beds of presumably the same age in the Wollongong district and the boulder-beds of the Talchirs was noticed by the late Dr. T. Oldham,[2] and when looking up the

[1] Notes on the occurrence of remarkable boulders in the Hawkesbury rocks by C. S. Wilkinson, F.L.S., G.S., Govt. Geologist. Trans. Roy. Soc. N. S. W., Vol. XIII, page 105 (1884).

[2] Mem. G. S. I., III, p. 209 (1863). As the volume is out of print and difficult to procure, I reprint the passage referred to, as it is of interest in the present juncture.

Speaking of a collection of specimens of Australian rocks procured and sent by Sir William Denison, he says,—

"And still further, many of the lower beds of the Australian group, there so abundantly rich in marine fossils, are very similar to many of the beds in the Indian *Talchir* series. There is the same mixture of pebbles, and large rolled masses in a matrix of fine silt; and much of this silt is of exactly the same peculiar bluish-green tint, so characteristic of these beds in this country, and which, once seen, can never be mistaken.

"I would not be misunderstood as desiring to give any great weight to a similarity in mineral texture or lithological aspect, in attempting to ascertain the true position of these rocks. But I am satisfied that this identity has a value, and by no means a light value, when, taken in connexion with every other point of evidence which is available, it is found in all cases tending to turn the balance in the same direction. And, basing my views on these considerations, I ventured to hold out a prospect in anticipation that future researches would enable a more accurate and detailed parallelism to be established between the rocks in both these countries, portions of which were now known to be synchronous, and that, while in all probability it would be found that starting from the common datum line of the coal-bearing rocks in either land, the sequence upwards would be established from Indian researches in this country, apparently supplying links wanting in Australia; on the other hand we should be enabled to supplement the evidences of the succession downwards (which is deficient in India) by a reference to Australian groups. As yet we have not been able to trace the existence of any marine deposits in this country of the same age as the 'Wollongong' sandstones of Australia, but there is nothing whatever in the few plants which occur in our Talchir beds which would militate against their being of the same general age (which I am disposed to think they are)."

literature of this controversy I was much struck by the passage. That any special stress should have been laid on the resemblance was not to be expected, for when the words were written the glacial origin of the Talchir boulder-bed had not been universally acknowledged, the very idea of a glacial epoch was still strange, and no one had yet dreamed of a palæozoic glacial epoch, still less of using such a conception in the correlation of distant deposits. But when I found that in Mr. W. T. Blanford's reply to Dr. Feistmantel[1] no notice was taken of this resemblance, although Mr. H. F. Blanford's suggestion, that the glacial beds of the Permian in England and the Talchirs in India were contemporaneous is quoted, I concluded that private information of later date had led to a modification of the views expressed as to the lithological resemblance of the beds.

Nevertheless I determined to pay special attention to this point, and was not surprised on visiting the section west of Newcastle, to find that the marine beds showed abundunt traces of glacial action. Blocks of slate, quartzite and crystalline rocks, for the most part sub-angular, are found scattered through a matrix of fine sand or shale, and these latter beds contain delicate *Fenestellæ* and bivalve shells with the valves still united, showing that they had lived, died and been tranquilly preserved where they are now found, and proving, as conclusively as the matrix in which they are preserved, that they could never have been exposed to any current of sufficient force and rapidity to transport the blocks of stone now found lying side by side with them. These included fragments of rock are of all sizes, from a few inches to several feet in diameter, the largest I saw being about 4 feet across in every direction as exposed in the cutting, and of unknown size in the third dimension; but I was informed by Mr. Wilkinson that in these same beds he has seen boulders of slate, &c., whose dimensions may be measured in yards.

It is impossible to account for these features except by the action of ice floating in large masses,[2] and I had the good fortune to discover, in the Railway cutting near Branxton, a fragment beautifully smoothed and striated in the manner characteristic of glacier action, besides at least two others which showed the same feature, though obscurely. This seems to show that the ice was of the nature of icebergs broken off from a glacier which descended to the sea-level.

Glacial evidence at Wollongong in Blue Mountains;

Beds of similar structure and indicating a similar mode of origin are also found at Wollongong, south of Sydney, and in the Blue Mountains. Though these have not been traced into connection with the marine beds west of Newcastle, the similarity of their position, fauna, and physical aspect, all leave little room for doubt that they are of the same age.

and in Queensland.

In Queensland beds of similar aspect have been described by Mr. R. L. Jack. These beds—also of marine origin, and indicating the presence of ice floating in the sea by which they were

[1] Rec. G. S. I., XI, page 148 (1878).

[2] Roughly speaking, it may be said to take 16 cubic feet of fresh-water ice floating in sea-water to float a cubic foot of granite, or 14 cubic yards to float 1 ton. It must be remembered that many of these fragments probably came from a distance, and that the ice was melting all the while. These figures must be reduced by two fifths if the rock is supposed to be immersed.

deposited [1]—contain 22 species of fossils, so far as the fauna is known, and of these 15 are common to the New South Wales marine carboniferous beds, the other 7 not having as yet been found to the southwards.

A "glacial epoch"

From what has been said it will be seen that over the greater part of Eastern Australia beds are found which, by their included fossils, are shown to be homotaxially equivalent, and which agree in indicating that during their deposition the seas under which they were formed bore on their surface masses of floating ice such as are now only seen in much higher latitudes than are occupied by any part of Australia. It would seem, then, that towards the close of the palæozoic era there was a widespread glacial epoch whose effects were felt over the whole of what is now Eastern Australia, and, if we allow this form of argument at all, it is to this period that we must refer the Bacchus Marsh beds rather than to that of the Hawkesbury sandstones which do not indicate so severe a climate.

to which the Bacchus Marsh beds should be referred.

Palæontological evidence.

The palæontology unfortunately does not help us to a definite conclusion. *Gangamopteris*, the only genus known from the Bacchus Marsh beds, has not so far been found either in the lower coal-measures of New South Wales, or in the Hawkesbury sandstones, but from the Newcastle beds, intermediate between them, two species of *Gangamopteris* have been obtained [2] of which one is identical with, and the other allied to, Bacchus Marsh forms.

The flora of the lower coal-measures comprises 7 species belonging to 4 genera; [3] of these, 4 belong to the genus *Glossopteris;* this latter is therefore, so far as the flora is at present known, the predominant genus, as is also the case in the Newcastle beds; further, one species, *G. browniana,* is also found in the latter; and besides this, the genera *Phyllotheca* and *Nöggerathiopsis* are common to the two floras, while *Annularia* is the only genus which has not, so far, been found in the Newcastle beds.

The Hawkesbury beds have only yielded two species of plants; in both cases the genus is represented in the Newcastle flora, and in one case, *Sphenopteris alata* Bgt., the Newcastle form, is only separated as a variety by Dr. Feistmantel. But if we group the Hawkesbury and Wianamatta beds together, as Dr. Feistmantel has done, we have 8 species belonging to 7 genera, of which but 1 species and 3 genera are common to the Newcastle flora.

From the facts just detailed it will be seen that so far as there is any balance

[1] Report on the Bowen River Coalfield, by R. L. Jack, F. G. S., Brisbane, 1879.

[2] *G. angustifolia,* McCoy, and *G. clarkeana,* Fstm., the latter being represented by *G. spathulata,* McCoy, in the Bacchus Marsh beds.

[3] The flora of the lower coal-measures according to Dr. Feistmantel (Tr. Roy. Soc. N. S. W., XIV, 103) 1880, is as follows:—

Phyllotheca australis Bgt.
Annularia australis Fstm.
Glossopteris browniana Bgt.
" " v. *præcursor* Fstm.
G. elegans, Fstm., *G. primæva,* Fstm., *G. clarkei,* Fstm.
Nöggerathiopsis prisca, Fstm.

at all, the Newcastle flora is more closely allied to that of the lower coal-measures than of the Hawkesbury beds.

Stratigraphical evidence.

As regards the stratigraphical relations of the beds, the Newcastle coal-measures are, so far as is known, perfectly conformable to the lower marine beds. On the other hand, though no unconformity has been traced between the Newcastle and Hawkesbury beds, yet if the coal-measures in the Blue Mountains are the representatives of the former,—and the balance of probability is very much in favour of this supposition,—there must be an unconformity and overlap of the Hawkesbury sandstones on the beds below them.

Turning to Victoria, we find that besides the Bacchus Marsh beds the only other rocks which can be referred to any portion of the carbonaceous series in New South Wales, are what are known as the *Tæniopteris* beds which are shown by their included fossils to be homotaxial with the Wianamatta beds or those immediately above them. Owing to the large surface covered by lava flows in Victoria, the exact relation of the *Tæniopteris* and *Gangamopteris* beds is not known, the two not having been found in contact; but there can be no doubt that the latter are completely overlapped by the former, which are frequently found in immediate contact with the silurian slates. I was further informed that the difference in the degree of induration of the two was such as to indicate a considerable difference of age, so that it would be improbable that they could belong to two periods so close to each other as the Hawkesbury and Wianamatta periods.

Contemporaneity of Talchir and carboniferous marine beds.

Seeing, then, that the palæontology and stratigraphy of the beds, so far from contradicting the conclusion derived from the evidence of glacial conditions, are either neutral or support it, we may conclude that the carboniferous marine beds of New South Wales are the most probable equivalents of the Bacchus Marsh beds of Victoria and hence of the Talchir beds of India.

Homotaxy of Damudas and Newcastle beds.

As a corrollary from the above conclusion, it would follow that the Damudas and Newcastle beds are far more nearly contemporaneous than Dr. Feistmantel will allow, and that the Gondwana series as a whole ranges from towards the latter end of the palæozoic into the secondary era, representing in part the interval between those two eras of European geology.

This was the opinion long held by the members of the Geological Survey of India, and was first seriously questioned by Dr. Feistmantel, who ascribed all the sub-divisions of the series to horizons in the mesozoic rocks of Europe, and also declared that the affinities between the Damuda and Newcastle floras were not such as to justify their being regarded as homotaxial.

It will be necessary therefore to see what these affinities are, and whether the difference between the two floras is such as to invalidate the conclusion I have formulated above.

Palæontological relations of Damuda and Newcastle beds.

To begin with, both floras are marked by the predominance of the genus *Glossopteris*, which, in the Newcastle flora, comprises 9 out of 26 species, or 35 per cent. of the total number of species, and 19 species out of 63, or 30 per cent. of the total number of species in the Damuda flora; of these, one species, *G. browniana*, is identical in both cases, and 3 Newcastle species, *G. linearis*, *G. ampla* and *G. parallela*, are represented by the allied Damuda species, *G. angustifolia*, *G. communis* and *G. damudica*. The genus *Phyllotheca* is represented in both floras, and the Australian form is allied to, and was long considered identical with, the *P. indica* of the Damudas. *Vertebraria* is found in both series, and is only known elsewhere from the "jurassic" beds of Siberia. *Sphenopteris alata* Bgt. is another species represented in allied forms in the Damuda flora, and the species *Gangamopteris angustifolia*, McCoy, is common to the two floras. Besides these the genus *Nöggerathiopsis* is represented in both floras, so that we have in all 6 genera and 2 species common to the two floras, besides 5 species represented by allied forms. On the other hand there are 3 genera out of a total of 9 not represented in the Damuda flora.

It may be that the resemblances between the two floras are not sufficient to justify our placing them, even approximately, on the same horizon; yet it cannot be said that the differences are such as to preclude the beds from which they were obtained from being regarded as, at any rate, approximately homotaxial with each other. I avoid the word contemporaneous for, though I think that probably it might be used in connection with the Talchirs and Bacchus Marsh beds, I see no possibility of arriving at any conclusion as to whether the Damudas and Newcastle beds were or were not contemporaneous; in all probability the two periods overlapped somewhat, and the affinities between the Hawkesbury and Wianamatta floras and that of the Damudas, as well as the internal evidence of that flora itself, would indicate that the latter was probably of somewhat later date than the Newcastle period.

In the foregoing pages I have not alluded, except incidentally, to the position of either the Australian or the Indian coal-measures in the European sequence. This question is altogether too large and complicated for me to take up at present when debarred from access to any scientific library, and has besides been already so well threshed out that I could hardly gather any grain by going over the straw again. This same cause—want of access to books—has prevented me from giving as full references to previous literature as I should have wished to do; but I have endeavored as far as possible to acknowledge my indebtedness to previous writers, and in conclusion I can but express my obligations to all those who assisted me in my enquiries during my visit to Australia, and in particular to Mr. C. S. Wilkinson, Government Geologist of New South Wales, and Mr. R. F. Murray, Government Geologist for Victoria, for the unreserved manner in which they placed all their information and experience at my disposal.

CAMP, AJMERE,
2nd November 1885.

3

Reprinted from *Geol. Mag.* **65:**323-327 (1928), by permission of Cambridge University Press

The Occurrence of *Glossopteris* in the Beacon Sandstone of Ferrar Glacier, South Victoria Land.

By W. N. Edwards.

(Published by permission of the Trustees of the British Museum.)

Introduction.

THE presence of members of the *Glossopteris* flora in Antarctica was first attested by the material collected from Mount Buckley by Dr. Wilson and Lieut. Bowers on Capt. Scott's last expedition to the South Pole. and the now famous specimens were described by Professor A. C. Seward (1914). Traces of fossil plants in Antarctica, were, however, first discovered by Mr. H. T. Ferrar, a member of the National Antarctic Expedition, 1901–4, and were briefly described by Arber (1907), who reported, on some specimens from the Ferrar Glacier, that " several of these show fair-sized carbonaceous impressions or markings, which in all probability are of vegetable origin". The evidence presented by this unpromising material did not "permit of any opinion as to the botanical nature or affinities of the fossils themselves, nor of the geological age of the beds in which they occur" (Arber, 1907, p. 48). Professor Seward examined the specimens and agreed with this conclusion (1914, p. 2).

GLOSSOPTERIS.

The material collected by Mr. Ferrar is in the Geological Department of the British Museum (Natural History), and, in the course of re-arranging part of the Museum collections, I recently had occasion to examine it. One of the specimens was a rather large block of hard shale with bands of carbonaceous matter. The impressions on the surface were unrecognizable, but it seemed obvious that such a block might prove to be of more value if it were

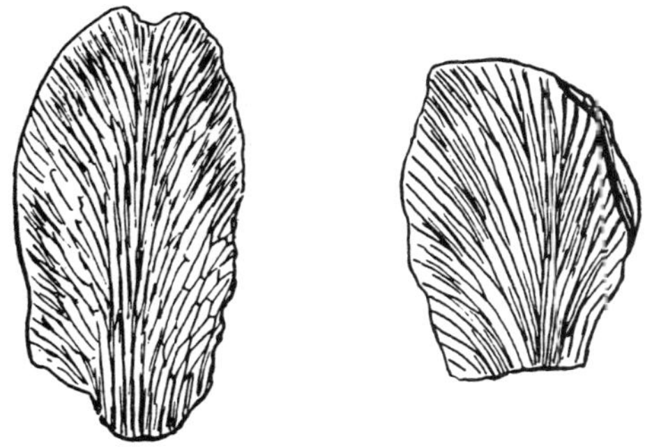

FIG. 1.—*Glossopteris indica* Sch., from Ferrar Glacier. A. Fragment with anastomosing venation clearly shown at one point on the right-hand side. × 2. V. 20445. B. Smaller fragment; anastomoses not clearly seen. × 2. V. 20447.

broken up, since it would be no great loss if it happened to be destroyed in the process. The shale was therefore carefully split open by Mr. F. M. Wonnacott, and we were at once rewarded by finding several undoubted fragments of *Glossopteris* fronds. The preservation is by no means good; the fronds are black impressions on a dark, often black shale; and the veins, when preserved at all, are faint silvery lines. However, as Professor Seward has said, *Glossopteris* is "one of the few genera that can be identified with confidence from fragmentary specimens" (1914, p. 6).

The specific identification is by no means simple, as will be apparent from the drawings of the best specimens. One of these is a small, almost spathulate leaf 2 cm. in length (incomplete), with a rather indistinct midrib, but with the anastomosing veins shown very clearly at one point (Fig. 1*a*); another is oval in form, 4 cm. in length (incomplete) and probably about 3·5 cm. in width

(Fig. 2). The venation varies considerably, especially in the angle made by the secondary veins with the midrib, but this variation is no more than might be expected in different leaves of a single species, and no more than is to be seen in the specimens from Mount Buckley figured by Seward (1914, pl. iii) as *Glossopteris indica* Schimper and *G. indica* var. *wilsoni* Seward. The latter were distinguished by the rarity of anastomoses between the secondary veins. Specimen V. 20446 (Fig. 2) from Ferrar Glacier resembles Seward's var. *wilsoni*, but even the secondary veins can only just be distinguished, and the apparent rarity of anastomoses may only be due to the poor preservation. This explanation may also apply to the specimens of *G. indica* var. *wilsoni* from Mount Buckley, which are by no means well preserved. Whether this is so or not, I think that the custom of applying varietal names to isolated fossil leaf impressions is to be deprecated. As Professor Seward himself has said (1910, p. 507), speaking of *Glossopteris*: "The

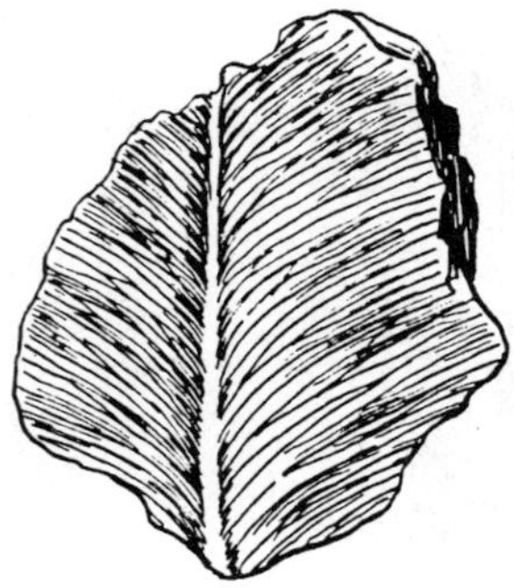

FIG. 2.—*Glossopteris indica* Sch. Ferrar Glacier. Incomplete leaf with nearly parallel veins and with apparently few anastomoses. Natural size. V. 20446.

arbitrary separation of sterile leaves, which differ by small degrees from one another in form and in the details of venation, by the application of specific names is a thankless task necessitated by custom and convenience; it is, however, idle to ignore the artificial basis of such separation." The use of trinominal nomenclature does not appear to add to the convenience of this artificial classification, and I think that in view of the imperfect preservation the Antarctic specimens of *Glossopteris*, both from Mount Buckley and Ferrar Glacier, should all be referred to a single form species. This may for the present be named *Glossopteris indica* Schimper. It is now usual to separate *G. indica* from *G. browniana*, though Professor Seward at one time united them. It is true that extreme forms may be distinguishable; but since neither Zeiller nor subsequent authors have clearly indicated any tangible line of demarcation between the two species, one should not attach undue importance to identifications from imperfect or poorly preserved material, for instance, in considerations of geographical distribution. The point to be emphasized here is that a *Glossopteris* of the *G. browniana* group occurs in Antarctica at Ferrar Glacier, some 500 miles north of the locality at Mount Buckley.

Two very imperfect fragments resemble the poorly preserved impressions from Mount Buckley, figured by Professor Seward (1914, pl. iii, fig. 17) as possible scale-leaves of *Glossopteris*. The Ferrar Glacier specimens are larger, but they also are unidentifiable. They may only be ordinary fronds poorly preserved.

Dadoxylon.

The flattened oval form of some of the carbonaceous specimens collected by Mr. Ferrar indicates the presence of plant stems, some several inches in diameter. Flakes of carbonaceous matter from some of these stems, examined by reflected light under the microscope, showed traces of the structure of gymnospermous wood. With the aid of collodion films it was possible to ascertain that the bordered pits on the radial walls of the tracheides were in one or

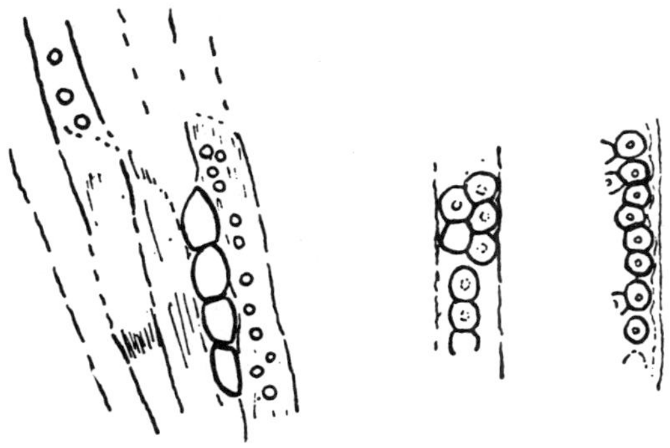

A B

Fig. 3.—*Dadoxylon* sp. Ferrar Glacier. Fragments of secondary wood detached on collodion films. Highly magnified. A. Tangential (V. 20450*a*). The medullary ray is only an impression on the film. B. Radial (V. 20451*a*).

two rows, and usually in contact; that the tangential walls were also pitted; and that the medullary rays were only a few cells in height. Wood of this type may be referred to the comprehensive form-genus *Dadoxylon*, but the possibility of identity with the wood from Priestley Glacier described by Professor Seward as *Antarcticoxylon priestleyi* is not excluded, since the radial pitting is similar. Seward's species has since been transferred by Walton (1923) to *Rhexoxylon* Bancroft, a genus founded on characters that cannot be fully determined from fragments of secondary wood only. Neither in *R. priestleyi* nor in the wood here described is the field pitting preserved, and so far the occurrence of tangential pitting has not been noted in any species of *Rhexoxylon*. The question of the possible relationship between *Rhexoxylon* and some of the Gondwanaland woods referred to *Dadoxylon* has already been discussed by Walton (1923).

Conclusions.

The following plants (not including some indeterminable remains) have previously been recorded from the Beacon Sandstone of Antarctica (Seward, 1914) :—

Glossopteris indica Schimper.
Glossopteris indica var. *wilsoni* Seward.
Vertebraria sp.
Rhexoxylon priestleyi (Seward).
Pityosporites antarcticus Seward.

I agree with Walton (1925, p. 11) in regarding the supposed spore, *Pityosporites antarcticus*, as merely a shrivelled pith cell. *Vertebraria* is a part of the plant which bore *Glossopteris*, so that the list may be reduced to the latter and a gymnospermous wood. The material collected by Mr. Ferrar has also yielded only *Glossopteris* and a gymnospermous wood. With the small amount of available material it is impossible to say whether this has any special significance, and the problem of whether or not *Glossopteris* was borne on a tree having wood of the *Dadoxylon* type will probably be settled in some other part of Gondwanaland.

The significance of the occurrence of *Glossopteris* in Antarctica has been fully dealt with by Professor Seward (1914), who also gives an account of the geology of the Beacon Sandstone, with references to the earlier literature. According to Ferrar, the Beacon Sandstone of South Victoria Land is about 2,000 feet thick, and the fossiliferous band with *Glossopteris* is near the top. The specimens here described are from a band 50 feet above the ice, below Hill B 1, in the south-west arm of Ferrar Glacier, about 7,000 feet above sea-level (Ferrar, 1907, pp. 41, 99). It is satisfactory to record that the material collected by Mr. Ferrar is of greater interest and importance than was formerly supposed.

WORKS QUOTED.

Arber, E. A. N. 1907. *Report on the Plant-Remains from the Beacon Sandstone.* In Ferrar, 1907, p. 48.

Ferrar, H. T. 1907. " Report on the Field Geology of the Region Explored during the *Discovery* Antarctic Expedition, 1901–4 " : *Brit. Mus. Nat. Hist., Nat. Antarctic Exped.*, vol. i, pp. 1–100.

Seward, A. C. 1910. *Fossil Plants*, vol. ii.

—— 1914. " Antarctic Fossil Plants " : *Brit. Mus. Nat. Hist.*, " *Terra Nova* " *Rep., Geol.*, vol. i, pp. 1–49.

Walton, J. 1923. " On *Rhexoxylon* Bancroft—a Triassic Genus of Plants exhibiting a Liane type of Vascular Organization " : *Phil. Trans. Roy. Soc.*, B, vol. ccxii, pp. 79–109.

—— 1925. " On Some South African Fossil Woods " : *Ann. S. Afr. Mus.*, vol. xxii, pp. 1–26.

4

Reprinted from pages 546-548 of *Geol. Soc. London Quart. Jour.* **27**:497-548 (1871)

ON SOME POINTS IN SOUTH-AFRICAN GEOLOGY

G. W. Stow

[*Editor's Note:* Only the Appendix to this article is reprinted here.]

Appendix: on the probable existence of an ancient Southern Continent.

The many similar forms of life, either fossil or recent, that are found scattered over various parts of different countries now so widely separated by the Indian and Pacific Oceans, seem to indicate that in very remote periods they must have been more intimately connected with each other than they are at present.

To those who believe that all the species of the same genus, and that in all probability all genera of the same family, have a common origin, it will appear almost self-evident that it must have been so. Thus remains of Dicynodont Reptiles and Labyrinthodont Amphibia are found both in India and South Africa, "affording," as Professor Huxley stated in a paper upon those from the former country, "new and interesting links with the fossil fauna of the Karoo beds of South Africa." In another paper he said, "There are two other forms of Labyrinthodonts which exhibit many similarities to the *Micropholis*. These are the *Brachyops laticeps* of Prof. Owen, from Central India, and a new form allied to the *Brachyops*, but distinct from it, from Australia, the *Bothriceps australis*."

It is not only in the fauna of the Dicynodon formation, but in the flora also that connecting links are found; thus a *Glossopteris* that has frequently been found in the Karoo and others from India and Australia are so nearly allied to each other that a high authority has stated that he "can find no specific distinction." With regard to this flora Mr. Tate, in his paper "On some Secondary Fossils from South Africa," says that it "presents close analogy with that of the coal formation of Eastern Australia, and the plant-bearing beds of Burdwan and Nagpur in India. The characteristic plant in each of these deposits is a *Glossopteris*; and it seems that the Indian, Australian, and South-African plants are specifically identical."

Amongst the fossil Mollusca of the succeeding Uitenhage formation we find many other such connecting genera. Thus the *Exogyra*, *Trigonia*, *Cucullæa*, *Pinna*, and *Crassatella* of the Sundays and Zwartkops strata find their representatives in Mauritius, India, and Australia, the *Crassatella* extending as far as New Zealand. With regard to the plant-bearing beds of this series, Mr. Tate writes*, —"Of the four species of *Pecopteris*, one is not satisfactorily distinct from *P. lobata* of India; and two others are closely allied to *P. indica*, also from the Jurassic plant-beds of the Rajmahal Hills. . . . . . *Asplenites lobata* is common to these Indian and African strata; and though, with one exception, they are distinct, yet, on the whole, the Jurassic plants of South Africa recall those of Scarborough and the Rajmahal Hills."

* Quart. Journ. Geol. Soc. vol. xxiii. p. 148.

The *Pectunculus* and *Perna* of the Zwartkops Pliocene limestone, the *Cardium*, the large *Natica*, *Loripes*, *Panopœa*, and *Akera* of the more recent formations, are found spread over the same extensive areas as those previously mentioned. Of the recent flora the 'Encyclopædia Britannica' gives the following:—"On the coast of Guinea and Congo the flora is intermediate between that of America and Asia. Species of *Sorghum*, *Sterculia acuminata*, the Kela-nut, and the Poison-bean of Calabar belong to this region." "In Chili there are many genera of Composites which are also represented in Australia and the Cape of Good Hope."

The most recent evidences of an ancient southern continent consist most probably in the wide-spread coral-reefs and islands in the Indian and Pacific Oceans, extending from the southern portions of the Red Sea and the northern part of the Mozambique Channel on the west, to the Island of Ducie in the extreme east, thus including the Seychelles, Madagascar, Mauritius, parts of the coast of Australia, New Caledonia, the Pacific Islands, to the before-mentioned Island of Ducie; and on the north of the Equator the Caroline Islands, the Marianne and Philippine Islands, the Chinese Seas, along the

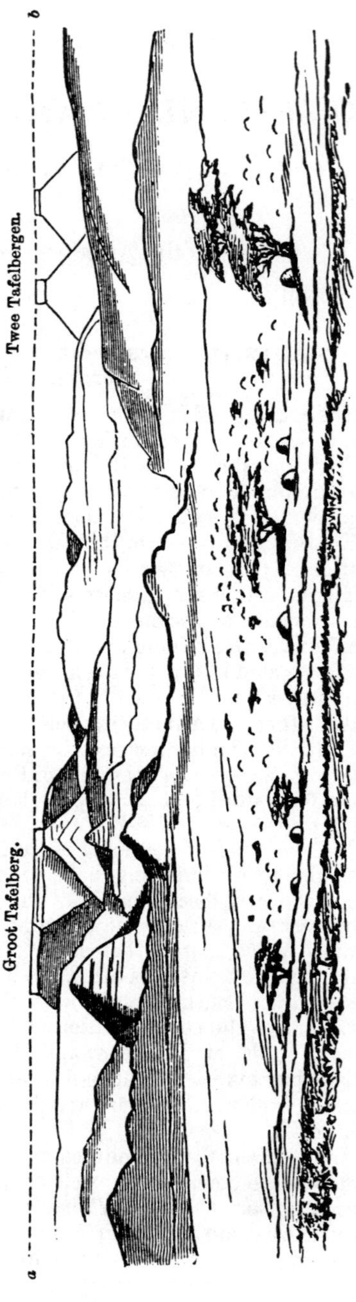

Fig. 17.—*Outline Sketch taken from Andrie's Neck.*

The dotted line *a b* indicates the level of the ancient plateau.

southern coast of Sumatra, and the Maldives and Laccadives, on the west of the Indian peninsula. As most of the species of the different corals forming these reefs cannot exist in water beyond a certain depth, the deep sea at present intervening between such widely separated groups of islands would have formed an impassable barrier to them; and thus the conviction seems forced upon us that most of these coral-reefs are now crowning the tops of the ancient mountains of a subsiding continent, along the shores of which the various genera of shells and other animals could have migrated—as much as we are impressed with the belief that the table tops of the Great Winterberg, the Groot Tafelberg, the Twee Tafelbergen, and others previously mentioned (see sketch, fig. 17) prove the existence of the elevated plateau that must at one time have occupied a large portion of Southern Africa.

Table of Elevations from Eland's Post (Kat River) to the Banks of the Orange River, near Aliwal. Taken by Mr. A. N. Ella, Government Inspector of Roads, and Mr. J. Graham, of Aliwal.

Locality.	Graham.	Ella.
	feet.	feet.
Eland's Post, Kat River	2200	
Katberg, highest point of road	......	5332
Ditto, flat on the north side, near road	4800	
Whittlesea	3100	
Queenstown	......	3432
Klopper's Fontein	4600	
Flat between Klopper's Fontein and foot of Pen Hoek	4800	
Foot of Pen Hoek, Stormberg	......	5068
Summit of ditto, ditto	......	5596
Burghersdorp	......	4382
Aliwal	4300	
Banks of Orange River, near Aliwal	......	4224
The flat around Bloemfontein, capital of the Orange River, Free State	5300	

Locality.	Hall.	Hall.
	feet.	feet.
The Great Winterberg is the culminating point of the Kat and Winterberg Mountains	7800	Gaika's Kop, 6800
Hangklip is the culminating point of the Andriesberg, an outlier of the Stormberg	6800	
	Lieut. Sherwell.	
Spitzkop or Compassberg is the culminating point of the Stormberg and Sneewbergen	10,259*.	

* Monts-des-Sources, in Basouta-land, is said to be the highest point in South Africa; but its exact height is not known.

5

Reprinted from *Geol. Soc. London Quart. Jour.* **26:**514–517 (1870)

2. *Notes on an* Ancient Boulder-clay *of* Natal. By Dr. Sutherland, Surveyor-General of the Colony. Contributed on Dr. Sutherland's behalf by Dr. Mann, F.R.G.S., F.R.A.S.

[Communicated by Prof. Ramsay, F.R.S., F.G.S.]

There is largely developed in the Colony of Natal a formation which seems to be in all essential particulars identical with Mr. Bain's claystone porphyry of the Cape of Good Hope. This formation flanks a long range of sandstone hills, which runs from the Tugela river frontier, at a distance of some six or eight miles from the sea, across the Umgeni river, and through the Berea hills to the mouth of the Umbilo. It also crops out extensively near Maritzburg, and stretches thence, in one direction, over the Umgeni and Umvoti to the Tugela valley between Greytown and the Biggersberg; and in the other direction over the Umlasi and Umkomasi towards the opposite frontier. In the latter course it goes onwards to the St. John's river and to the further districts of the Cape.

The deposit itself consists of a greyish-blue argillaceous matrix, containing fragments of granite, gneiss, graphite, quartzite, greenstone, and clay-slate. These imbedded fragments are of various size, from the minute dimensions of sand-grains, up to vast blocks measuring 6 feet across, and weighing from 5 to 10 tons. They are smoothed, as if they had been subject to a certain amount of attrition in a muddy sediment; but they are not rounded like boulders that have been subjected to sea-breakers. The fracture of the rock is not conchoidal, and there is manifest in its substance a rude disposition towards wavy stratification. The general appearance is that of a clay which has been deposited by aqueous action, and afterwards metamorphosed by heat, pressure, and chemical action. In places the deposit exhibits the most unmistakable ripple-markings, which seem to indicate that the wind, or other force, which agitated the water during the period of deposit, acted at different times in different directions. The thickness of the deposit varies in different situations, but in some places amounts to as much as 1200 feet.

The clay is compact and tenacious, and unfavourable to abundant supply of water. Borings into its substance remain dry to almost any depth. The surface-soil, which is associated with the formation, is stiff and non-fertile until it has been loosened to a considerable depth and thoroughly worked.

The boulder-bearing clay rests generally upon old Silurian sandstones, which in their turn are based upon granite and gneiss, and which in that connexion constitute the peculiar South-African table mountains. Upwards it passes first into newer shales, and through them into the sandstones and shales, which are associated in Natal with Carboniferous deposits. The transition is very gradual, without any distinct line of demarcation, and often stretches through a quarter of a mile of debatable ground. The transition is very well seen in the Town hill directly above Maritzburg. The only real difference between the shale and Boulder-clay seems to be, that in one of them the matrix is of uniform homogeneity and fineness, while in the other extraneous fragments of rock are brecciated in the argillaceous substance. Near the Umpambinyone and Umzinto rivers the boulder-bearing clay passes into beds which very nearly simulate the condition of true slate, but have their lines of cleavage in the direction of, instead of transverse to, the general stratification. In one instance, in this situation, there is a fine flagstone very closely resembling the Caithness Sandstone. Ripple-markings are plentifully developed where the Boulder-clay passes into these fine slates and flagstones.

The old sandstones which lie immediately beneath the Boulder-clay have their upper surface, in many instances, deeply grooved and striated, as if a semiplastic substance, containing hard and angular fragments, had been passed over it with considerable pressure.

The rocky fragments imbedded in the Boulder-clay formation are all of them of the character of the rocks that are contemporaneous with, or inferior to, the Silurian Sandstone. Fragments of a higher series are never found in the matrix. In some instances very ponderous rock-fragments are found as much as fifty or sixty miles away from the sources of their derivation.

Mr. Bain, in his communication to the 'Journal of the Geological Society' in the year 1856, has very exactly expressed the condition and relations in which this boulder-formation is found in Natal, flanking the sandstone-beds which cap the granite and gneiss. It is, however, certainly very questionable whether this so-called Claystone Porphyry of Mr. Bain is really of the nature of porphyry.

Dr. Sutherland holds that it is obviously a brecciated conglomerate, derived from aqueous deposit, rather than an igneous porphyry, and he urges the following reasons in justification of this view.

In the first place, he has sought for years for the vent through which such a vast mass of erupted rock could have been thrown. Where such a mighty mass is concerned its eruption from beneath could not have been effected without leaving very obvious marks of volcanic disturbance. There are no such indications anywhere. Mr. Bain has suggested the possibility of some such vent having existence about the sources of the Orange river. In that district there is certainly no trace of such rock-convulsion.

Then the abundant ripple-markings, and the gradual transition into fine shales and sandstones of unquestionably aqueous origin, are incompatible with the theory of volcanic or igneous formation; as is also the rubbed condition of the fragments, and the absence in them of all signs of fire-action or fusion.

And yet, again, there is the transport of the vast ponderous blocks to long distances, and the scorings and groovings of the sandstone upon which the Boulder-clay formation rests, to be accounted for.

On the whole, these circumstances seem to indicate that the constituents of this Boulder-clay have been derived from the superficial denudation of older rocks by aqueous agency, and that it has assumed its existing relations to the other rocks while in the condition of a moist and plastic mass. In all probability the finer shales, containing impress of the ripple-marks, were formed during periods of approximate repose; and cataclysmic violence and disturbance had ceased altogether when the deposit of such beds as the Maritzburg shales began.

Dr. Sutherland inclines to think that the transport of vast massive blocks of several tons' weight, the scoring of the subjacent surfaces of sandstone, and the simultaneous deposition of minute sand-grains and large boulders in the same matrix, all point to one agency as the only one which can be rationally admitted to account satisfactorily for the presence of this remarkable formation in the situations in which it is found. He believes that the boulder-bearing clay of Natal is of analogous nature to the great Scandinavian drift, to which it is certainly intimately allied in intrinsic mineralogical character; that it is virtually a vast moraine of olden time; and that ice, in some form or other, has had to do with its formation, at least so far as the deposition of the imbedded fragments in the amorphous matrix are concerned. He dwells particularly upon the fact that Prof. Ramsay has already assigned certain breccias of Permian age to glacial periods and agency, and that there is good reason for referring the coal-bearing shale of Natal, into which this boulder-bearing clay passes almost imperceptibly, to the Permian system.

For these various reasons, Dr. Sutherland submits that the Boulder-clay formation of Natal should be classed with the Permian glacial breccias of Prof. Ramsay.

Discussion.

Dr. Mann called attention to some specimens which he had brought in illustration of Dr. Sutherland's paper, showing the gradual transition from a state closely resembling that of an igneous rock, into that of a truly stratified deposit.

Mr. T. M'K. Hughes, while admitting a recurrence of glacial periods, disputed the evidence in the cases hitherto brought forward. He adduced instances of similar accumulation of similar fragments in recent and ancient deposits which could not be referred to glacial origin; and showed how striation of the included fragments was frequently produced by movements in the mass subsequent to consolidation, pointing out that the supposed glacial deposits of Old Red age were formed under conditions similar to those under which such irregular accumulations were formed at the present day; that all the striated fragments occurred close to faults; that the Permian in the north of England, where it was deposited between high mountains, and might therefore be expected to be supplied in part by glaciers, contained no fragments from the higher ground of the Pennine range. He urged that the Natal deposit was not like a Boulder-clay, being ripple-marked and distinctly bedded, and that it had been subjected to pressure so great as to have produced distinct cleavage.

Prof. Ramsay combated the views of Mr. Hughes, and maintained that there was no necessity for supposing that all the pre-Carboniferous rocks were above water at the time of the deposit of the Permian beds of the north of England. Still he had never maintained that any of those deposits were due to glacial action, but only more southern beds. He pointed out that in the Natal beds under discussion enormous blocks of rock occurred, which were 60 or 80 miles from their original home, and still remained angular; and there was a difficulty in accounting for the phenomena on any other hypothesis than that suggested. He still maintained the probability of the occurrence of glacial episodes, not only in Permian but in other ages, as he had done now fifteen years ago. A Boulder-clay might be recognized, not only by the striæ on the stones, which were not always present, but by their shape and the general character of the agglomeration of the beds.

6

Reprinted from pages 188-190 and 218-227 of *Geol. Soc. South Africa Trans.* **24:**188-227 (1921)

The Carboniferous Glaciation of South Africa.

By Alex. L. du Toit, D.Sc., F.G.S.

(Read 10th October, 1921.)

[Plate XI.]

Chapter I.—Introduction.

South Africa has been extraordinarily favoured with geological problems of more than uncommon attraction, but the most fascinating is probably that of the Dwyka Glaciation, its study promising to become even yet more engrossing with increase of knowledge.

The Carboniferous Ice Age of the Southern Hemisphere—with which this paper deals in part—has for long formed a theme much debated not only by geologists, but by geographers and meteorologists as well, and in spite of the great strides that have been taken in the accumulation of geological data, the causes of this extremely ancient climatic revolution are candidly admitted by most authorities to remain almost as obscure, complex and elusive as ever.

As regards this episode in South Africa much that is illuminating has been written, and that too by geologists of eminence, nevertheless the immense unstudied areas undoubtedly have tended to cripple all their attempts to solve the puzzle. Eighteen years of field work have indeed caused me to realise only too clearly both the magnitude and the complexity of the problem, and only the unlikelihood of being able to continue this investigation has led to the setting down of the following outlines of this revolutionary period of earth history.

Since the interpretation of present-day glacial phenomena is by no means straightforward, that of so remote a period and with such a fragmentary record is obviously beset with exceptional difficulties.

The subject is one with several important side issues impossible to handle save in the briefest manner; for example, the comparison of the respective glaciated units of the Southern Hemisphere or the causes of this widespread refrigeration. As it happens. the information that we possess concerning the Indian and South American sections is disappointingly meagre, and the amount still to be learnt about them very considerable.

It has been assumed that the general outlines of the distribution and nature of the Dwyka glacial deposits and of the current views of the glaciation itself are so universally known as to require no detailing; sufficient will it be merely to state that south of latitude 21° S. the whole of South Africa must towards the close of the Carboniferous

Epoch have been temporarily covered by ice, which upon melting left behind a mantle of morainic matter that in turn became buried beneath younger—Permian and Triassic—deposits. At the present day it happens that over very wide areas this glacial mantle is either still buried beneath such formations or has been entirely removed by erosion; it is indeed only over extremely restricted strips of territory, where the requisite stage of denudation has been attained, that the glacial deposits are adequately exposed and striated surfaces can be sought for—a great contrast to the conditions in the Pleistocene.

Again, ignorance of the geology of certain and by no means limited parts of the country, particularly of the northern and north-eastern confines of the glaciated region, hampers interpretation.

Wherever striae are preserved on the glaciated floor, the movements of the vanished ice-mass are precisely indicated, but at the same time it is most important to note that they only furnish a record of the direction of movement at a *particular stage* in the glacial invasion; they belong to the time when the ground-moraine began to form a protecting layer between the ice and the underlying rock or when the retreat commenced. The boulder-beds are of enormous thickness in places, dwarfing those of the Pleistocene, and, even granted that they were produced by one and the same ice-sheet, there are strictly no grounds for presuming that the ice-body maintained throughout its life a course of movement unchanged in its direction.

The general course can be inferred, if not always with accuracy, at least approximately, from the inclusions of certain rocks of distinctive lithological characters referable to formations of known and restricted distribution. Among such "indicators" are the Red granite of the Bushveld, the felsites, basic lavas and sandstones of the Waterberg System of the Transvaal, the Ventersdorp volcanics, the Campbell Rand Dolomite of the Cape and Transvaal, the brown Griquatown jaspers and the purplish Matsap quartzites of Griqualand West—especially the conglomeratic variety of the last-named with its brilliant red jaspery pebbles. Over limited areas "trails" of boulders make their appearance on the "lee-side" of some localised formation, *e.g.*, the Koras lavas from near Upington, the Klokfontein rhyolites south of Kimberley and the Nondweni "bankets" of Zululand. Now and then some unique rock-type has been recognised in the tillite many hundreds of miles away from the only known exposures of that particular formation.

From the evidence thus acquired the annexed maps, Plate XI. and Fig. 1, have been drawn, the arrows marking the direction of the groovings upon observed striated surfaces. Such common types as granite, gneiss, diabase quartzite, etc., are of practically no value in this connection, because, first, they are not as a rule sufficiently distinctive; and, secondly, they are of wide distribution, underlying not improbably most of the Karroo basin and doubtless also composing the region now buried beneath the Indian Ocean, a quarter that certainly contributed to the eastern and southern Dwyka.

In certain areas there is great latitude in our conjectures owing to the scantiness of the data, so that the maps have to be regarded as first attempts and subject to revision with advances in knowledge; that of the eastern side of the Union, Fig. 1, is believed to be on the whole accurate.

At the period of maximum glaciation the ice apparently formed a continuous body across the African Peninsula fully a thousand miles from east to west and perhaps but little less than that from north to south.

Up to now the glacial phenomena in South Africa have been ascribed to a single ice-body whose centre lay in the Transvaal, a sheet comparable indeed with that of Greenland at the present day. Evidence, however, has gradually been accumulating to show that it must have arisen through the union or conflict of several apparently distinct ice-bodies of which those of the Transvaal and Natal were the most important with subsidiary centres of dispersal in Griqualand West and in the S.-W. African Protectorate.[1] To the north lay land; to the south the ice moved into deep water, the conditions in that direction forming perhaps a parallel to the Great Barrier of the Antarctic.

While the tillite is in the north-east usually less than 100 feet in thickness and is occasionally missing, over most of the area south of the twenty-ninth parallel it is as a rule several hundreds of feet thick and in the extreme south from 1,000 to 1,500; indeed in the Southern Karroo 2,000 feet is probably not far over the mark.

[*Editor's Note:* Material has been omitted at this point.]

[1] Formerly known as German South-West Africa.

CHAPTER VI.—SOME CONSIDERATIONS OF THE GONDWANALAND GLACIATION.

While no account could be considered as complete without reference to the causes that gave rise to the Carboniferous Glaciation, it must frankly be admitted that a good deal of what can be advanced partakes in a great degree of speculation.

The several elements of the ancient Continent "Gondwanaland" are, for example, now separated by such vast stretches of ocean that attempts towards restoring the Carboniferous geography are bound to be merely approximate; the present investigation indeed reveals the complexity pertaining to only a small portion of the region involved. Reference might be made to the palaeographical maps of De Lapparent and Koken, for instance; it will be sufficient to point out that the attempt by the latter is vitally affected by the younger (Lower Permian) age adopted for the glaciation.

There has been a good deal of misconception as to the nature and probable extent of this palaeozoic glaciation, for certain geologists have casually regarded the regions affected as having embraced the tropics; actually the position is vastly different.

In Madagascar and in South Africa the northerly limits yet known are about 23° S. and 22° S. latitude respectively, though possibly it may have attained to 21½° S. in Africa; in the opposite direction the Dwyka is in its full development still in 34° S.

It must be observed that there are several instances of tillites known from the Belgian Congo, for example in the Lufira Valley in Katanga and between Matadi and Leopoldville. Though ascribed to the Dwyka, there is some doubt upon this correlation; the formation is in one case certainly more ancient, and the same is not improbably true of the Lower Congo glacials. Admitting that certain of the Katanga deposits are of Dwyka age, this area would have to be regarded as a glacial "outlier," playing a part analogous to the Alpine region in the glaciation of Europe. While the latitude is only 10° S., it should be recollected that during the Pleistocene refrigeration glaciers actually descended the slopes of Ruwenzori and Kenia, which are situated right upon the Equator. Further information will be awaited with interest.

In the Australian region the northerly limit is in Western Australia 23° S. and in New South Wales 30° S., but the effect of (northerly) drifting ice is manifested on the Bowen River in Queensland in lat. 20° S., though seemingly at a somewhat later date than the main phase; the southernmost exposures are present in 43° S. in Tasmania. In the Falkland Islands the southerly limit is 51° S., while in the Argentine and Brazil the tillite is represented between latitudes 38° S. and 21°, perhaps 20° S. Across the Equator the range is from 30° N. to 18° N. in India, while the Squantum Tillite of North America belongs to 42° N. latitude.

It is consequently manifest that the glaciation during this epoch can with justice be described *as nowhere equatorial, but almost wholly extra-tropical.*

This statement by no means disposes of the extreme difficulties introduced by latitude, for they remain well nigh insuperable under any of the older views, which one and all presuppose that the spacing upon the globe of the several units remained practically the same from the Carboniferous down to the present day.

On the contrary, the solution follows in the simplest manner with the perception of the idea that *the several sections represent portions of Gondwanaland actually disrupted and forcibly torn apart from one another subsequently to the Triassic.* This revolutionary idea of continental and oceanic evolution, ably championed by Wegener, was recently applied to this problem by the writer,[17] and not only has a most remarkable mass of evidence in its favour hard to account for otherwise, but has proved extraordinarily stimulating in its application and indeed can be looked upon as a master-key to the Past.

Involving as it does the conception of the drifting apart of the several portions of Gondwanaland upon a yielding sub-crustal "sole" over vast distances, the existence of such past glacial conditions in present day low latitudes receives a simple explanation. The arguments physical and stratigraphical for this daring hypothesis cannot be presented here; sufficient it will be to observe that this idea is accepted as supplying the sole reasonable explanation yet put forward of the various facts.

In Fig. 4 are given three suggested restorations of the land during the Carboniferous and Permian, when the fragments of this ancient continent were supposedly much closer together and were surrounded by an ocean that persisted to a very much later date. Gondwanaland it will be noticed is pictured as a continent situated almost wholly

---

[17] A. L. du Toit. *S.A.A.A. Sci.*, vol. xviii., 1921 (in the press).

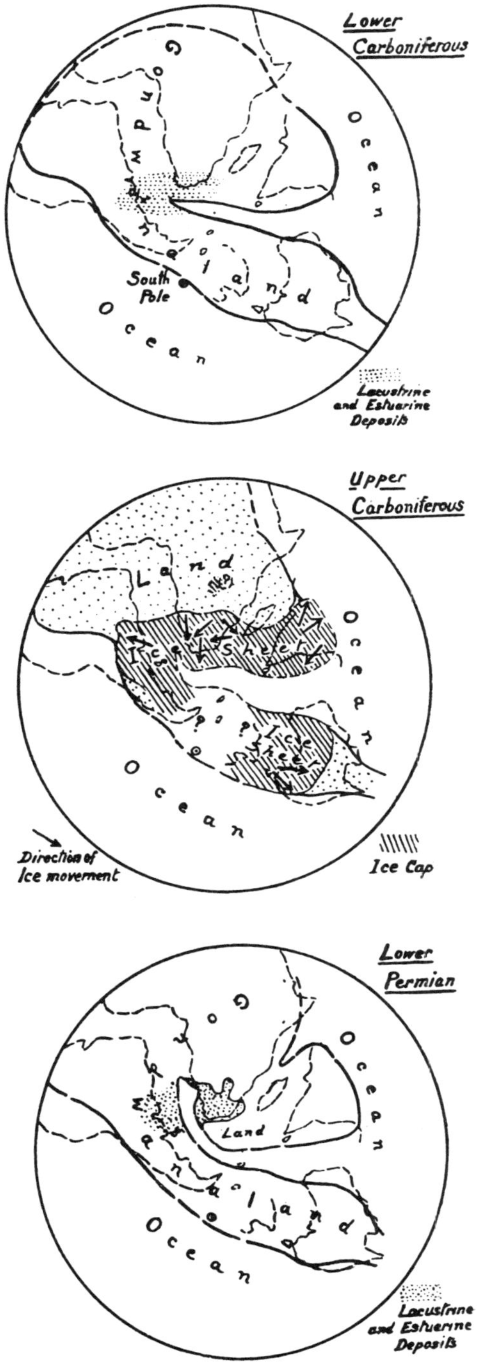

FIG. 4.—Hypothetical arrangement of Land and Water in the Southern Hemisphere during the Carboniferous and Permian.

within the Southern Hemisphere and deeply indented on the north-east by a gulf which is situated between India and Australia. This was a relic of the Devonian Sea and persisted as such until the close of the Cretaceous, when it widened out immensely to form a great ocean. Its presence would have led to a bilobate character of the ice-cap instead of a symmetrical one.

One of these lobes may have been a continuous body stretching from India across to Brazil, manifestly compound, shedding ice both equatorwards and polewards. In northern Peninsular India it is known that the ice moved north-westwards into the ocean—in the Salt Range region—and also to the north-east apparently. Information is meagre regarding the source of the southerly-lying Talchir boulder-bed, but some observations made by Ball in the Narbada Valley and in Orissa (Bengal) point to a north-easterly origin. Probably, therefore, the ice failed to reach so far to the east as Burma, but terminated in that direction in ocean.

The Southern Brazilian tillite was, according to Coleman,[18] derived from a sheet probably having its centre to the south-west, off the present coast-line. Both he and Woodworth also record certain erratics of a peculiar quartzite or grit with banded jasper pebbles, which from their accounts are just like those collected by the Transvaal ice from the ranges of Matsap beds in Griqualand West and transported so far westwards at least as the 18th meridian. With the continental disruption hypothesis in mind, could they not possibly have been carried much further westwards still?

In the north-west of Parana State marine conditions appear to have prevailed, but the boulder-beds are known near to or on the Chilian border and also well to the west and south-west of Buenos Aires, so that this territory may perhaps constitute the western end of the lobe, for ocean probably lay further to the west.

Whether the tillite of the Falkland Isles fell within this extensive east-west trending ice-region is, however, uncertain; it might equally well have formed part of the ice-capped ground linking South America, Antarctica and Australia. It must be noted that tillite has not yet been discovered beneath the *Glossopteris*-bearing strata of South Victoria Land, but for the present it may be assumed that that area lay within the region of glaciation.

In Australia a great ice-sheet radiated northwards across Tasmania, Victoria and South Australia, travelling for the most part over land, but discharging into the ocean in the Hunters River and Macleay areas in New South Wales and also in the extreme west of Western Australia in the Irwin River and Gascoigne regions. This territory constitutes the extremity of the second great lobe, which may have stretched across to South America; the two lobes were separated by a gulf.

Under the hypothesis adopted all the areas known to have been glaciated, even Peninsular India, can be brought together inside the

[18] A. P. Coleman, *Jour. Geol.*, vol. xxvi., p. 310, 1918.

forty-fifth parallel south, which we may note is close to the existing northerly limit of drift ice, so that the Carboniferous glaciation of this continent on these assumptions would present no greater difficulties than that of the Northern Hemisphere in the Pleistocene. So many points of similarity with the latter obtrude themselves, it might be remarked, as to convince me that the conditions during the much earlier epoch must have been very closely comparable indeed.

There is no doubt, for example, that over all this glaciated territory the land probably nowhere stood much higher than a few thousands of feet above sea-level, while furthermore in certain places the ice actually moved over the ocean floor, as in Tasmania. As regards the Southern Karroo it was remarked that the floor must have dipped far below sea-level until ultimately the ice floated, and it is not unlikely that the regions where such conditions prevailed at the time of maximum glaciation were quite extensive. Edgeworth David has indeed discussed the possibility of an ice-cap having commenced to form upon an archipelago and, as the Lower Carboniferous was a period of general positive (downward) movement, the continents were doubtless fringed with islands when the refrigeration set in.

The evidence to hand effectually disposes of the view that the cooling was due to extreme elevation of the various centres of Ice-dispersal, one still held in certain quarters.

Another obscuring idea is that of the *Double Ice-cap, i.e.*, the simultaneous glaciation of the two polar areas with approximately equal intensity The well-established fact that the Pleistocene Glaciation was represented in but attenuated degree south of the Equator at once throws doubt upon this hypothesis, as usually formulated. Reviewing the Carboniferous episode such doubt becomes greatly strengthened by the limited development of tillites of that age in the Northern Hemisphere, an opinion that is furthered, if we come to regard the Indian occurrences as belonging to an area originally south of the Equator, but since transferred across it.

All this points to very unequal refrigeration in the two halves of the globe in both of these particular periods, and it seems as though this may be a very important generalisation. Such a conclusion would not be altogether unexpected, for the initiation of an ice-cap in a single hemisphere would upset the normal atmospheric and oceanic circulations of the entire globe much more intensely than were each half possessed of an ice-mass, displacement of the heat-equator being greater in the former case.

Another concept pervading most glacial thought is that of the *Symmetrical Polar Ice-cap.* Now of all the characters attending the Pleistocene glaciation the most important one in my mind is the *markedly eccentric position of the ice-masses in the Northern Hemisphere,* the significance of which somehow seems to have been almost entirely overlooked.

Whereas parts of Alaska and almost the whole of Siberia escaped being over-ridden by the ice, though entering well within the Arctic

Circle, the southerly limits of the great ice-fields were 52° N. in Europe and actually as much as 37½° N. latitude in North America. On the hypothesis of Wegener that these two continents were closer together at this date, the intervening ocean would not have been such a serious obstacle, so that at one time the ice not improbably formed a continuous mass across the greater part of the regions referred to, thus linking up the New World to the Old. The centre of the cap would thereupon have fallen somewhere in Greenland, far away indeed from the pole of rotation.

In North America, though there can be made out several oscillations of the southerly moving ice-lobes, the advances and retreats of the margin progressed in a regular manner from *west to east*, so that the newly-formed moraine was always being laid down upon and to the eastward of the previous deposit. Though not so clearly marked, the same sequence is present in Europe, the Baltic Glacier being the latest over-riding body and overlapping towards the north-east and north.

The centre or "pole" of glaciation would appear to have repeatedly shifted its position, so that the ice continually encroached upon virgin ground to the east and abandoned moraine-covered country to the west. This suggests that the centre—or centres, if the cap were bi-lobar—slowly, perhaps spasmodically, described a curve over ground situated wholly away from the North Pole, presumably within the area bounded by Baffin Land on the west and Spitzbergen on the east. On this view the angular distance from the polar axis of the American centre would have been 15° at least or 1,000 miles and may possibly have exceeded 20°. Even at the present day the centre of the Arctic Ice lies fully 5° from the pole—on the Canadian side—, but the Antarctic body is, on the contrary, nearly symmetrically arranged.

In instituting comparisons with the Carboniferous occurrences it is most satisfactory to obtain from South Africa evidence that points to a migration of the ice-centre in a manner precisely similar to that of the Pleistocene. The overlapping took place also from *west to east*, which is just what one might anticipate on *a priori* grounds, seeing that such an easterly movement accords with the direction of travel of cyclonic disturbances.

Recent studies of the large ice-masses of Greenland and Antarctica have proved most definitely that over such bodies the prevalent atmospheric condition is *anticyclonic*. Again it has been discovered that above a height of ten kilometres an anticyclone is colder than a cyclone and that there are rapid changes in temperature across the vertical plane in crossing from one to the other, by which precipitation of moisture is facilitated. It has also been noted that the fall of snow is greater over the infra-marginal belt of an ice-cap than at the centre, so that the ice in the interior may be nearly stagnant, whereas that on the margin is moving rapidly; this certainly helps to explain the low surface gradients of such bodies and yet the vast distances to which they were enabled to extend.

Now it will be conceded that the initiation of a cold anticyclonic area of this character some distance from the polar axis would lead to more marked disturbances—both in temperature and pressure distribution—in the particular polar circle and adjacent temperate girdle than were the cap located in a truly axial position. Its encroachment upon the temperate girdle would lead to its rapid accretion and extension; to quote Wright—" A great ice-sheet once established sets up machinery of its own, which is constant in its action and practically independent of external causes."[19]

Granted the validity of the disruption hypothesis, the several glaciated elements of Gondwanaland moved away from the original centre of glaciation, though at different rates, and, when their present situations are compared and the enclosing ring of Tertiary foldings is considered, it is difficult not to conclude that the pole of glaciation failed to coincide with the South Polar axis and that the ice-mass was *eccentric thereto* and consequently like that of the Pleistocene. We are thus enabled to eliminate the fanciful hypothesis of " pole-wandering " as an explanation of the Carboniferous Ice Age, the application of which by Koken to this problem unfortunately leads to the introduction of still greater difficulties.

The *theory of eccentric ice-capping*, implying, first, a slow movement of the ice-centre around the polar axis and, secondly, the progressive development of the glacial maxima along its shifting margin, receives in the case of the Carboniferous epoch certain support from the palaeontological side.

A recent review by Edgeworth David[20] of the Carboniferous and Permian of the several elements of Gondwanaland points strongly, despite the uncertainties of exact correlation, to a lack in synchronism in the glaciations of the regions involved.

So far back indeed as the Middle Carboniferous refrigeration had attained some magnitude in the Hunters River area of New South Wales, glacial deposits associated with *Rhacopteris* flora being found in the Kuttung Series, while the movement of the ice was a little west of north; this is the earliest record anywhere in the Carboniferous epoch. After an interval glacial conditions returned and became more widespread, as shown by the pre-Greta tillites of New South Wales (Hunters River), Victoria (Bacchus Marsh), Tasmania (Wynward) and South Australia (Inman Valley), good evidence for climatic pulsations and for oscillations of the ice-front being forthcoming in addition.

The boulder-beds of South America and South Africa came into being presumably at the same time, that is to say during the Upper Carboniferous (Uralian), their polyphase character being inferable for the former through Woodworth's discovery of a two- or three-fold tillite in south-western Brazil, and detailed above for the latter. The palaeontological evidence leans rather towards the Western

[19] W. B. Wright. *The Quaternary Ice Age*, p. 26, 1914.
[20] C. A. Sussmilch and T. W. E. David. *Proc. Roy. Soc., N.S.W.*, liii., p. 293, 1920.

Australian and the Indian deposits being slightly younger, approaching the beginning of the Permian in fact; contemporaneous with them appear to have been the minor glacial horizons of the *Eurydesma* zone of the Lower Marine beds of the Hunters River area.

A new cycle of refrigeration is marked by the two glacial horizons in the Upper Marine beds (Lower Permian) of New South Wales and Tasmania, and possibly this is reflected in the peculiar arkoses of the Middle Ecca beds of Natal.

When it is recollected that not until the close of the Triassic did Gondwanaland commence to break up, and further that this continent was built up out of masses most of which date back to the pre-Cambrian and which had but rarely, and then only in small degree, been submerged by the ocean since that date, the explanation is forthcoming for the numerous evidences of *past glaciations* represented within its confines other than those of the "Permo-Carboniferous." Indeed the occurrences themselves go far to indicate that much of the territories which ultimately constituted Gondwanaland were in closer community throughout this vast period of time and probably occupied an area that lay either in moderate or in high latitudes.

For example, we need only quote the (probably) pre-Cambrian tillites of South Africa in at least two distinct periods (Namaqualand, Griqualand West and Heidelberg (Transvaal)) and those of the Congo, the Devonian glacials in the Table Mountain series of the Western Cape, the Cambrian tillite of South Australia and the Blaini "Boulder-bed" near Simla, in India, of similar or of pre-Cambrian age. In addition, Ball, Shaler and Passau[21] have described Triassic boulder-clays in Central Africa, while certain phenomena in the equivalent Hawkesbury Series of New South Wales have been ascribed to a cold cycle during this epoch. Further discoveries of glacial beds within the limits of "Gondwanaland," either in new formations or as extensions of those already known, are not only likely, but can safely be predicted accordingly.

To discuss the actual causes dominant or contributory of the refrigeration would be out of place here. With the recognition that a particular latitude or longitude in a previous geological epoch not only may, put probably does, not have a close relationship to that of the present, marks, I venture to say, an enormous advance in geo-analysis.

Not improbably the solution, astronomical, physical or climatological, will be perfectly simple; in regard to the first-named aspect what is greatly to be desired is a definite pronouncement from astronomers as to the validity of Drayson's theory of the varying obliquity of the earth's axis and the consequences thereof.

Chapter VII.—Summary.

(1) In pre-Dwyka (Lower Carboniferous) times South Africa was a country of low relief, a peneplain in the south-west, rising steadily

[21] S. H. Ball and M. K. Shaler. *Jour. Geol.*, vol. xviii., p. 681, 1910; G. Passau. *Ann. Soc. Geol., Belg.* Publ. N.C.G., vol. xl., p. 152, 1913.

towards the north-east, where the surface was diversified by low ranges; in the Transvaal the country was in parts somewhat broken. Several nearly parallel ranges formed the Griqualand West "highlands," on the south-east side of which lay the "Kaap Valley," an important trough conducting the drainage south-westwards from the Transvaal region; in the extreme north-west in the S.-W. African Protectorate rose another mountainous block.

The flatter parts of the region could not have stood high above sea-level, while a large body of water lay to the south; the higher ground attained altitudes of from 3,000 to possibly 5,000 feet above sea-level.

(2) The Dwyka Ice-sheet, which came into being during the Upper Carboniferous (Uralian), was a compound body or "multiple unit" formed by the coalescing of ice radiating from several distinct centres, namely, Namaland, Griqualand West, Transvaal and Natal.

(3) Over the three main centres the ice attained to a maximum at a slightly different date and the successive overlapping of the resulting glacial deposits took place from west to east. In the east the conflict between the Transvaal and the Natal Ice can be clearly made out and an important set of fluvio-glacial sediments resulted therefrom.

(4) The general direction of the ice was in a southerly or poleward direction, the northern margin of the compound body, where presumably the motion was equatorwards, having been neither located nor studied yet. In the south the ice probably merged into an unbroken front discharging into standing water, probably an estuary across that section included within the Cape Province.

(5) The boulder-beds represent for the most part ground-moraine and north of latitude 32½° S. rest upon an uneven, polished or striated floor, but to the south of that parallel they pass downwards without a break into deep-water shales and were largely the product of the occasionally buoyed-up ice-sheet. Just as in North America, the tillite is thin or missing at or about the centres of radiation, while thick away therefrom, sometimes enormously so.

(6) The motion of the ice can be made out from the striae on the floor and from the boulders in the tillite, many of which are far-travelled, some having been found distances of as much as 800 miles from their sources. Oscillations of the ice margins occurred, but truly interglacial periods are as yet doubtful.

(7) With the retreat of the ice the sea invaded the till-covered region of the north-west and by its further encroachment led to the development of a great estuary covering practically all the Cape Province, along the north-eastern side of which ice-rafting occurred, as indicated by boulder-shale and boulder-mudstone, such deposits being particularly characteristic of the old Kaap Valley. Deep water followed up the retreating ice tongues and green and black muds were laid down upon the strictly glacial deposits.

(8) In the north-east either the land rose owing to isostatic readjustments as the ice-cap melted or else the country lay too high to be submerged immediately and for a considerable period this region with its unevenly spread mantle of till lay exposed to the weather, as the result of which the morainic covering became bleached and denuded, while secondary deposits of conglomerates, sands and clays were spread out along the valleys.

Nevertheless ultimately the greater part of this region experienced subsidence and became covered by deltaic deposits—the Transvaal "Coal Measures"—and by still higher zones of the Karroo System.

(9) Gondwanaland in the Carboniferous is pictured as a much smaller continent than as usually conceived. Subsequently to the Triassic it was forcibly disrupted, the several sections of the crust proceeding to move apart from one another until checked in the Tertiary by the ring of earth-folds pushed up in the paths of the drifting masses by the latter. The ocean came to occupy the spaces between the land areas: whence the present positions of South America, Africa, Madagascar, India, Australia and Antarctica.

(10) The areas known to have been glaciated are under this hypothesis brought together in the Southern Hemisphere and could be all made to fall between the South Pole and the forty-fifth parallel. The difficulties as regards the development of ice-caps in low latitudes, left unsolved under the older hypotheses, is thus simply disposed of. Between pre-Cambrian times and the close of the Triassic Gondwanaland or its primitive elements experienced repeated glaciation.

(11) In the several sections of this Carboniferous land the glaciation seems to have culminated at slightly different dates, appearing the earliest and enduring the longest in New South Wales.

(12) The asymmetrical position of the vast ice-fields of the Arctic, and more so those of the Pleistocene, as regards the polar axis, becomes the basis of what is being called the *Eccentric Ice-cap Hypothesis*, from which it follows that the effects of refrigeration can be and are very unequal in the two hemispheres; to the Carboniferous Ice Age does this apply very markedly.

Part II

GEOPHYSICS

Editor's Comments on Papers 7 Through 12

During the same period that certain fossil and glacial evidence was strongly lending its support to the theory of continental drift, evidence of another type altogether was being gathered by geophysical techniques—evidence that would eventually lead to the solution of the riddle of global tectonics. One of these lines of evidence involved the fossil magnetism that is locked into certain kinds of rocks at the time of their formation. An early worker in the field, the Italian Macedonio Melloni, reported in 1853 (Paper 7) that his studies of volcanic rocks from Sicily and southern Italy demonstrate clearly that lavas cooled in the presence of the earth's magnetic field take on the magnetic orientation of the earth's magnetic field. Fifty years later, Pierre David (Paper 8) and B. Brunhes (1906) further demonstrated the stability of such thermo-remanent magnetism by observations on lava flows and their underlying contact-metamorphosed clays and on volcanic building stone. These two French paleomagnetists also were the first to notice that some rocks show a reversed polarity and to suggest that the earth's magnetic field must have been reversed at the time those rocks formed (see also Folgerhaiter, 1899; Mercanton, 1910; Chevallier, 1925; Matuyama, 1929; Minakami, 1941; Roche, 1956). Then, another fifty years elapsed before rock magnetic studies were used to track continental drift through time and thereby reopen the continental drift debate after World War II (see Graham, 1949; Clegg, Almond, and Stubbs, 1954; Runcorn, 1956). Polarity reversal waited still longer for acceptance but eventually proved to be the key that revealed the secrets of global plate movement (see Roche, 1951; Opdyke and Runcorn, 1956; Nagata, et al., 1957; Rutten, 1959; Cox, Doell, and Dalrymple, 1963).

Considering the fact that earthquakes are the most obvious direct result of contemporary tectonic activity, it is not surprising that seismology played an important role in gradually increasing our understanding of crustal movement. What is surprising is that the meaning of the most fundamental of all seismic observations—that earthquakes occur mostly in relatively narrow global belts—first spelled out in detail by F. de Montessus de Ballore about 1900 (Paper 9), was not adequately explained until the mid-1960s when the theory of plate tectonics was developed. Prior to that time, Ballore's observations and the more precise epicenter location studies of later workers remained a puzzle (see Barazangi and Dorman, 1969; National Earthquake Information Center, 1970).

Even more puzzling were the observations of H. H. Turner (1922) who discovered deep-focus earthquakes and K. Wadati (Paper 10) who studied the spatial arrangement of earthquake foci near the Japanese Islands. Wadati found that the foci tend to occur along a well-defined plane that dips away from the Japan trench and beneath the Japanese island arc to depths of 700 kilometers and that the associated volcanoes tend to cluster above an intermediate depth zone on the plane. The problem was that any material below about 50 kilometers should have too little strength to accumulate sufficient stress to result in an earthquake. Furthermore, the geometrical arrangement was a complete puzzle (see also Gutenberg and Richter, 1938, 1939; Benioff, 1949). Neither of these observations was explained in any real sense until sea-floor spreading and plate tectonics theory suggested that a cooler, rigid slab of material is sinking into the mantle at such locations.

Papers 11 and 12 describe seismic refraction observations that, along with studies of seismic surface waves, finally established the structure of the oceanic crust. Prior to this work, some geophysicists were still holding that at least some ocean basins are underlain by substantial thicknesses of continental crust (see, for example, Gutenberg, 1936, p.1600). However, the seismic refraction studies carried out by workers from Lamont-Doherty Geological Observatory, Cambridge University (Hill, 1952; Gaskell, Hill, and Swallow, 1958), and Scripps Institution of Oceanography (Raitt, 1949, 1956) soon demonstrated that the ocean basins are typically underlain by a few hundred meters of sediment above an oceanic crust only about 6 kilometers thick and that no continental crust is present. This finding meant that, in some respects at least, Alfred Wegener's model was correct. It did not, however, end the mechanical problems of continental drift. Furthermore, many workers came to regard the newly discovered thin oceanic crust as very old, original crust, a conclusion exactly the opposite of what is now known to be the case.

REFERENCES

Barazangi, M., and J. Dorman, 1969, World Seismicity Maps Compiled from ESSA, Coast and Geodetic Survey, Epicenter Data, 1961-1967, *Seismol. Soc. America Bull.* **59:**369-380.

Benioff, H., 1949, Seismic Evidence for the Fault Origin of Ocean Deeps, *Geol. Soc. America Bull.* **60:**1837-1866.

Brunhes, B., 1906, Recherches sur la direction d'aimantation des roches volcaniques, *Jour. Physique,* 4th ser, **5:**705-724.

Chevallier, R., 1925, L'aimantation des laves de l'Etna et l'orientation du champ terrestre en Sicile du XIIe au XVIIe siecle, *Ann. Phys.* **4:**5-162.

Clegg, J. A., M. Almond, and P. H. S. Stubbs, 1954, The Remanent Magnetization of Some Sedimentary Rocks in Britain, *Philos. Mag.* **45:**583-598.

Cox, A., R. R. Doell, and G. B. Dalrymple, 1963, Geomagnetic Polarity Epochs and Pleistocene Geochronometry, *Nature* **198:**1049-1051.

Folgerhaiter, G., 1899, Sur le variations seculaires de l'inclinaison magnetique dans antiquité, *Jour. Phys.,* 3rd ser. **8:**5-16.

Gaskell, T. F., M. N. Hill, and J. C. Swallow, 1958, Seismic Measurements Made by H.M.S. *Challenger* in the Atlantic, Pacific and Indian Oceans and in the Mediterranean Sea, 1950-1953, *Royal Soc. London Philos. Trans.,* ser. A, **251:**23-83.

Graham, J. W., 1949, The Stability and Significance of Magnetism in Sedimentary Rocks: *Jour. Geophys. Research* **54:**131-167.

Gutenberg, B., 1936, Structure of the Earth's Crust and the Spreading of the Continents, *Geol. Soc. America Bull.* **47:**1587-1610.

Gutenberg, B., and C. F. Richter, 1938, Depth and Geographical Distribution of Deep-Focus Earthquakes, *Geol. Soc. America Bull.* **49:**249-288.

Gutenberg, B., and C. F. Richter, 1939, Depth and Geographical Distribution of Deep-Focus Earthquakes (2nd paper), *Geol. Soc. America Bull.* **50:**1511-1528.

Hill, M. N., 1952, Seismic Refraction Shooting in an Area of the Eastern Atlantic, *Royal Soc. London Philos. Trans.* ser. A, **244:**561-596.

Matuyama, M. 1929, On the Direction of Magnetization of Basalt in Japan, Tyosen and Manchuria, *Imp. Acad. Japan Proc.* **5:**203-205.

Mercanton, P. L., 1910, Etat magnetique des diabases de l'Isfjord au Spitzbergen, *Acad. Sci. C. R.* **151:**1092–1094.

Minakami, T., 1941, Magnetization of the New Lava Flows of Miyakesima Island, *Tokyo Univ. Earthquake Research Inst. Bull.* **19:**612–618.

Nagata, T., S. Akimoto, S. Uyeda, Y. Shimizu, M. Ozima, and K. Kobayashi, 1957, Paleomagnetic Study on a Quaternary Volcanic Region in Japan, *Philos. Mag. Supp. Adv. Physics* **6:**255–263.

National Earthquake Information Center, 1970, Seismicity of the Earth, *Map NEIC 300,* NEIC, Rockville, Md.

Opdyke, N. D., and S. K. Runcorn, 1956, New Evidence for Reversal of the Geomagnetic Field near the Pliocene-Pleistocene Boundary, *Science* **123:**1126–1127.

Raitt, R. W., 1949, Studies of Ocean-Bottom Structure off Southern California with Explosive Waves (abs.), *Geol. Soc. America Bull.* **60:**1915.

Raitt, R. W., 1956, Seismic-Refraction Studies of the Pacific Ocean Basin, Part I; Crustal Thickness of the Central Equatorial Pacific, *Geol. Soc. America Bull.* **67:**1623–1640.

Roche, A., 1951, Sur les inversions de l'aimantation remante des roches volcaniques dans les monts d'Auvergne, *Acad. Sci. C. R.* **233:**1132–1134.

Roche, A., 1956, Sur la date de la derniere inversion du champ magnetique terrestre, *Acad. Sci. C. R.* **243:**812–814.

Runcorn, S. K., 1956, Paleomagnetic Comparisons between Europe and North America, *Geol. Assoc. Canada Proc.* **8:**77–85.

Rutten, M. G., 1959, Paleomagnetic Reconnaissance of Mid-Italian Volcanoes: Geologie en Mijnbouw **21:**373–374.

Turner, H. H., 1922, On the Arrival of Earthquake Waves at the Antipodes, and the Measurement of the Focal Depth of an Earthquake, *Royal Astron. Soc. Monthly Notices, Geophys. Supplement* **1:**1–13.

7

ON THE MAGNETIZATION OF VOLCANIC ROCKS

Macedonio Melloni

This article was translated expressly for this Benchmark volume by Orpheus L. Johnson of the University of Wisconsin-Parkside, from "Sur l'aimantation des roches volcaniques," in Acad. Sci. C. R. **37:**229–231 (1853)

I am sending you a report on the permanent magnetization of lava (1), about which Mr. von Humboldt must have spoken to you some time ago in one of his last letters. The numerous obstacles that I am presently encountering in my efforts to acquire newspapers and scientific news have not yet permitted me to learn whether this communication took place and, if it did, what new developments it included from our very good friend, who is certainly one of the most competent authorities in such matters. Furthermore, since you now have the original before you, you can judge perfectly well for yourself what consequences there can be in all of this with respect to the magnetism of the earth. I will add, however, that I continue to be concerned with the magnetism of rocks and that such studies are not entirely without value for scientific progress. Thus, I have just recently realized that if the general law of the permanent magnetization of lava has escaped the attention of observers until the present, it is because of the faulty reading of the instruments designed for this sort of research.

One normally measures the degree of magnetism in rocks by the attractive and repulsive influence that they, in their natural state, exercise on a pivoted magnetized bar. While it is true that repulsion is, as cannot be doubted, one of the distinctive qualities of magnetization, it has been an error to conclude that a piece of rock is not magnetized because its various parts attract both magnetic poles. In fact, the lavas from Vesuvius and the Phlegrian Fields presently being used to renew the surface of the principal streets of Naples are certainly magnetized, since they repulse the needle of my magnetoscope from 10 to 100 and even 120 degrees. At the same time they also attract in all directions both poles of a mineralogist's magnetized rod that I borrowed from professor Scacchi.

From this lack of repulsion in individual speciments, one ought not to conclude that the large masses of lava from which they come do not repulse, in certain directions, either one or the other pole of a magnetized needle. It is certainly probable, as I wrote to Mr. von Humboldt in the letter mentioned above, that the enormous differences observed by Mr. Peters between the slopes of the various areas located around Etna (from 2.5 to 31 degrees) are due to the combined repulsive and attractive forces of the lava erupted by this giant of European volcanoes. And I do not doubt that I will succeed in showing the attraction of the southern pole and the repulsion of the northern pole of a magnetized needle all around Vesuvius, if the

Academy of Science of Naples furnishes me the means that its permanent Secretary has requested for the completion of my experiment plans.

A piece of glowing steel plunged into cold water becomes hardened and acquires at the same time permanent magnetic polarity through the earth's action, but if it is reheated and let to cool slowly in the air, it loses both properties. Things do not occur in the same fashion when one repeats the same observations on a fragment of lava because magnetization takes place in both experiments, although the intensity is a bit greater in the first; and in addition, the lava reheated to a glow and then placed in a fixed position still retains, after cooling, a certain amount of southerly magnetism in its lower part and an equal amount of northerly magnetism in its upper part.

This magnetization is stable and not dependent on any position given to the fragment of cooled lava. One can, however, easily reverse its poles or give them any direction one chooses by again heating the piece (to red) and letting it cool in a suitable position with respect to the earth's magnetic axis.

We know that all combinations of iron, nickel, and cobalt are not magnetic. It is, therefore, natural that a magnetized needle does not react in the vicinity of certain rocks, in which analysis reveals the existence of a certain amount of iron or other substances that are magnetic in their isolated state. Now I find that a few of these rocks, inert relative to weak external magnetic force, react to a much greater force and then become true permanent magnets, in turn capable of acting upon all manner of magnetized needles.

I can prove, furthermore, by means of natural and artificial magnetic substances having a coercive force and more or less magnetized, that the quantity of material capable of acting, by attraction, on the two poles of the needle is very small compared to the effect resulting from the magnetization. All of this appears to me to be obviously contrary to the rather strange idea that knowing the proportions of the material sensitive to a very powerful magnet contained in the rocks of a given country would furnish the elements of a calculation necessary to determine the deviations of the magnetized needle with respect to the magnetic meridian. These deviations come in no way, in my opinion, from the absolute quantity of the material that influences the magnetized needle and even less from the quantity that became magnetic through the artificial action of the magnet used, but rather from the degree of direct force—in other words, from the degree of magnetization the rocks naturally possess. If the circumstances are even slightly favorable for me, I will most certainly include in a second report the details on the observations and experiments that have led me to this last conclusion.

(1) Memoir on the magnetic polarity of lavas and rocks of volcanic origin, read at the Naples Academy of Science on 21 January 1853.

8

ON THE STABILITY OF THE DIRECTION OF MAGNETIZATION IN SOME VOLCANIC ROCKS

Pierre David

This article was translated expressly for this Benchmark volume by Orpheus L. Johnson of the University of Wisconsin-Parkside, from "Sur la stabilité de la direction d'aimantation dans quelques roches volcaniques," Acad. Sci. C. R. **138:**41–42 (1904)

The measurements taken during more than two years, in collaboration with Mr. Brunhes, and referred to the Academy (Comptes Rendus, 15 July 1901 and 7 December 1903), had shown us that all volcanic rocks of the region have a stable remanent magnetization that is probably in the direction of the earth's field at the time the rock solidified. It might henceforth be interesting to establish whether, during a known period of sufficiently long duration, the direction of magnetization of these rocks had varied.

I began by studying different stones, taken on location, from a low wall built during the Gallo-Roman period.These stones, of different origin and construction (scoria, basalt, domite), which remained in place for about 2,000 years, are all permanently magnetized, but the direction of magnetization is extremely varied from one to the other. This, in fact, seemed to indicate that the direction of magnetization had not been modified during that period of 2,000 years. It was, however, difficult to orient these stones with enough precision to cut them into cubes and make declinometer measurements. I then thought of cutting a few cubes from a very well-preserved pavement of the former temple to Mercury located at the top of the Puy-de-Dome, which dates manifestly from the Gallo-Roman period. These cubes were taken from four domite paving stones; from three of them we took cubes separated by a distance of more than a meter in order to verify the measurements. The results obtained are the following, taking as a point of departure the present declination:

Paving No. 1	Piece No. 1	Declination: 144 degrees east Inclination: −55 degrees
	Piece No. 2	Declination: 143 degrees east Inclination: −58 degrees
Paving No. 2	Piece No. 1	Declination: 86 degrees east Inclination: −52 degrees
	Piece No. 2	Declination: 85 degrees east Inclination: −54 degrees
Paving No. 3	Piece No. 1	Declination: 74 degrees west Inclination: 53 degrees
	Piece No. 2	Declination: 71 degrees west Inclination: 53 degrees

Paving No. 4	Piece No. 1	Declination: 141 degrees east Inclination: −56.5 degrees

For all the paving studies, the inclination has the same absolute value, with the approximation that measurements permit, negative for three and positive for the other. The declination values, on the contrary, are less accurate. The conclusion seems to be then that since the installation of these pavings, their magnetic direction has not varied through the influence of the earth's field, in spite of the variations and perturbation that could have taken place since that time.

Furthermore, the constant absolute value of the inclination seems to indicate that all these pavings were taken from the same quarry in parallel slabs as is done frequently in quarries at present. Upon laying, some of them were probably placed in opposite directions, which would explain the change in the sign of inclination. As for the declination, the cutting of the paving stones could have modified it in some slight way. We note, furthermore, that two cubes of the same dimensions have the same magnetic moment. Here there would perhaps be a means of resolving the often discussed question on the origin of the enormous blocks used in building a temple whose ruins are still imposing after 2,000 years.

9

ON THE EXISTENCE OF TWO GREAT CIRCLES OF MAXIMUM SEISMIC INSTABILITY

F. de Montessus de Ballore

This article was translated expressly for this Benchmark volume by Orpheus L. Johnson of the University of Wisconsin-Parkside, from "Sur l'existence de deux grands cercles d'instabilité sismique maxima," Acad. Sci. C. R. **136:**1707–1709 (1903)

A long time ago the periphery of the Pacific Ocean was named Circle of Fire because of the numerous volcanoes that surround it. Noted also was the frequency of earthquakes along the approaches to the mid-ocean depression, stretching ideally as far as the Antilles and to the Indo-Gangetic plain.With particular regard to the unstable seismic regions, it is possible to give substance to this concept through the considerable number of quakes recorded to date, of which at least 156,878 are known to me. Of course, the purely instrumental tremors, or microseisms, are eliminated because their epicenters are too often unknown. We have only considered macroseisms.

If we consider the great circle of the terrestrial sphere whose pole is at 35 degrees 40 minutes north and 20 degrees 40 minutes east ([prime meridian at] Paris), which we will call Ando-Japanese-Malayasian or the Pacific, one sees that a zone of about 15 degrees on all sides encompasses epicenters of 64,426 seisms, including as one might expect seismic regions varying greatly in instability.

Second western quadrant

North: Northern Andes, Antilles (from Puerto Rico to Granada), Panama, and Darien: 3,494 seisms.
South: Southern Andes, central Andes, Guayaquil: 6,368 seisms.

First western quadrant

North: Atlantic slope of Costa Rica, Great Antilles, Honduras and Belize, British Columbia, Alaska: 897 seisms.
South: Pacific slope of Costa Rica, Nicaragua, Salvador, Guatemala, Mexico, California and the Rocky Mountains, the Aleutians: 13,018 seisms.

First eastern quadrant

North: Kamchatka and the coast of the Sea of Okhotsk, Mongolia and Manchuria, Sakhalin Island, western part of Hokkaido Island, Lu-Shun and Korea, China coast, Hainan Island, Malaccan Isthmus, northern Sumatra (to the north of the Bengkulu [Sumatra]-Singapore line): 2,482 seisms.

South: Kuriles, balance of Japan, Taiwan and Liou-Kiou, Philippines, Mindanao, Borneo, Moluccas, southern Sumatra, Java: 38,167 seisms.

The other great circle, Alpino-Caucasian-Himalayan-New Zealand, or Mediterranean, has as its pole 40 degrees 45 minutes north and 152 degrees 30 minutes west (Paris). Its corresponding zone of the same size includes epicenters of 84,023 seisms. The seismic regions are located as follows:

Second western quadrant

North: Azores, Galicia, northern Portugal, northern France, Switzerland, continental Italy, central Europe, eastern Alps, Hungary, Caucasus: 23,962 seisms.

South: Cape Verde Islands, Canary Islands, Madeira, Barbary Coast, Spain and the balance of Portugal, south of France (Maritime Alps, Provence, Pyrenees), Baleares, Corsica and Sardinia, Sicily, peninsular Italy, Greece, Balkan peninsula, Syria, Asia Minor, Armenia: 50,453 seisms.

First western quadrant

North: Eastern Caspian, Syr-Dar'ya, Issyk-Kul, Afganistan, southern flank of the Himalayas, Assam, Burma, Tibet, Yunnan: 6,612 seisms.

South: Persia, Baluchistan, Oude and Lahore, Punjab, Industan, Ceylon, Andaman and Nicobar: 812 seisms.

First eastern quadrant

North: New Zealand (North Island and Cooks Straits): 1,665 seisms.

South: Australia, Tasmania, remainder of New Zealand: 360 seisms.

Second eastern quadrant

South: The Guianas (as a remainder): 25 seisms.

The balance of the earth's surface, outside these two narrow belts, produces only 8,429 seisms as opposed to 148, 449, only one-twentieth. There are only three truly unstable regions: The Lake Baikal region, the Ka-shih region, and the basin of the lower Wei-Ho.

The distribution of seismic regions is not identical along the two zones. In the Pacific zone they are spread without interruption along an immense arc of 225 degrees (25,600 km), from the Gulf of Ancud (southern Chile) to Engano Island (southern Sumatra). In the central Pacific zone, on the contrary, they are massed together and overflow the circle more to the north and the south on an arc that is more than half as short, only 114 degrees (12,700 km) from the Azores to Yunnan with the exception of the independent and isolated center of instability of New Zealand.

The two great circles cross to the south and off Poeloe-Nias Island [west of Sumatra] and between the Galapagos and Darien. The angle between them is 67 degrees 5 minutes.

The preceding firgures attribute 41.07 percent of the seisms to the Pacific circle as opposed to 53.56 percent in the Mediterranean circle with a remainder of 5.37 percent. The difference of 12.49 percent in favor of the latter is practically negligible because it could very well simply result from the fact that these observations are much older than those for the Pacific circle.

We can then say that the surface of *the earth quakes almost uniquely and almost equally along its two lines of relief* (the word *relief* being understood to mean the difference in level between the bottom of the oceans and the mountain crests) and that *these two lines of instability are developed along two great circles forming between them an angle of 67 degrees.* Chance obviously has little to do with such results.

10

Reprinted from *Geophys. Mag.* **8**:305-325 (1935)

On the Activity of Deep-focus Earthquakes in the Japan Islands and Neighbourhoods.

K. Wadati.

1. *Introduction.*

The existence of deep-focus earthquakes has become to give a general interest of geophysicists and it is now necessary to get a precise catalogue of these earthquakes of all the world. This laborious work has been carried on by the International Seismological Summary since long ago. Recently V. Conrad[1] gives a catalogue of deep-focus earthquakes from these data. It is considered, however, that his results cannot completely fulfil the demand of investigators of the present time because their accuracy is somewhat insufficient for the positions of epicenters as well as the focal depths. Indeed, this work must be carried out by ample data of precise observations, particularly those made at places near the epicenter, and at the same time it is indispensable to have at hand correct tables of travel times for seismic waves of all kinds.

The author tried to get a catalogue of deep-focus earthquakes[2] occurred during 1905~1934 (IV) in the region near the Japanese Islands where is noted for their frequent occurrence. This is given in the first part of this paper. Then in the second part are given some statistics taken by the above results to see the general state of activity of deep-focus earthquakes. And in the last part some investigations are made about the seismic as well as volcanic activity in the region under consideration.

---

(1) V. Conrad: Gerl. Beitr. z. Geophys. Bd. 40. S. 113.

(2) "Deep-focus earthquakes" is here used in a wide sense, that is for such earthquakes whose foci lie at a deeper place in the crust compared with those in the case of ordinary shallow-focus ones. But afterwards in this investigation, earthquakes are classified into three according to their focal depeths, „deep-focus," "intermediate" and "shallow-focus." The first is used when the focal depth is larger than 200 km. below the surface, the second when it lies between 100 and 200 km. and the last when it is less than 100 km. As is well known shallow-focus earthquakes occur in almost all cases at the depth less than 60 km, most frequently at about 30 km, and also very frequently just near the Earth's surface.

Part 1. Catalogue[1] of Deep-focus Earthquakes.

2. The present catalogue of deep-focus earthquakes consists of three parts. In the first part are collected deep-focus earthquakes occurred near the Japanese Islands during 1905～1923, and in the second those occurred during 1924～1934 (IV). These two groups of earthquakes are given in the same table (Table 1. A and B). The reason why these deep-focus earthquakes are divided according to their times of occurrence is that the seismological observations of our country were not so complete in earlier years and consequently the results shown in Table 1 (A) may be inaccurate in some cases. While those obtained for earthquakes occurred after the year 1924 are considered to be much reliable. Of course, when the energy of earthquake is so small as to be observed only by few stations or when the epicenter situated in the sea bearing far away from the land, the accuracy of the results is greatly diminished.

Table 1 (A). Catalogue of Deep-focus Earthquakes 1905～1923. (H≧200 km.)

No.	Time of Occurrence*	Epicenter		Depth of Focus	(E)	Notes
1	1905, Aug. 25th 18^h 48^m	North Japan Sea?	—	Deep	B?	
	〃 Oct. 10th 5^h 10^m	——*	—	—	D?	*These earthquakes may perhaps be shallow ones.
2	1906, Jan. 21st 22^h 49^m	The Ise Bay?	—	320^{km}	A?	
	1907, Feb. 6th 17^h 38^m	——*	—	—	D?	
3	〃 Mar, 26th 20^h 22^m	South Japan Sea.	$\lambda=136°E$ $\varphi=38$ N	450	A?	
4	〃 May 4th 17^h 37^m	NE-ern off to the Is. of Ogasawara.	$\lambda=154°E$ $\varphi=31°N$	400	B	Epicenter of No. 4 is questionable.
5	〃 May 25th 23^h 04^m	Neighbourhood of Vladivostock.		more 500	A	
6	1908, Apr. 19th 17^h 01^m	Coast of Maritime Prov. of Siberia.	$\lambda=134°E$ $\varphi=43°N$	ditto	A	
7	〃 June 17th 10^h 18^m	Upper valley of the Tama river.	$\lambda=139.3°E$ $\varphi=35.7°N$	200?	C?	
8	1909, Mar. 18th 7^h 28^m	Far off the Coast of Kii.	$\lambda=137°E$ $\varphi=30°N$	350?	C	

* Times are all refered to the Central Standard Time of Japan, 135°E.

(1) A more detailed one was published in Japanese in the Seismological Magazine Vol. 8 No. 2

No.	Time of Occurrence	Epicenter		Depth of Focus	(E)	Notes
9	1909, Oct. 3rd 23^h 2^m	Southern off to Hatidyô Is.	λ=140°E φ=31°N	300?km	D	
10	1910, May 5th 0^h 19^m	Western off to the Is. of Ogasawara.	λ=139.0°E φ=27.8°N	350?	D	
11	〃 June 27th 0^h 59^m	WSW-ern off to Hatidyô Is.	λ=142.5°E φ=32.0°N	300?	D	
12	1911, Sep. 6th 9^h 55^m	North Japan Sea	λ=138.5°E φ=47.0°N	400?	B?	Gött: $iS-iP=9^m5^s$ Not so deep?
13	1912, Mar. 10th 20^h 10^m	Eastern part of the Sôya Straits	λ=145°E φ=46°N	250?	C?	Pul: $S-iP=8^m13^s$
14	〃 May 20th 16^h 54^m	NW-ern off to the Is. of Ogasawara.	λ=142.4°E φ=28.0°N	400	B?	Pul: $iS-iP=9^m6^s$
15	〃 July 25th 8^h 24^m	SW-ern off to Hatidyô Is.	λ=137.6°E φ=31.0°N	450?	C?	
16	1913, Jan. 20th 8^h 51^m	Northern part of the Kuril Is.	λ=154°E φ=49°N	200	A	
17	〃 Nov. 2nd 16^h 51^m	Neighbourhood of Oosima Is.	—	within 400?	D	
	1914, Oct. 17th 7^h 9^m	——	—	—	C?	Shallow?
18	1915, Mar. 18th 3^h 46^m	Southern off to the cape of Erimo.	λ=142.5°E φ=41.5°N	360	B?	Graz: $S-iP=9^m57^s$
19	〃 June 28th 0^h 29^m	SE-ern off to Etorô Is.	λ=150.0°E φ=42.5°N	250?	A	
20	〃 Oct. 9th 0^h 37^m	Neighbourhood of Hatidyô Is.	λ=140.0°E φ=33.0°N	250	B	
21	1916, Jan. 25th 20^h 38^m	Straits of Sôya.	λ=143.1°E φ=45.75°N	240	B	
	〃 Feb 1st 16^h 39^m	——	—	—	—	Shallow?
22	1917, Feb. 22nd 0^h 47^m	South Japan Sea	λ=135.3°E φ=36.1°N	350?	C	Epicenter conld not be accurately located.
23	〃 July 31st 12^h 25^m	North to Vladivostock	λ=130.0°E φ=45.0°N	more 500	A	
24	1918, Jan. 31st 6^h 20^m	Coast of Maritime Prov. of Siberia.	λ=135.0°E φ=44.5°N	255	A	
25	〃 Apr. 10th 11^h 6^m	North to Vladivostock	λ=130.3°E φ=43.5°N	350	B	I.S.S. $H=0.07$
26	〃 Dec. 31st 14^h 31^m	——	—	—	D	Certainly deep.
	1919, Jan. 4th 18^h 23^m	——	—	—	D	Shallow?
27	〃 Mar. 26th 22^h 37^m	SE-ern off to the Hatidyô Is.	λ=140°E φ=31°N	350?	D	Epicenter inaccurate.
28	〃 June 1st 15^h 53^m	Northern off to Isigaki Is.	λ=124.0°E φ=25.0°N	240	B	I.S.S. $H=0.04$
29	1920, Feb. 23rd 2^h 32^m	Okhotsk Sea.	λ=146.5°E φ=47.0°N	360	B	I.S.S. $H=0.05$

No.	Time of Occurrence	Epicenter		Depth of Focus	(E)	Notes
29	1920, Apr. 12th 8^h 5^m	Southern off to Kamchatka.	$\lambda=159°E$ $\varphi=48°N$	—	B	Not so deep?
30	〃 Apr. 15th 21^h 14^m	Southern off to the cape of Omaezaki.	$\lambda=138.3°E$ $\varphi=34.1°N$	400^{km}	C	
31	〃 May 6th 18^h 40^m	Neighbourhood of Vladivostock.	$\lambda=132.0°E$ $\varphi=43.4°N$	500?	C	I.S.S. 0.07
32	1921, Mar. 4th 21^h 52^m	SW-ern off to Hatidyô Is.	$\lambda=139.0°E$ $\varphi=29.8°N$	450	C	I.S.S. 0.06
33	〃 Mar. 15th 13^h 32^m	Western off to Hatidyô Is.	$\lambda=138.2°E$ $\varphi=33.6°N$	350	C	
34	〃 July 4th 23^h 20^m	Far western off to the Is. of Ogasawara	$\lambda=136.5°E$ $\varphi=25.5°N$	200	B	Tukubasan: $S-P=111^s$ $ScS-P=13^m$ 28.5^s
35	〃 Sep. 8th 7^h 29^m	Eastern off to Hatidyô Is.	$\lambda=141.9°E$ $\varphi=33.4°N$	300?	D	
36	〃 Sep. 27th 6^h 16^m	Southern off to Hatidyô Is.	$\lambda=140.0E$ $\varphi=31.0°N$	320	D	
37	1922, July 14th 20^h 10^m	Manchuria?	—	—	D	Probably deep.
38	〃 Oct. 25th 6^h —	Kuril Islands.	—	—	D	ditto.
39	1923, Feb. 12th 21^h 28^m	WNW-ern off to Hatidyô Is.	$\lambda=138.5°E$ $\varphi=33.4°N$	360?	C	
40	〃 Mar. 21st 17^h 28^m	NW-ern off to Etorô Is.?	—	200?	C	
41	〃 June 29th 19^h 47^m	Western off to the Is. of Ogasawara	$\lambda=140.9°E$ $\varphi=27.3°N$	more 400?	C	I.S.S. 0.06
42	〃 July 21st 1^h 50^m	Western off to Hatidyô Is.	—	300?	C	Epicenter and depth are questionable.

Table 1 (B). Catalogue of Deep-focus Earthquakes.
1924～1934 (IV). (H≧200 km.)

No.	Time of Occurrence	Epicenter		Depth of Focus	(E)	Notes
43	1924, Feb. 22nd 7^h 43^m	Middle part of the Japan Sea	$\lambda=137.0°E$ $\varphi=38.3°N$	$200?^{km}$	D	inaccurate.
44	〃 Apr. 3rd 11^h 30^m	SW-ern off to Hatidyô Is.	$\lambda=139.0°E$ $\varphi=32.0°N$	320	C	I.S.S. 0.05
45	〃 Apr. 13th 4^h 8^m	Ensyû-nada	$\lambda=137.3°E$ $\varphi=33.6°N$	300?	D	
46	〃 Apr. 26th 3^h 07^m	Southern off to the Is. of Ogasawara	$\lambda=143.0°E$ $\varphi=25.5°N$	240?	C	inaccurate.
47	〃 May 28th 18^h 53^m	Okhotsk Sea.	$\lambda=147.6°E$ $\varphi=47.8°N$	400	A	
48	〃 June 3rd 11^h 42^m	Southern part of the Idu peninsula	$\lambda=138.9°E$ $\varphi=34.6°N$	210	D	

No.	Time of Occurrence	Epicenter		Depth of Focus	(E)	Notes
49	1924, Dec. 14th 5^h 24^m	Niigata Prefecture, Japan Sea Coast.	$\lambda=137.7^\circ$E $\varphi=37.0^\circ$N	200?^{km}	B	
50	1925, Mar. 16th 17^h 54^m	—	—	350?	D	Epicenter could not be accurately located.
51	〃 Mar. 27th 13^h 18^m	SW-ern off to Hatidyô Is.	$\lambda=138.7^\circ$E $\varphi=29.7^\circ$N ?	400?	D	
52	〃 Apr. 20th 0^h 50^m	Western off to Hatidyô Is.	$\lambda=138.0^\circ$E $\varphi=33.0^\circ$N	350	B	I.S.S. 0.045
53	〃 May 16th 3^h 27^m	SW-ern off to Hatidyô Is	$\lambda=138.5^\circ$E $\varphi=30.5^\circ$N	400	C	〃 0.05
54	〃 May 27th 11^h 30^m	South Japan Sea	$\lambda=134.2^\circ$E $\varphi=37.7^\circ$N	370	A	〃 0.05
55	〃 Oct. 20th 18^h 42^m	Southern off to Hatidyô Is.	$\lambda=140.0^\circ$E $\varphi=29.0^\circ$N	400	C	〃 0.05
56	1926, Jan. 15th 23^h 53^m	Eastern part of the Sôya Straits.	$\lambda=142.6^\circ$E $\varphi=45.2^\circ$N	360	C	〃 0.06
57	〃 Feb. 11th 14^h 6^m	Far NW-ern off to the Is of Ogasawara.	$\lambda=139.0^\circ$E $\varphi=29.0^\circ$N	200?	D	Not so deep?
58	〃 Apr. 2nd 1^h 04^m	SES-ern off to the coast of the Sima Peninsula	$\lambda=137.0^\circ$E $\varphi=34.2^\circ$N	350	A	I.S.S. 0.045
59	〃 July 27th 3^h 55^m	Neighbourhood of Hikone.	$\lambda=136.4^\circ$E $\varphi=35.4^\circ$N	350	B	〃 0.05
60	1927, Jan. 15th 23^h 31^m	Eastern off to the Is. of Oki.	$\lambda=134.5^\circ$E $\varphi=36.2^\circ$N	420	C	〃 0.06
61	〃 Apr. 8th 3^h 33^m	Etigo or Ettyû district.	$\lambda=138^\circ$E $\varphi=37^\circ$N ?	400?	E	
62	〃 May 18th 6^h 46^m	NW-ern off to Vladivostock.	$\lambda=131.0^\circ$E $\varphi=44.0^\circ$N	500?	B	
63	〃 June 18th 11^h 27^m	Kumano-nada.	$\lambda=138.5^\circ$E $\varphi=33.6^\circ$N	300	C	
64	〃 Aug. 12th 9^h 34^m	Western off to the Is. of Ogasawara.	$\lambda=140.0^\circ$E $\varphi=27.5^\circ$N	500	B	I.S.S. 0.06
65	〃 Aug. 21st 7^h 13^m	SW-ern off to Hatidyô Is.	$\lambda=139.0^\circ$E $\varphi=32.1^\circ$N	300	D	〃 0.05
66	〃 Aug. 24th 14^h 30^m	ditto.	$\lambda=139.0^\circ$E $\varphi=32.0^\circ$N	300	D	〃 0.09
67	〃 Sep. 13th 0^h 29^m	Ensyû-nada	$\lambda=137.9^\circ$E $\varphi=31.2^\circ$N	280	D	
68	〃 Dec. 19th 4^h 50^m	North Japan Sea.	$\lambda=132.0^\circ$E $\varphi=41.5^\circ$N	350	C	
69	1928, Jan. 18th 22^h 11^m	Southern part of Ensyû-nada	$\lambda=137.9^\circ$E $\varphi=33.4^\circ$N	400?	D	
70	〃 Mar. 29th 14^h 07^m	SW-ern off to the Is. of Ogasawara.	$\lambda=138.2^\circ$E $\varphi=31.7^\circ$N	410	A	I.S.S. 0.06
71	〃 May 8th 13^h 46^m	Okhotsk Sea.	$\lambda=149.0^\circ$E $\varphi=50.0^\circ$N	450	A	〃 0.07
72	〃 Aug. 16th 12^h 50^m	NW-ern off to the Is. of Ogasawara.	$\lambda=140.0^\circ$E $\varphi=28.3^\circ$N	450	C	〃 0.07

No.	Time of Occurrence	Epicenter		Depth of Focus	(E)	Notes
73	1928, Aug. 23rd 10^h 18^m	Western part of the Sôya Straits.	$\lambda=140.0°E$ $\varphi=51.0°N$	300?km	B	I.S.S. 0.07
74	〃 Aug. 28th 2^h 59^m	SW-ern off to Hatidyô Is.	$\lambda=139.0°E$ $\varphi=32.5°N$	300	D	After Honda and Takehana.
75	1929, Mar. 17th 21^h 15^m	Sôya Straits.	$\lambda=144.0°E$ $\varphi=48.5°N$	200	C	
76	〃 June 3rd 6^h 39^m	Off the coast of the Sima peninsula	$\lambda=137.5°E$ $\varphi=34.0°N$	320	A	I.S.S. 0.05
77	〃 Sep. 28th 23^h 57^m	Okhotsk Sea.	$\lambda=146.0°E$ $\varphi=47.0°N$	250?	D	Not so deep?
78	1930, Mar. 6th 12^h 32^m	NW-ern off to the Is. of Ogasawara.	$\lambda=139.5°E$ $\varphi=28.7°N$	250	C	After H. & T.
79	〃 Mar. 11th 1^h 27^m	Okhotsk Sea.	$\lambda=148.0°E$ $\varphi=50.0°N$	660	C	After Mi amoto.
80	〃 Sep. 29th 13^h 53^m	Kagosima Bay	$\lambda=130.6°E$ $\varphi=31.6°N$	260?	D	After H. & T.
81	1931, Feb. 20th 14^h 34^m	North Japan Sea.	$\lambda=135.7°E$ $\varphi=44.5°N$	350	A	
82	〃 Mar. 1st 23^h 23^m	Straits of Sôya.	$\lambda=143.7°E$ $\varphi=46.0°N$	250	C	After H. & T.
83	〃 Apr. 21st 9^h 03^m	Middle part of the Japan Sea.	$\lambda=134.2°E$ $\varphi=38.5°N$	350	C	
84	〃 June 2nd 11^h 38^m	Valley of the Masuda River.	$\lambda=137.°30'E$ $\varphi=35°58'N$	240	C	
85	〃 June 30th 1^h 43^m	Kumano-nada	$\lambda=136.8°E$ $\varphi=33.9°N$	350	B	
86	1932, Feb. 3rd 16^h 35^m	NW-ern off to the Is. of Ogasawara.	$\lambda=140.1°E$ $\varphi=29.0°N$	400	B	
87	〃 Apr. 5th 4^h 17^m	Southern off to Hatidyô Is.	$\lambda=139.14°E$ $\varphi=30.54°N$	410	A	After H. & T.
88	〃 Apr. 28th 12^h 43^m	Kumano-nada.	$\lambda=136.77°E$ $\varphi=34.0°N$	320	C	〃
89	〃 May 5th 13^h 11^m	Oosaka-Bay	$\lambda=135.3°E$ $\varphi=34.6°N$	360	C	〃
90	〃 July 25th 17^h 25^m	Neighbourhood of the Biwa Lake	$\lambda=135°52'E$ $\varphi=35°13'N$	360	C	After Takehana.
91	〃 July 27th 9^h 31^m	SWS-ern off to Hatidyô Is.	$\lambda=139.0°E$ $\varphi=31.2°N$	300	C	After H. & T.
	〃 Sep. 2nd 21^h 57^m	Southern off to the Is. of Ogasawara.	—	—	—	
92	〃 Sep. 23rd 23^h 23^m	North Japan Sea.	$\lambda=139.0°E$ $\varphi=44.7°N$	330	A	〃
93	〃 Oct. 2nd 0^h 9^m	NW-ern off to the Is. of Ogasawara.	$\lambda=140.5°E$ $\varphi=29.5°N$	300	C	〃
94	〃 Oct. 14th 21^h 37^m	SW-ern off to Hatidyô Is.	$\lambda=138.8°E$ $\varphi=31.6°N$	300	C	〃
95	〃 Oct. 26th 2^h 3^m	Eastern off to the Nakasiretoko peninsula.	$\lambda=144.6°E$ $\varphi=46.2°N$	370	B	After Miyamoto.

No.	Time of Occurrence	Epicenter		Depth of Focus	(E)	Notes
96	1932, Nov. 13th 13^h 48^m	North Japan Sea.	$\lambda=137.25°E$ $\varphi=43.57°N$	320^{km}	A	After H. & T.
97	〃 Dec. 5th 9^h 20^m	Kumano-nada.	$\lambda=137.0°E$ $\varphi=33.7°N$	350	C	〃
98	1933, Feb. 9th 12^h 57^m	SW-ern off to Hatidyo Is.	$\lambda=138.8°E$ $\varphi=31.7°N$	250	C	〃
99	〃 Mar. 12th 4^h 33^m	Western off to the Is. of Ogasawara	$\lambda=140.2°E$ $\varphi=26.4°N$	300	A	〃
100	〃 May 24th 13^h 36^m	Okhotsk Sea.	$\lambda=145.6°E$ $\varphi=46.7°N$	450	C	〃
101	〃 May 29th 8^h 40^m	SW-ern off to Hatidyô Is.	$\lambda=138.0°E$ $\varphi=32.4°N$	300	D	〃
102	〃 Sep. 3rd 1^h 42^m	Southern off to Hatidyô Is.	$\lambda=139.4°E$ $\varphi=30.3°N$	380	A	
103	〃 Sep. 6th 23^h 05^m	Ensyû-nada	$\lambda=137.8°E$ $\varphi=34.4°N$	250	C	〃
104	〃 Sep. 20th 12^h 57^m	Kumano-nada	$\lambda=136.6°E$ $\varphi=34.1°N$	330	C	〃
105	〃 Nov. 19th 10^h 34^m	SW-ern off to Hatidyô Is.	$\lambda=139.0°E$ $\varphi=32.6°N$	250	E	〃
106	〃 Dec. 5th 4^h 34^m	NE-ern off to the Nakasiretoko Peninsula	$\lambda=144.0°E$ $\varphi=46.6°N$	340	A	After Miyamoto.
107	1934, Apr. 20th 1^h 14^m	Southern off to Hatidyô Is.	$\lambda=139.5°E$ $\varphi=30.0°N$	350	B	

The last part consists of a catalogue containing intermediate earthquakes occurred during 1924～1934 (IV), (Table 2). To know an accurate depth of an intermediate earthquake is very difficult when its epicenter lies far off the coast, and therefore some of such earthquakes may be probably omitted mistaking as shallow-focus ones. Earthquakes having a focal depth of about 80 or 90 km. are not here included though they are by no means regarded to belong to the same kind of earthquake as that of ordinary shallow focus ones.

Table 2. Catalogue of Intermediate Earthquakes.
1924～1934 (IV). (200 km. $>H\geqq$100 km.)

No.	Time of Occurrence	Epicenter		Depth of Focus	(E)	Notes
1	1924, Jan. 27th 3^h 15^m	Bôsô-Peninsula	$\lambda=140°6.7'E$ $\varphi=35°19.6'N$	140^{km}	E	by Eqk. Res. Ins. Tôkyô.
2	〃 July 1st 0^h 40^m	Neighbourhood of Etrô Is.	$\lambda=147.6°E$ $\varphi=44.7°N$	about 160	A	
3	1925, Jan. 28th 13^h 6^m	Eastern off to Nemuro.	$\lambda=147.2°E$ $\varphi=43.2°N$	about 120	A	
4	1926, June 5th 18^h 10^m	Off the coast of the Satuma Peninsula	$\lambda=130.0E$ $\varphi=31.5°N$	160	B	I.S.S. 0.035

No.	Time of Occurrence	Epicenter		Depth of Focus	(E)	Notes
5	1926, June 29th 23^h 26^m	NW-ern off to the Is. of Okinawa.	λ=126.8°E φ=27.3°N	140?^{km}	A	I.S.S. 0.02
6	1927, Nov. 11th 4^h 48^m	Valley of the Hime River, Nagano Pref.	λ=137.9°E φ=36.6°N	190	D	
7	〃 Nov. 18th 5^h 22^m	Sagami-Bay	λ=139°19.9′E φ=35°18.6′N	110	D	by E. R. I. T.
8	〃 Dec. 10th 11^h 45^m	Western part to Yonezawa.	λ=139.8°E φ=37.8°N	160	D	
9	〃 Dec. 31st 14^h 50^m	Upper valley of the Ara-river, Saitama Pref.	λ=139.2°E φ=36.1°N	140	D	
10	1929, Jan. 13th 9^h 3^m	Neighb. of Kuril Is.	λ=1[illegible]4.8°E φ=49.7°N	120	A	I.S.S. 0.015
11	〃 June 4th 12^h 39^m	Neighbourhood of Mt. Hakone.	λ=139°8.5′E φ=35°14.9′N	120	E	by E. R. I. T.
12	〃 Oct. 6th 4^h 1^m	Kunisiri Is.	λ=146.0°E? φ=44.2°N ?	150?	C	
13	〃 Oct. 10th 4^h 45^m	Neighb. of Mt. Aso.	λ=131.1°E φ=32.9°N	120	D	
14	1930, Apr. 18th 11^h 20^m	Valley of the Tama-river, Kanagawa Pref.	λ=139°30.0′E φ=35°32.2′N	115	E	by E. R. I. T.
15	〃 Aug. 30th 5^h 2^m	SE-ern off to Kunisiri Is.	λ=146.5°E φ=44.2°N	150	C	After Miyamoto.
16	〃 Dec. 21st 23^h 51^m	SE-ern off to Formosa	λ=122.2°E φ=20.4°N	150	A	
17	1931, Jan. 6th 12^h 23^m	Neighbourhood of Ura kawa, Erimo Peninsula.	λ=142.8°E φ=42.4°N	100	C	After Oka.
18	〃 Jan. 9th 10^h 46^m	Neighbourhood of the Tazawa Lake, Akita Pref.	λ=140.6°E φ=39.8°N	130	C	
19	1931, Nov. 12th 22^h 11^m	Neighbourhood of Oosima Is.	λ=139.4°E φ=34.8°N	100	D	
20	1932, Feb. 19th 22^h 25^m	Neighb. of Hatidyô Is.	λ=140.3°E φ=32.9°N	150	C	
21	〃 Aug. 21st 18^h 04^m	Neighb. of Kumagaya.	λ=139.4°E φ=36.2°N	120	E	
22	〃 Nov. 18th 5^h 11^m	Southern off to the Satuma Peninsula.	λ=130.4°E φ=31.1°N	100	C	
23	〃 Dec. 27th 6^h 15^m	Western off to the coast of Naha.	λ=125.5°E φ=23.0°N	about 150?	B	
24	1933, Mar. 19th 0^h 51^m	Southern off to Hatidyô Is.	λ=140.0°E φ=32.2°N	120	B	After H. & T.
25	〃 Apr. 28th 10^h 31^m	Valley of the Ara-river, Saitama Pref.	λ=139.4°E φ=36.1°N	about 100	D	
26	〃 Oct. 2nd 4^h 24^m	Neighb. of Mt. Aso.	λ=131.4°E φ=32.9°E	〃	D	
27	1934, Feb. 1st 9^h 16^m	Neighb. of Ootahara, coast of the Sagami-Bay.	λ=139°20′E φ=35°20′N	〃	C	
28	〃 Feb. 24th 15^h 24^m	SES-ern off to the Is. Ogasawara.	λ=145.0°E φ=21.0°N	150?	A	a little deeper?

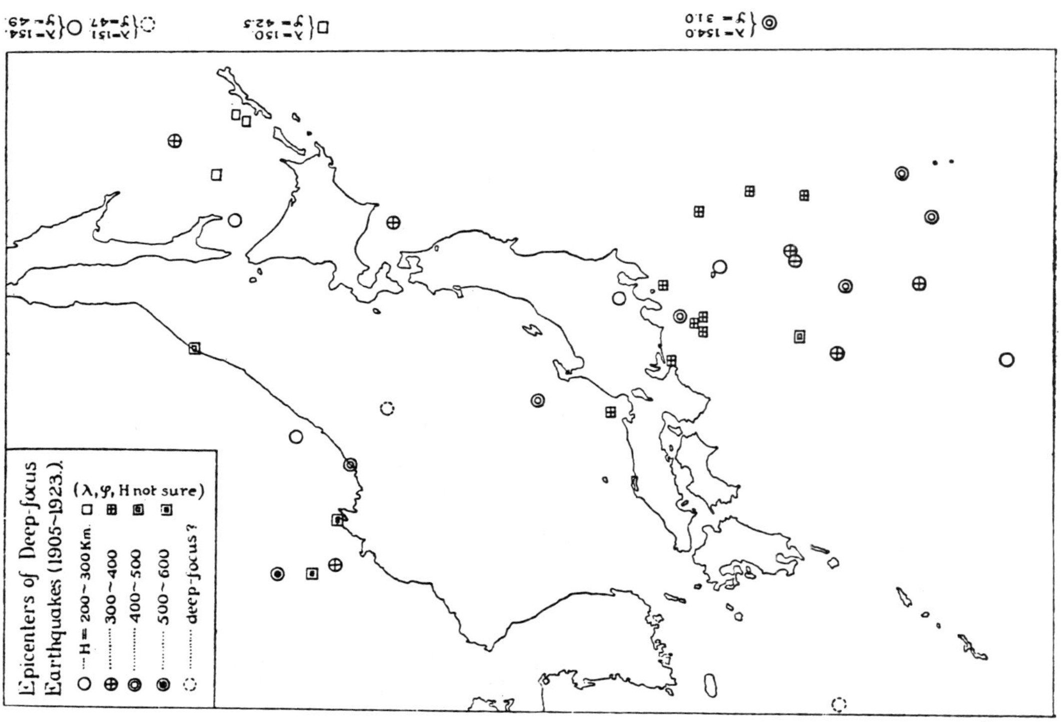
Epicenters of Deep-focus Earthquakes (1905~1923.)
○ ---H = 200~300 Km.
⊕ .... 300~400
◎ .... 400~500
◉ .... 500~600
◌ .... deep-focus?
(λ, φ, H not sure)

Fig. 1.

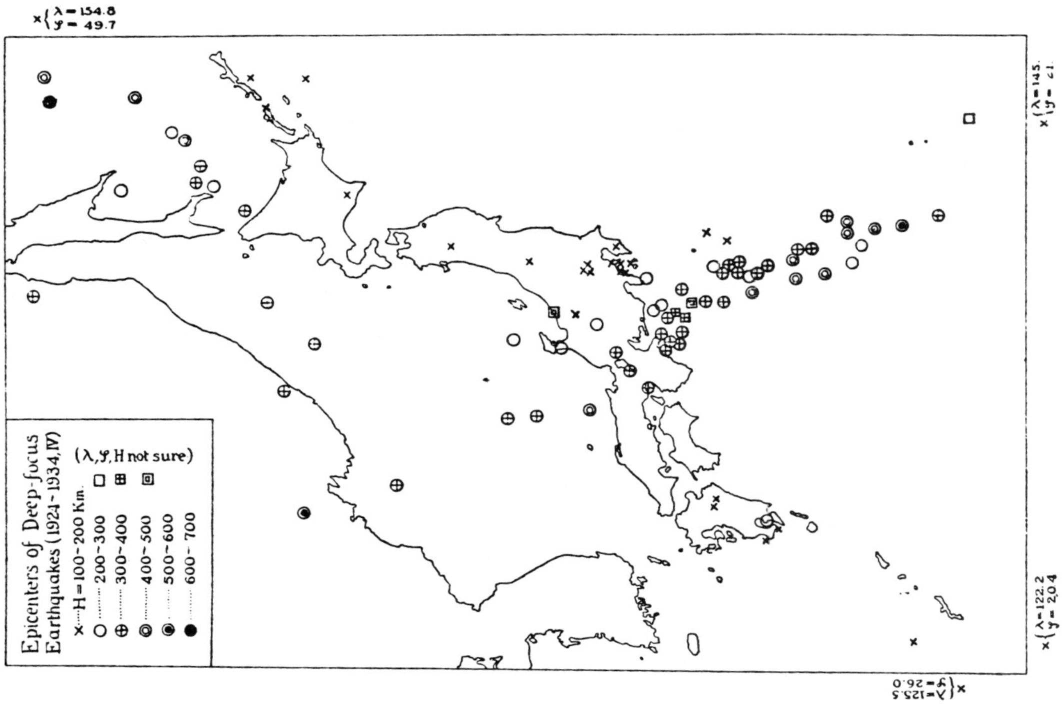
Epicenters of Deep-focus Earthquakes (1924~1934, IV)
× ---H = 100~200 Km.
○ .... 200~300
⊕ .... 300~400
◎ .... 400~500
◉ .... 500~600
● .... 600~700
(λ, φ, H not sure)
λ = 154.8
φ = 49.7

Fig. 2.

3. *Remarks.*

1° Illustration of the Distribution of Deep-focus Earthquakes.

The positions of epicenters of deep-focus earthquakes are exhibited in the map of Fig. 1 and Fig. 2. These epicenters are indicated by different marks respectively according to their depth of focus. In Fig. 1 are shown those given in Table 1 (A) and in Fig. 2 those in Table 1 (B).

2° Illustration of the Distribution of Intermediate Earthquakes.

The positions of epicenters of intermediate earthquakes are shown in Fig. 2, the same map containing those of deep-focus earthquakes occurred in the same period.

3° Observational Materials.

For the present investigation observational materials[(1)] are taken from following publications.

1. Annual Report of the Central Meteorological Observatory of Japan, Tôkyô. (Seismological Part) 1905∼19 , in Japanese.
2. Monthly Geophysical Review (氣象要覽) and Seismological Magazine (驗震時報). published by the C. M. O. of Japan, Tôkyô.
3. Seismological Reports of Meteorological and Seismological Observatories of our and foreign countries.
4. Publication of the Imperial Earthquake Investigation Committee.
5. Seismological Report of the Earthquake Resarch Institute, Tôkyô.
6. International Seismological Summary.
7. Papers on deep-focus earthquakes and etc.[(2)]

4° The method of Investigation.

It must be unnecessary to mention here precisely about the method used for the present investigation to determine the position of epicenter and the focal depth of earthquakes. First a rough estimation of the focal depth was done examining the figure of seismogram and the distribution of seismic intensity. In old observations made at distant places *iP*, *iS* and poor surface waves are always found in the case of deep-focus earthquakes. Now and then some phases denoted by *i* are reported to appear after the phase *iP*, and it

(1) Principal data are exhibited in the catalogue written in Japanese loc. cit.

(2) Old seismograms recorded at Tukuba are specially investigated by the author for this object. He thanks very much to Mr. Takayama and Mr. Oki for their kindness in giving him facilities for it.

proves that these are sometimes nothing but the phases *pP* or *sP* etc. which are now generally known. Of course these phases are particularly useful for the depth determination. If the focal depth can be estimated in any way, then the epicenter is easily located. For the present purpose, the tables[1] of travel times obtained by us were mainly used.

5° *Time of Occurrence.*

The time of occurrence[2] is given in the second column of the tables. This shows a rough time when the first wave of that earthquake arrived to the nearest station. Since the position of focus is known, the accurate time when the disturbance first occurred at the focus may be calculated in an order of second by the aid of the travel time tables though it is not easy. This is, however, not done in the present case.

6° *Epicenter and Focal Depth.*

To represent the positions of epicenters as well as the focal depths of these deep-focus earthquakes is a principal object of this catalogue. These are examined anew for all earthquakes and there are given in Table 1 and 2 the values considered by the author as the most reliable. But it must be noticed that those of Table 1, i. e. the values for earthquakes occurred before the year 1923, could not be so accurately obtained in most cases.[3]

7° *Largeness of Earthquakes (E).*

The forth column in the catalogues shows a rough estimation concerning the largeness of earthquakes which is denoted by (E). It is classified into five, A, B, C, D and E according to the area of seismic disturbance. A and B are used for earthquakes whose waves were widely observed by seismographs installed at all stations of the world, but the former stands for a particularly remarkable one. C shows such an earthquake that its waves were only observed in the region not so far from the Japanese Islands. While D and E are used for comparatively small earthquakes whose waves were only observed in a con-

(1) Wadati and others: "On the Travel Time of Earthquake Waves" Part I. Geophysical Magazine Vol. 7 No. 1 and Part V. Vol. 7 No. 3, Part VI. Vol. 8 No. 2.

(2) In this investigation times are all refered to the Central Standard Time of Japan (135°E).

(3) For example, the deep-focus earthquake of No. 18 (1915, III 18, H=360 km.) is obtained to have its epicenter at λ=142.5°E, φ=41.5°N, where is southern off to the coast of the Erimo Peninsula, Hokkaidô. While since then there has never been observed such a deep-focus one in this region. It is supposed that the epicenter of this earthquake should be located in the northern part of the Japan Sea.

fined part of our country and the latter for particularly small one. Of course, this classification merely shows a rough order of magnitude of seismic energy issued from the origin of respective earthquakes.

4. *Distribution of Deep-focus and Intermediate Earthquakes near the Japanese Islands.*

1° Deep-focus Earthquakes.

The general appearance of the distribution of deep-focus earthquakes can be seen in Fig. 2 and also in Fig. 1 though it is not so regularly observed in the latter. First of all, epicenters of most deep-focus earthquakes lie quite regularly along two zones which are intersected perpendicularly with each other near Vladivostock. The one is named as *traversing deep-focus earthquake zone* and the other as *Sôya deep-earthquake zone.* It is the former one that the author first found in the investigation of deep-focus earthquakes in this region, which extends over more than 2000 km. from the neighbourhood of Vladivostock to the Islands of Ogasawara, traversing the Japan Proper almost perpendicularly through its narrow part. And the other one lies in the northern part of the Japan Sea and extends to the north-western direction along the Tisima (Kuril) Islands, passing through the Straits of Sôya. These two zones are also seen by the distribution of epicenter shown in Fig. 1, but not so regularly.

The total number of deep-focus earthquakes shown in Fig. 1 is not so many as that shown in Fig. 2, though the period of statistics is rather longer in the former case. This is due to the omission of deep-focus ones when they were found, since it is very difficult to find them from the observational data of old days.

Besides these two deep-focus earthquake zones, there exists another one along the Ryûkyû Islands starting from the middle part of Kyûsyû Island. This is named as *Kyûsyû deep-focus earthquake zone.* Fig. 3 represents these three zones and also contains some geographical names which are used in this investigation.

2° Intermediate Earthquakes.

As was already noticed in previous investigations, it is very remarkable that the distribution of intermediate earthquakes is quite different from those of deep as well as shallow-focus ones, because earthquakes of this kind are apt to occur in inland places in general. It will be, however, described in

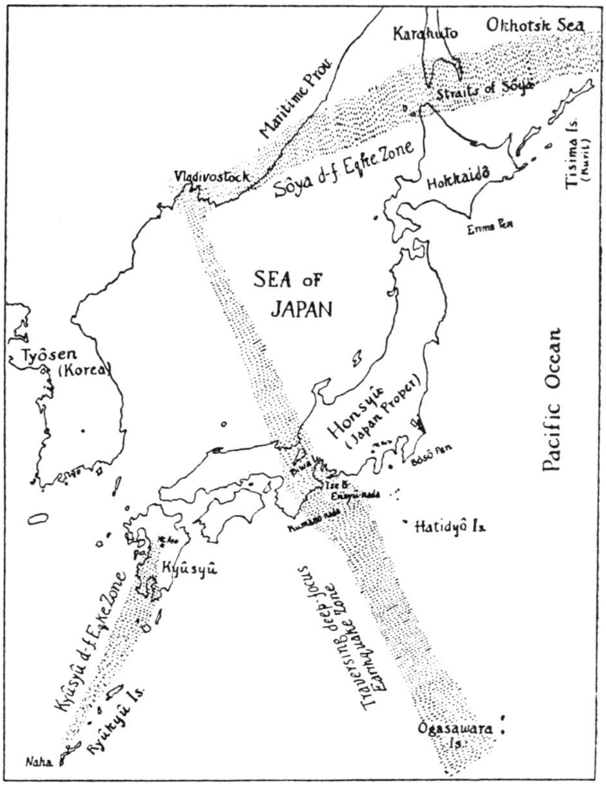

Fig. 3. Deep-focus Earthquake Zones near the Japanese Islands.

the later part that the distribution of intermediate earthquakes has an intimate connection with those of deep-focus ones and moreover with that of active volcanoes in this region.

By the way it is noticeable that there scarcely occurred such an extraordinarily large intermediate earthquake in recent years as was sometimes experienced in the cases of shallow as well as deep-focus earthquakes.

Part II. Some Statistics concerning the Seismic Activity.

5. By the results shown in Tables 1 and 2, some statistics were taken to see the general state of seismic activity in the region near the Japanese Islands in the recent ages. For the sake of comparison are used in this investigation similar results[1] already obtained for shallow-focus earthquakes.

(1) These are taken from the seismological section given in the "Monthly Geophysical Review" published by C. M. O. of Japan.

6. Frequency Distribution of Focal Depth.

It must be interesting to know the frequency distribution of focal depth by the results shown in the tables now obtained. This is tabulated in Table 3, specially taking into account of the largeness of earthquake.

Table 3. Number of deep-focus and intermediate earthquakes occurred during 1905~1934, IV.

Focal Depth	Number of Earthquakes					
	A	B	C	D	E	Total
H>600 km.	—	—	1	—	—	1
500~599	3	3	1	—	—	7
400~499	5	3	9	—	—	21
300~399	10	9	18	12	—	49
200~299	3	5	10	6	1	25
unknown	—	1	—	3	—	4
Total	21 (13)	21 (10)	39 (26)	24 (14)	2 (2)	107 (65)
100~199 km.	(6)	(3)	(7)	(7)	(4)	(27)

* Figures put in parentheses show numbers of earthquakes occurred during 1924~1934, IV.

We can see in this table that those having a focal depth of 300~400 km. take place most frequently. Thus in the region under consideration 300~400 km. is considered as the most usual depth of focus for deep-focus earthquakes, while there occur rather seldom at such a depth of about 200 km. This tendency can be also noticed in Dr. Conrad's investigation(1) on deep-focus earthquakes in the world, though his result is not entirely coincident with that of Table 3 on the point of their absolute values of focal depths.

There are 8 earthquakes in the table whose foci lie at so deep place as more than 500 km., and the deepest of them even attains to 660 km.(2) This figure must be astonishing for the focal depth of an earthquake.

In the investigation of earthquakes occurred in old times, there must be many omissions of deep-focus ones having a small energy. Thus the frequency

(1) Gerl. Beit. Geo. Bd. 40. S 113. His investigation is based on the data given in the I.S.S. and then the focal depth of so-called normal earthquakes is assumed as 100 km. This makes the focal depths very large compared with those obtained by us.

(2) No. 79, occurred in the Okhotsk Sea.

distribution according to their larginess of earthquake shown in Table 4 cannot be directly accepted. We can only see its general tendeney by the figures put in parentheses since they are obtained by recent earthquakes only.

The number of shallow-focus earthquakes is of course far larger than those of deep-focus and intermediate ones. But if we take large earthquakes only and compare their frequencies with one another, they are comparable. The statistics taken during 1923~1934, IV are as follows.

Deep-focus earthquakes	(A, B) ..	42
	(C, D, E)	65
Intermediate earthquakes	(A, B)	9
	(C, D, E)	18
Shallow-focus earthquakes	(large[1])	152
	(small)	numerous.

7. *Geographical Distribution of Deep-focus Earthquakes.*

The number of occurrence of deep-focus earthquakes are geographically distributed as follows:

Traversing deep-focus earthquake zone.	Neighbourhood of the Ogasawara Is. (South of 30°N)	15
	Neighbourhood of Hatidyô Is. (South of 34°N, East of 138°E)	27
	The Ensyû-nada, The Ise Bay.	11
	Neighbourhood of the Lake Biwa.	5
	Southern part of the Japan Sea	8
	Neighbourhood of Vladivostock	9
Sôya deep-focus earthquake zone.	North-eastern part of the Japan Sea	4
	The Sôya Straits	12
	The Okhotsk Sea	5
	Others	8

8. *Annual Distribution of Frequency of Deep-focus Earthquakes.*

It was already investigated by Conrad[2] that the frequency of deep-focus earthquakes shows a special distribution in a year. In the present investiga-

(1) We take all shallow-focus earthquakes which belong to "remarkable" earthquakes (顯著地震) in the Monthly Geophysical Review.

(2) loc. cit.

tion also the same character is observed,[1] as we can see clearly in Table 5 that the frequency curve of deep-focus earthquakes has a maximum in May and another small maximum in October. Generally speaking, they are liable to occur frequently in spring while comparatively rare in winter. Statistics of this kind were occasionally taken for shallow-focus earthquakes by some authors but we cannot see in these cases such a remarkable annual distribution generally as is obtained for deep-focus earthquakes.

Table 5.
Annual distribution of frequency of deep-focus earthquakes. (1905~1933)

Month	frequency	%
Jan.	7	6.5
Feb.	8	7.5
Mar.	16	15.0
Apr.	15	14.0
May	11	10.3
June	10	9.3
July	8	7.5
Aug.	7	6.5
Sep.	10	9.3
Oct.	7	6.5
Nov.	3	2.8
Dec.	5	4.7

Part III. On the Distribution of Seismic Foci and the Volcanic Activity.

9. *Lines of Equal Depth of Earthquake Foci.* As is known by Fig. 2, epicenters of deep-focus earthquakes lie very regularly along two zones respectively. But if we examine precisely their focal depths one by one, we must notice that the focal depths are also distributed in a regular manner. Thus it is possible to draw *lines of equal focal depths (contour lines)* for deep-focus earthquakes occurred in recent years in the region near Japan. Further it is interesting that foci of intermediate earthquakes, whose distribution seems to have no relation to that of deep-focus ones at first sight in Fig. 2, are also quite regularly distributed if we draw contour lines by them.

Thus contour lines can be drawn for depths of every 50 km. by deep-focus as well as intermediate earthquakes and they are shown in Fig. 4. Earthquakes used for this object are those of recent years given in the catalogue. Strictly speaking, some earthquakes show abnormal focal depths to draw contour lines finely, but it may be allowable if we remember that positions of epicenter and focal depths of some earthquakes are not always obtainable so accurately. At any rate, the fact that contour lines can be drawn in general by earthquake foci is itself important.

(1) Of course, Conrad's investigation contains many of deep-focus earthquakes given in the present catalogue. But this tendency of annual frequency distribution can be observed for deep-focus earthquakes occurred in another region.

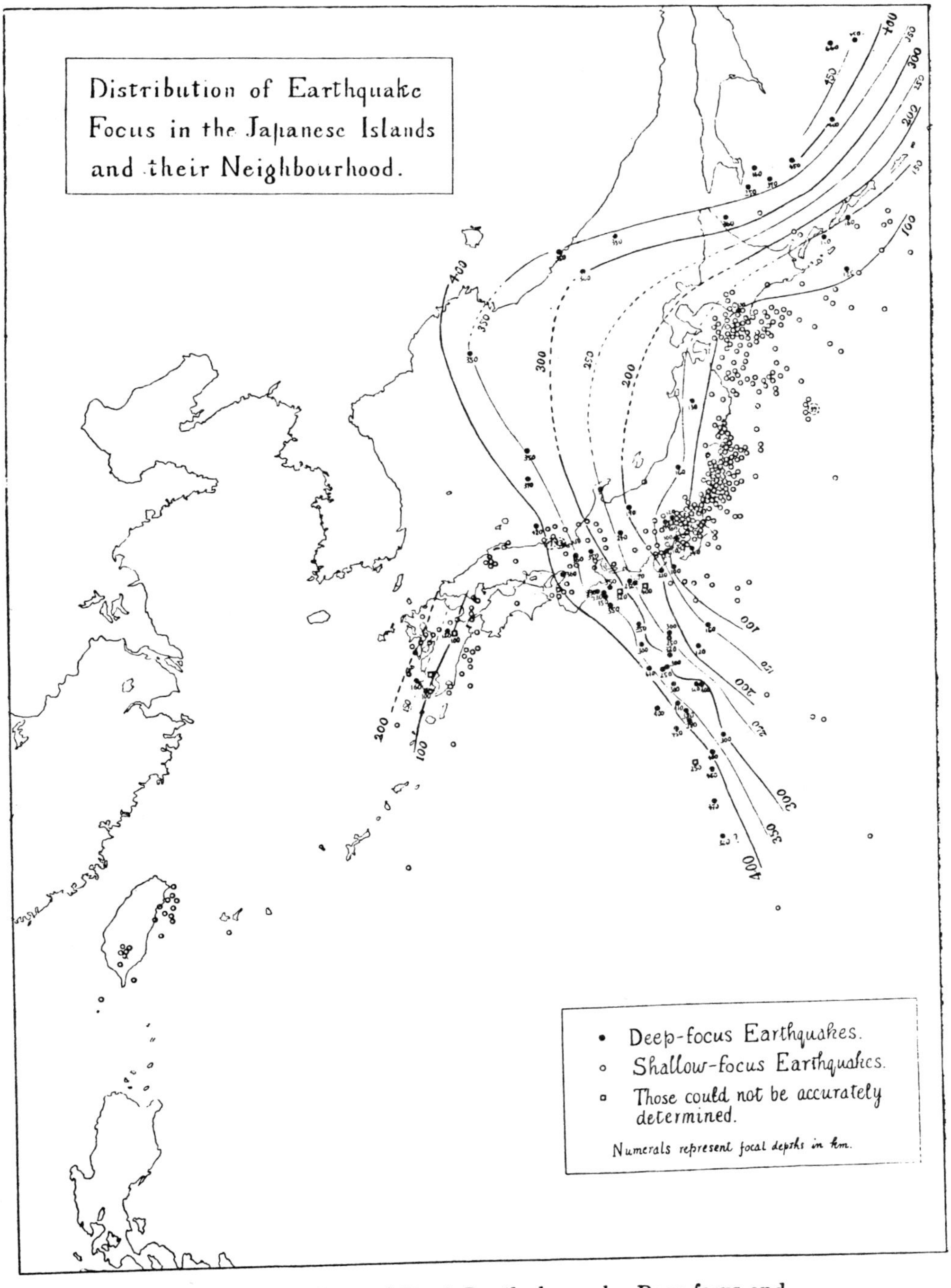

Fig. 4. Lines of Equal Focal Depth drawn by Deep-focus and Intermediate Earthquakes.

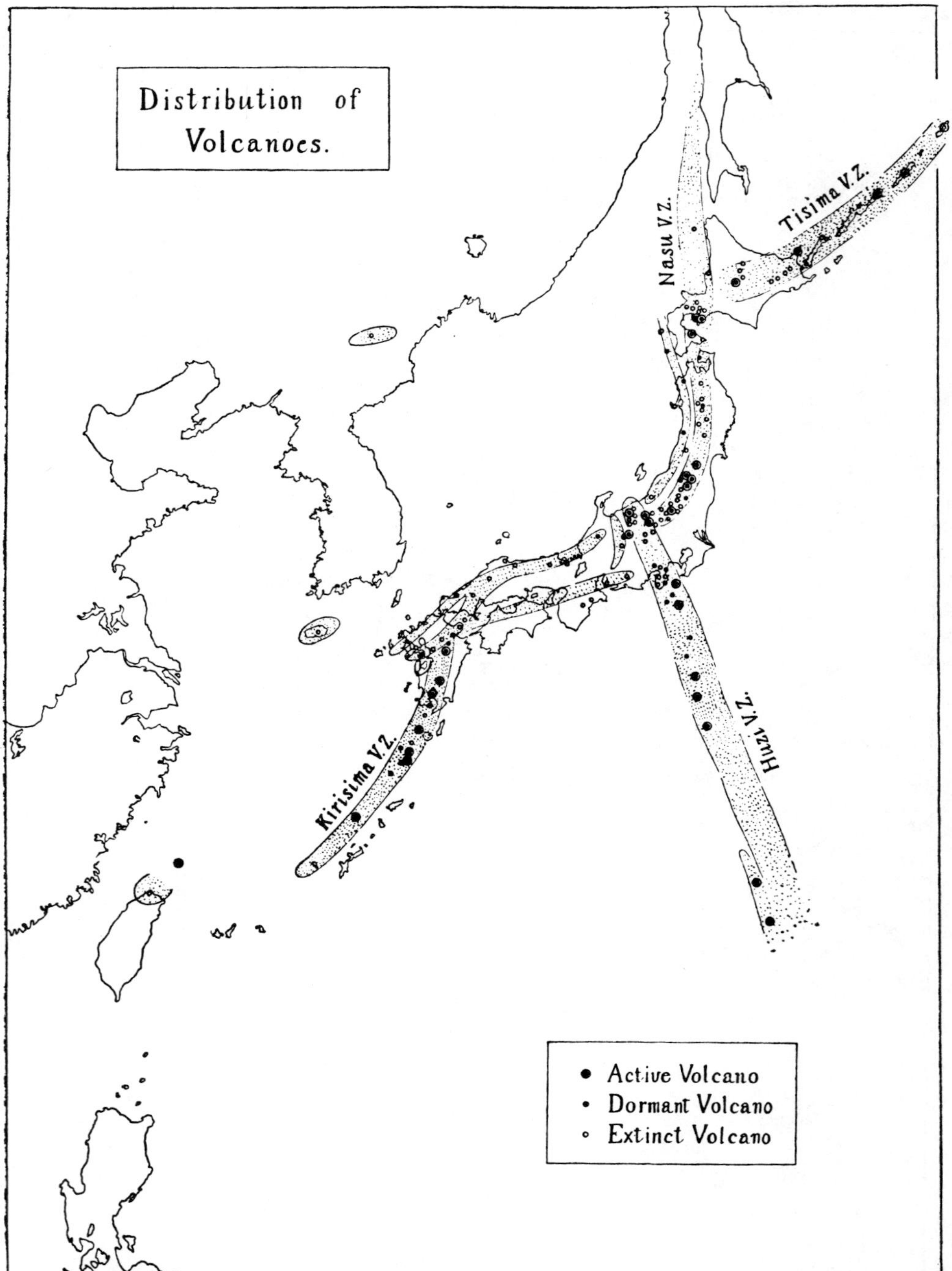

Fig. 5. Distribution of Volcanoes and Volcanic Chains in Japan.

In Fig. 4 are also shown ·epicenters of large shallow-focus earthquakes occurred during 1924~1933. Though the contour lines were drawn neglecting the existence of these shallow-focus earthquakes, most of them seem to have some relations to the contour lines. From this view-point we can know that there are different kinds in shallow-focus earthquaks, but this problem is left to another investigation.

10. The contour lines representing the focal depth of more than 300 km. run naturally along the deep-focus zones and those of the larger depth situate on the western side in the "traversing deep-focus earthquake zone" and on the northern side in the "Sôya zone". As we can imagine that these two remarkable deep-focus earthquake zones are connected in the neighbourhood of Vladivostock and make a single winding zone, these contour lines are so drawn as to lie those of the larger depth on the outer side.

Some contour lines are also drawn in Fig. 4 along the Kyûsyû deep-focus earthquake zone. But since there occurred only one deep-focus and few intermediate earthquakes in this period, and moreover the former was a very small shock, it seems to be rather unreasonable to drawn contour lines in such a way as shown in Fig. 4. It is only done by a supposition which can be presumed by the distribution of contour lines obtained in other regions.

11. *Relation between the Distribution of Contour Lines and Volcanoes.*

The relation between the activities of earthquakes and volcanoes has been considered since long ago, but a definite conclusion seems to have not yet been given. Indeed, earthquakes occur frequently where there are many volcanoes. This tendency is true when it is examined in a very wide region, for example such as numerous volcanoes and earthquakes which are found along the coast of the Pacific Ocean. It is, however, not always true if we examine precisely in a comparatively small region. In the neighbourhood of Japanese Islands rather the reverse has been considered as the case.

In Fig. 5 are shown all eminent volcanoes in Japan. It is true that earthquakes[1] occurred rather seldom in the neighbourhood of the volcanoes as is seen by Fig. 4 and Fig. 5, *so far as the discussion be made about all earthquakes in general.* But if we take intermediate earthquakes only, it is not so. Intermediate earthquakes occurred now and then just near volcanoes

(1) Of course, so called volcanic earthquakes are excepted, which are ordinarily of small shocks and considered to be caused by the volcanic action.

as if there exist some relations between them.

12. It was already noticed that the traversing deep-focus earthquake zone runs quite parallel to the famous volcanic chain of Huzi (Fuji). And now the Sôya deep-focus earthquake zone is obtained to be parallel to the volcanic chain of Nasu in its north-eastern part along the Tisima (Kuril) Islands.

Now comparing Fig. 4 and 5 we can find that contour lines and the distribution of *active and dormant* volcanoes have a close connection with each other. This is clearly seen because volcanoes shown in Fig. 5. are classified into three kinds, active, dormant and dead volcanoes, and there are given by different marks respectively. Wherein "active" is used for those erupted in the recent about 60 years, while "dormant" for those having a history of eruptions in old ages.

Generally speaking, the possibility of drawing contour lines of the focal depth suggests that there exists in the crust something like *a weak surface* at where the earthquake outburst is liable to occur. This surface extends slopewise in the crust near the Japanese Islands having some connections with the volcanic activity. Contour lines corresponding to larger depths lie on one side of the active volcanic chains and those of small depth on the other side. It is interesting that the volcanic activity seems to have no connection with the occurrence of shallow-focus earthquakes, but closely connected with that of earthquakes having rather deep foci. Thus we can probably say that the volcanic and seismic activities are of the same phenomenon so far as they are considered at deep place in the crust.

The mechanism of earthquake outburst at the focus has been precisely investigated by seismologists from the initial movements of earthquakes observed near the epicenter, especially for deep-focus ones. The contour lines now obtained are considered to play an important rôle for the mechanism of occurrence of deep-focus earthquakes. Indeed recent investigation made by H. Honda and M. Takehana(1) shows that almost all deep-focus earthquakes occurred near Japan are caused in like manner by a certain state of stress stored in the crust. This seems to be favourable to the present result, but we cannot give a concluded discussion about this problem at present until more sufficient materials are prepared.

---

(1) H. Honda and M. Takehana: Seismological Magazine Vol. 8 No. 1 (In Japanese.)

14. Generally speaking, volcanoes are found in chain running nearly parallel to the continental margin and earthquakes occur also in a zone along a volcanic chain. But deep-focus earthquakes are apt to take place on one side nearer to the continent and shallow focus ones on the other side where is in most cases bordered on a very deep sea. This tendency seems to be observable in many volcanic regions in the world. Of course, we cannot say decisively but if the theory of continental drift suggested by A. Wegener be true, we may perhaps be able to see its traces of the continental displacement in the neighbourhood of Japan consulting the figures of contour lines as well as the distribution of volcanoes shown in Fig. 4 and Fig. 5 respectively.

In conclusion, author wishes to express his sincere thanks to the directors of the meteorological observatories of our country for their kindness to give him seismological data of old times, and also he does the same to Miss. Masuda for her assistance in making the catalogue of deep-focus earthquakes.

September 1934, at the Central Meteorological
Observatory of Japan, Tôkyô.

When this paper was in the press, an investigation of Mr. Honda (This Magazine Vol. VIII, No. 2 pp. 179~186) was published, which dealt with the similar problem about the geographical distribution of deep-focus earthquakes.

11

Reprinted from *Geol. Soc. America Bull.* **60:**1915 (1949), courtesy of the Geological Society of America

STUDIES OF OCEAN-BOTTOM STRUCTURE OFF SOUTHERN CALIFORNIA WITH EXPLOSIVE WAVES

Russell W. Raitt
University of California, Marine Physical Laboratory, San Diego, Calif.

Beginning in March 1948 seismic reflection and refraction techniques have been used to study the sediment thickness and basement velocity of the Pacific Ocean within an area between Point Eugenia in the south to Santa Cruz Island in the north and extending 800 nautical miles west of San Diego.

TNT bombs of 1 to 4 pounds for the reflection studies and ranging up to 100 pounds for the refraction studies are used as sound sources. Reflected and refracted waves are received with a pressure-actuated crystal hydrophone. Optimum results are achieved when source and receiver depths are equal to one-quarter wave length of sound of the explosion bubble pulse frequency.

In a few exceptional areas, strong echoes were observed which are believed to originate at the top of the crystalline basement beneath the sediment. However, greatest success was achieved by refraction studies.

Beyond the continental slope, in water of the order of 2000 fathoms depth, the sediment thickness was of the order of a few hundred meters, and the basement velocity was approximately 6.5 km sec^{-1}. One reversed profile of 60 km length indicates that the velocity increases to 8 km sec^{-1} at a depth of about 5 km beneath the basement surface. A strong second arrival has a velocity of 3.6 km sec^{-1} and may be a basement shear wave.

Inside the continental slope the greatest sediment thickness observed was about 3 km in Santa Cruz Basin, whose water depth is 1050 fathoms. Basement velocities range from about 4.5 km sec^{-1} near land, where metamorphic rocks outcrop, to 6.3 km sec^{-1} beneath Santa Cruz Basin.

12

Reprinted from *Seismol. Soc. America Bull.* **40:**233–242 (1950)

SEISMIC REFRACTION MEASUREMENTS IN THE ATLANTIC OCEAN BASIN*

(PART ONE)

By MAURICE EWING, J. L. WORZEL, J. B. HERSEY, FRANK PRESS, and G. R. HAMILTON

ABSTRACT

A REVERSED seismic refraction measurement was made 120 miles northwest of Bermuda (400 miles east of Cape Hatteras) in 2,800 fathoms of water. A velocity of 24,800 feet per second (7.58 km/sec.) for the second layer was identified with the ultrabasic layer of earthquake seismology. Assuming a velocity of 5,600 feet per second (1.70 km/sec.), clearly indicated by earlier measurements, a thickness of 4,500 feet was obtained for the sedimentary layer. The granitic and intermediate layers were absent.

I. INTRODUCTION

A RECENT passage of the research vessels ATLANTIS and CARYN from Woods Hole to Bermuda gave the first opportunity to use two ships in an offshore seismic refraction profile, all previous work of this kind having been done with

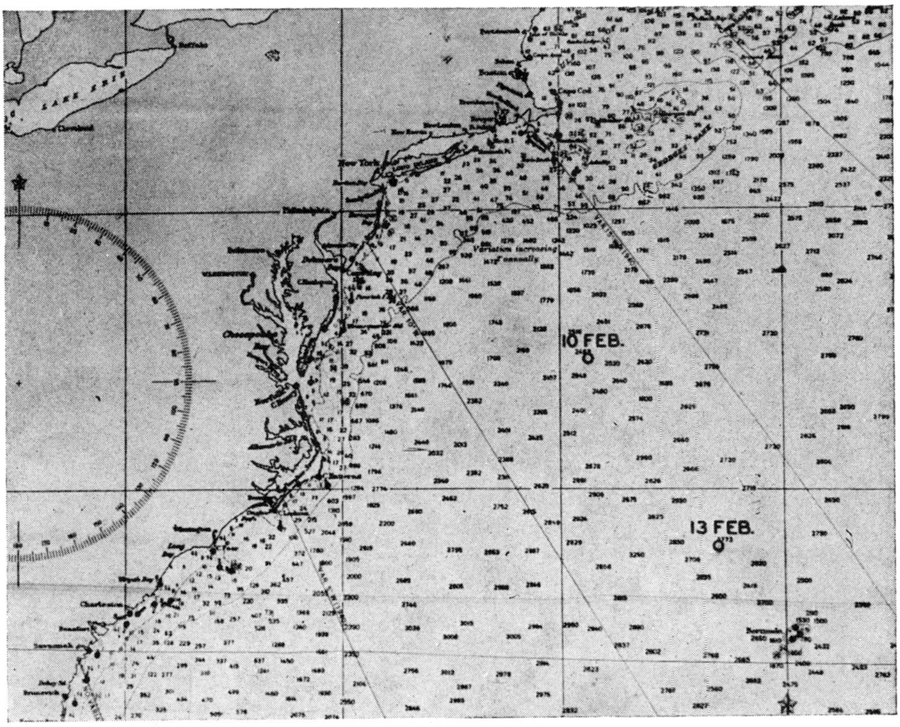

Fig. 1. Station locations.

* Manuscript received for publication July 22, 1949.

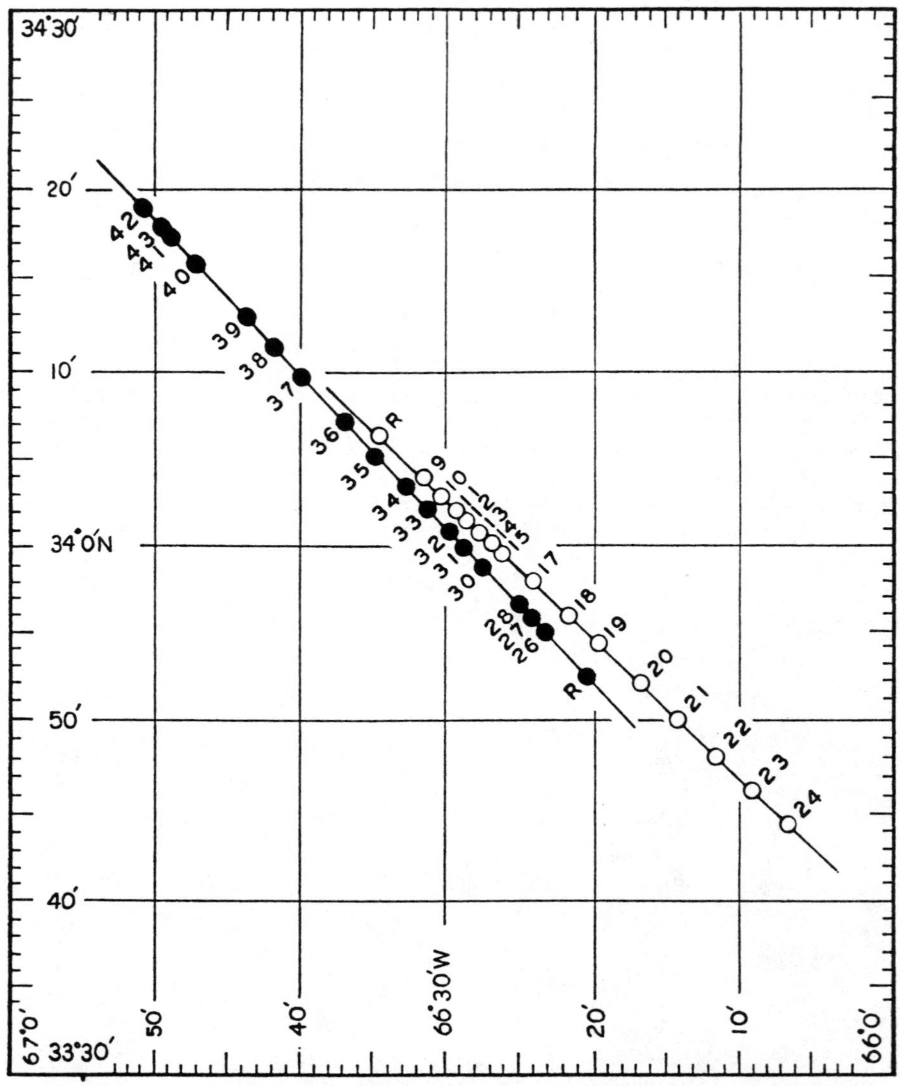

Fig. 2. Station chart, February 13, 1949. Legend: ○, Profile I; ●, Profile II; ●R, receiving vessel; ○R, receiving vessel.

a single ship and a twenty-foot whaleboat. Time was allotted for two reversed profiles, but stormy weather caused cancellation of the first one after shots up to seven miles distance had been made at latitude 37° 23′ N and longitude 69° 23′ W on February 10, 1949. On February 13, in an area about 125 miles northwest of Bermuda, in 2,800 fathoms, at latitude 34° N and longitude 66° 30′ W (fig. 1), a complete reversed refraction profile was made.

II. Acknowledgments

This work was done as part of a longer cruise during February and March, 1949, by the research vessels Atlantis and Caryn of the Woods Hole Oceanographic Institution under its contract NObs-2083 with the Navy Department, Bureau of Ships. Additional equipment, material, and personnel for this part of the cruise were supplied by the following agencies: Penrose Fund of the Geological Society of America; Office of Naval Research, contract M6-onr-271; and the Geophysical Research Directorate, Cambridge Field Station, A.M.C., U. S. Air Force, Contract W-28-099-ac-396.

III. Method of Measurement

Figure 2 shows the locations of shots 9 to 24 made by the Atlantis while running to about 35 miles on course 140° T from the first receiving position of the Caryn. This constitutes Profile I. Profile II, consisting of shots 26 to 43, was made with the Caryn about 18 miles 140° T from her first position and with the Atlantis running about 35 miles on course 310° T. The charges ranged up to 300 lbs., and were all fired under way. The operation required 11 hours, and produced the most complete refraction profile beyond the continental margins known to us.

The travel time of the seismic signals was determined by recording on the seismogram a radio signal sent from the shot point at the instant of the explosion. As a safeguard against missed radio signals, events at both vessels were related to absolute time by the aid of WWV time signals. Loran coverage was good and the positions of both vessels were checked frequently by Loran. The actual distance between vessels for each shot was determined from the travel time of the direct sound transmitted through the water.

The seismic detector was a hydrophone floated at a depth of about 50 feet about 100 feet from the Caryn. The amplifying and recording system had several channels for recording both the high-frequency water waves and the lower-frequency ground waves.

IV. Results

Figure 3 shows the seismograms for shots 11, 12, 13, 15, and 20. The ground waves are marked with G, the direct water waves by D, and the first reflection R. The numerals represent travel time in seconds. The first two traces on shot 11 are a high-frequency rectified channel, the third and fourth are dead, the fifth is an intermediate-frequency channel, and the sixth to ninth are low-frequency channels. It will be noted that the first four traces are covered, to conserve space, on all other shots, and that the paper speed was reduced by a factor of about 4 on shot 20. This series of records illustrates the many characteristic features of the phases R, G, and D, which distinguish them, and also

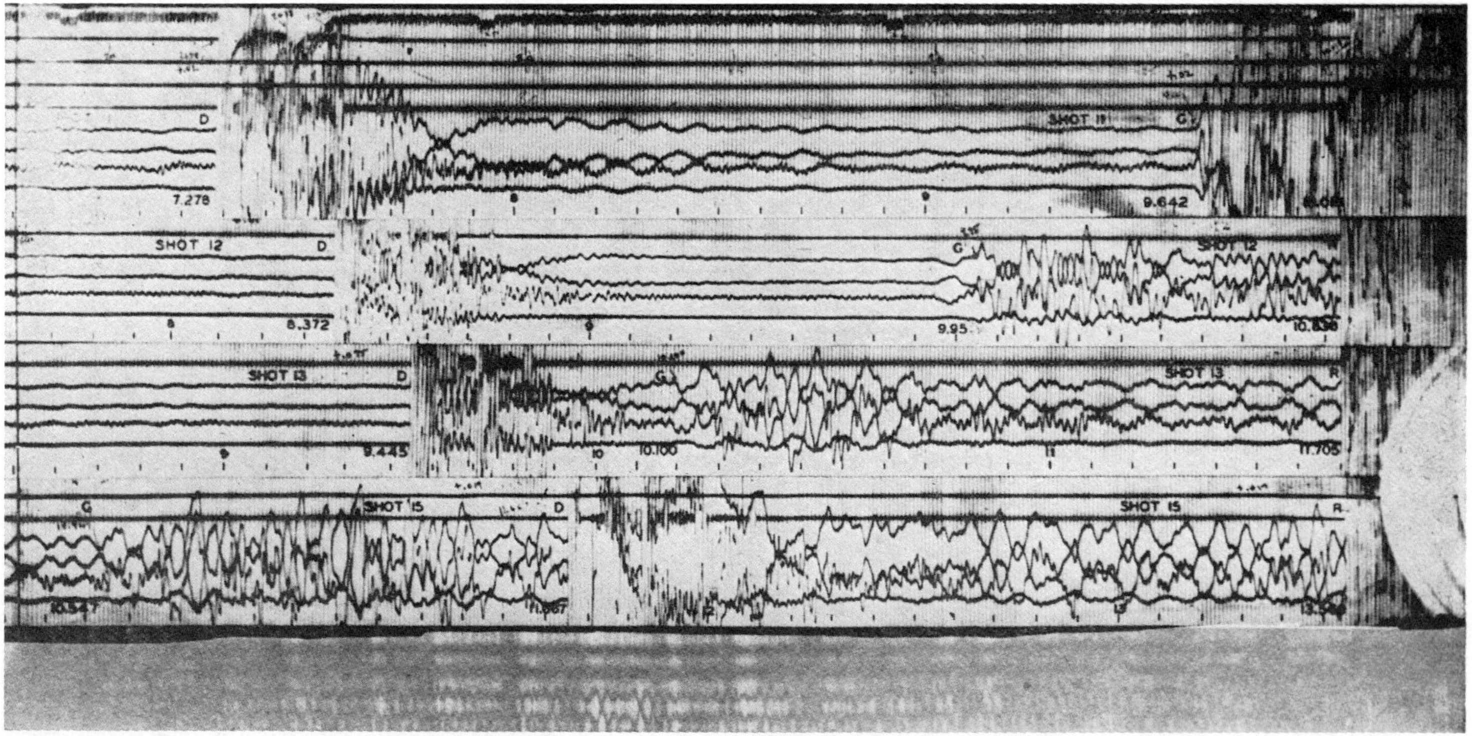

Fig. 3. Representative seismograms where G is travel time of ground wave, D is travel time of direct water transmission, and R is travel time of first bottom reflection. Travel time is given in seconds.

TABLE 1

Shot no.	Charge	Explosion instant	Water travel time	Ground wave	Bottom reflections			Ground-wave residual[a]
					R_1	R_2	R_3	
	lbs.		*sec.*	*sec.*	*sec.*	*sec.*	*sec.*	*sec.*
9	4.5	095523.56	4.16	......	8.032	14.365	21.09	......
10	4.5	100842.97	5.84	......	9.028	14.966	21.48	......
11	25	101907.53	7.28	9.642	10.018	15.560	21.960	−.069
12	25	102713.14	8.37	9.95	10.838	16.077?	22.27	+.024
13	25	103422.86	9.45	10.100	11.705	16.675	22.735	−.045
15	25	105052.95	11.67	10.547	13.569	18.002	23.667	−.026
17	55	112058.22	.....	......	......	......	......	......
18	300	113924.27	.....	......	......	......	......	......
19	300	115921.75	20.67	......	21.83	24.80	29.27	+.150
20	300	121949.76	24.62	13.13	25.61	28.23	32.11	+.015
21	300	124011.96	28.06	13.82	28.93	31.27	34.90	+.030
22	300	130007.85	31.64	14.50	32.45	34.57	37.76	+.007
23	300	132002.33	35.08	15.16	35.83	37.76	40.76	−.008
24	300	134032.42	38.37	15.78	39.00	40.81	43.55	−.074
26	4.5	161937.62	4.185	......	8.019	14.337	20.967	......
27	4.5	162623.62	5.465	......	8.781	14.765	21.44	......
28	25	163346.33	6.80	......	9.662	15.250	......	......
30	25	165540.30	10.27	10.090	12.38	17.10	......	−.146
31	25	170540.50	12.14	10.580	13.970	18.308	......	+.126
32	25	171311.88	13.59	10.760	15.253	19.340	24.77	−.002
33	55	172426.32	15.72	11.180	17.182	20.900	25.92	−.020
34	55	173548.52	17.94	11.765	19.270	22.634	27.32	+.106
35	55	175043.21	20.65[b]	12.290	......	......	......	......
36	55	180716.34	23.94	12.958	24.990	27.673	31.64	+.062
37	300	183028.30	28.17	13.80	29.06	31.40	34.93	+.031
38	300	184522.70	30.95	14.40	31.77	33.92	37.19	+.057
39	300	190030.15	33.66	14.83	34.42	36.43	......	−.076
40	300	193050.69	38.72	15.97	39.37	41.19	43.92	+.025
41	300	194521.42	41.28	16.30	41.91	43.36	46.26	−.172
42	300	195934.60	.....	......	......	......	......	......
43	300	201527.54	.....	......	......	......	......	......

[a] Residuals are observed times less times calculated from $T = 8.283 + X/25{,}500$ for Profile I and from $T = 7.959 + X/24{,}200$ for Profile II.
[b] Water wave not recorded—interpolated from neighboring shots.

the advantageous fact that D is brief, permitting G to be read in the interval between D and R.

The list of data taken from each shot is shown in table 1 and the travel-time curves for both profiles are plotted from a common origin in figure 4, where the squares represent Profile I and the circles represent Profile II. The abscissas are water-wave travel times in seconds, which may be considered as the distance between the two ships in units of 5,000 feet. The ground waves lie

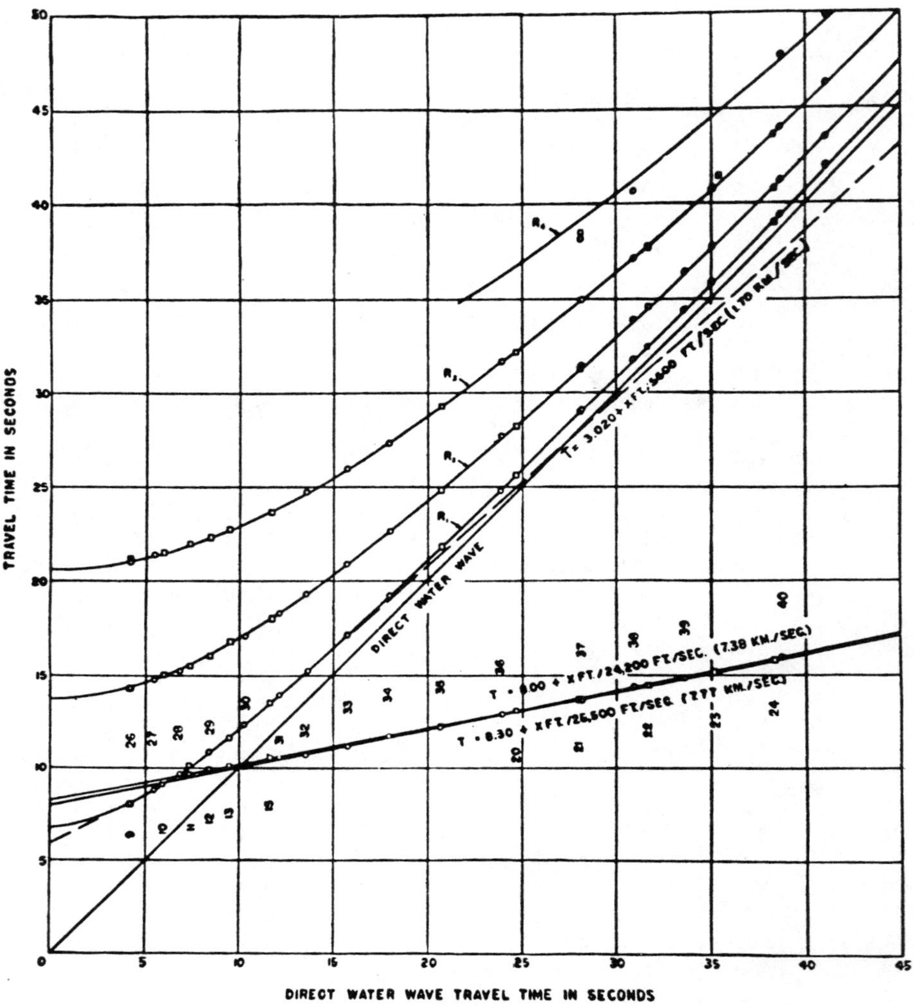

Fig. 4. Travel-time curves. Legend: ○, receiving station, 34° 06′.5 N, 66° 35′ W, shot station azimuth, 140° T; ● receiving station, 35° 52′.5 N, 66° 20′.5 W, shot station azimuth, 310° T.

on two slightly different straight lines which intersect at about 27 seconds or about 22.5 miles, which compares well with the distance of 18 miles separating the two positions of the CARYN. The close agreement of the two "apparent velocities," namely, 24,200 ft/sec. or 7.38 km/sec. and 25,500 ft/sec. or 7.77 km/sec., indicates that only a very slight slope is present in the basement surface and that the true velocity in the basement rocks is 24,800 ft/sec. or 7.58 km/sec.

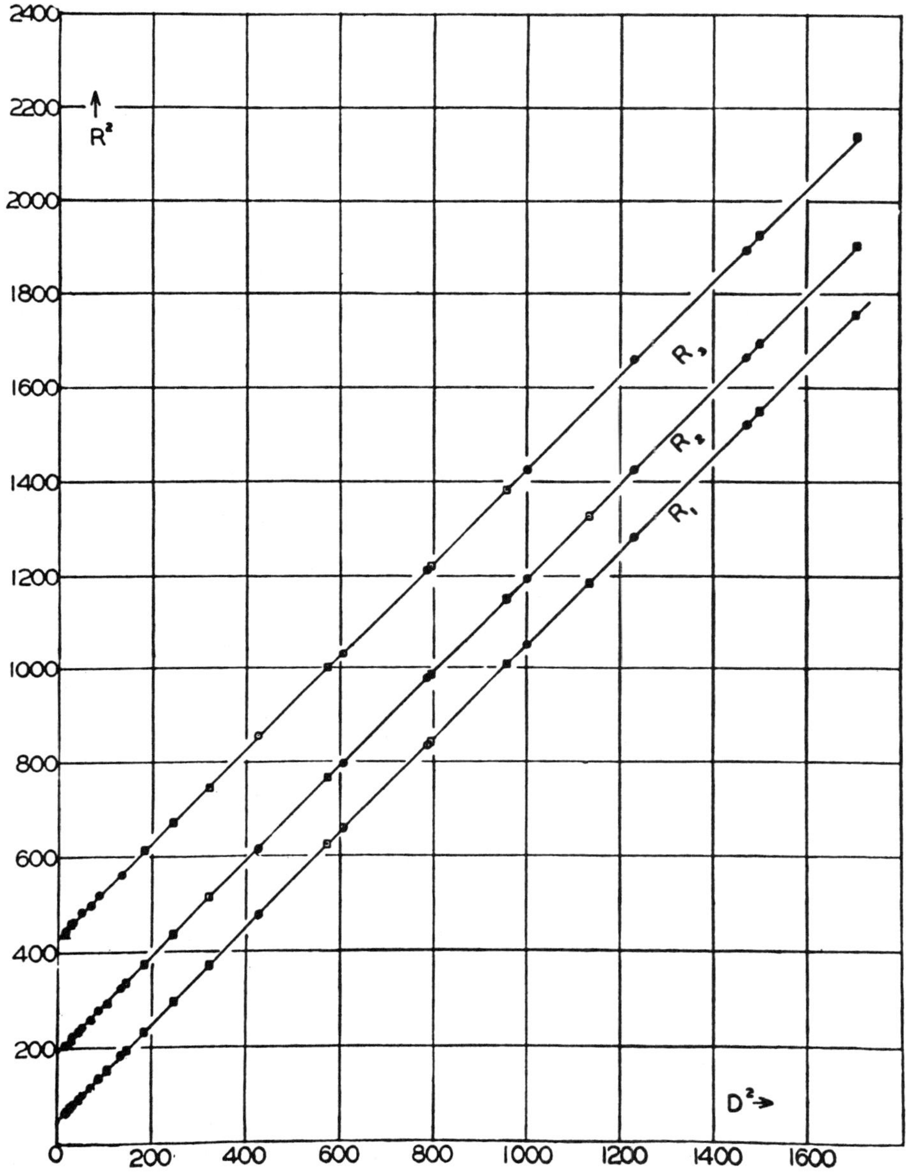

Fig. 5. Graph of R^2 vs. D^2 where R is the travel time of the first bottom reflection and D is the travel time of direct water transmission

The curves R_1, R_2, and R_3 are the reflections of various orders from the ocean floor, and squares and circles lie on common curves. The fact that the ground-wave curves cut across the reflection hyperbola R_1, instead of coming tangent to it, shows the presence of a layer of low-velocity medium (presumably unconsolidated sediment) between the ocean floor and the basement surface. In our other work, mostly unpublished, the velocity of sound in this sedimentary layer has been given as 5,600 ft/sec. or 1.70 km/sec. by a strong masked phase emerging from the R_1 hyperbola at about 14 seconds. In the present profile this phase is absent, possibly because the velocity is even less than 5,600 ft/sec. Despite this absence, we have used the velocity 5,600 ft/sec. shown by the dashed line in figure 3, in computing the thickness of the sedimentary layer, finding it to be 4,500 feet, with a slope of 20 minutes. The absence of this phase will be investigated in subsequent work. The theory that a variation in absorption coefficient is responsible is one which will be tested.

Some question might be raised about the identification of the direct water wave and some of the bottom reflections at longer distances. The validity of treating the refraction paths through the water as straight lines might also be questioned. Consequently, in figure 4, the squares of the water-wave times (distances) have been plotted as abscissas, and the squares of the reflection times R_1, R_2, R_3, etc., as ordinates. These quantities are related by the equations

$$R_1^2 = D^2 + H^2$$
$$R_2^2 = D^2 + 4H^2$$
$$R_3^2 = D^2 + 9H^2$$

where D is the direct water-wave travel time and H is the time which would be required for a vertical reflection from the ocean floor.

In figure 5 it is seen that the lines for the various orders of reflection are straight and have intercepts in the ratios 1 : 4 : 9, etc. Thus identification and treatment of the direct and bottom reflected waves is on a firm basis.

V. Discussion

The velocity 7.58 km/sec., found here under 2,800 fm. (5.13 km.) of water and 4,500 ft. (1.37 km.) of sediment, compares well with the velocity 7.78 km/sec. at a depth of 38 km. near Albany, N. Y., and 7.76 km/sec. at a depth of 25 km. near New Haven, Conn., as found by Slichter* from studies of quarry blasts. It also compares well with the value 7.747 km/sec. at a depth of 33 km. given by Jeffreys (1939) as a world average of P_n, particularly when we note that extrapolation of this velocity back to zero depth by use of the rate of increase of .003 km/sec/km. deduced from his velocity-depth table gives 7.65 km/sec. Byerly (1946) found 7.72 km/sec. at a depth of about 14

* Personal communication, in which Professor Slichter kindly made available to us his preliminary results.

km. from the Port Chicago explosion, but he has found 8.0 km/sec. for the speed of P_n in central California. He suggests that the speed of P_n may be lower in northern California. Gutenberg (1944) gave 6.91 km/sec. for the second intermediate layer at depths of 22 to 23 km.

Leet (1938 and 1941) gave for P_n 8.43 km/sec. at a depth of 36 km. in New England.

Hodgson (1947) examined a profile extending from the Canadian Shield to New England and found for P_n 8.2 km/sec. at a depth of 36 to 47.5 km. Tuve *et al.* (1948) found 7.05 km/sec. at a depth of 24 km. and 8.15 km/sec. at a depth of 42 km. in the Washington, D.C., and Maryland area. Reich *et al.* (1948) found 6.55 km/sec. at a depth of 21 km. and 8.2 km/sec. at a depth of 31 km. in the southern part of the Black Forest in Germany. Rothé *et al.* (1948) found 6.41 km/sec. at 16 km. and 8.13 km/sec. at 31 km. from the same region. The British National Committee for Geodesy and Geophysics (1948) reports 8.10 km/sec. at a depth of 26½ km. in northern Germany. Gutenberg (1946) found 7.99 km/sec. and 8.04 km/sec. for P_n in the New Mexico–Arizona–southern California area.

We had originally decided to identify our velocity 7.58 km/sec. with P_n on the basis of Jeffreys' world average, but as the result of advice generously given by Byerly and Gutenberg we have decided that there must be kept open the possibility that it is the lower part of the intermediate layer. The decision between these possibilities can be made from longer refraction profiles in the oceans, from refraction studies over the continental margins, or from a more concordant value of P_n from seismologists.

The significance of the present result will be increased if in subsequent work it is found to be typical of the entire ocean basin. A companion paper on the implications of this result for earthquake surface wave dispersion has been submitted for publication.

Summary

A reversed seismic refraction profile was made a latitude 34° N, longitude 66° 30′ W, in 2,800 fathoms of water, 400 miles east of Cape Hatteras and 120 miles northwest of Bermuda. The basement rock was found to have a velocity of 24,800 ft/sec. (7.58 km/sec.). No velocity was determined for the sedimentary cover. Using the value 5,600 ft/sec. (1.70 km.) for this layer, as indicated by earlier measurements, a thickness of 4,500 ft. (1.37 km/sec.) was obtained.

(Contribution 483 from the Woods Hole Oceanographic Institution).
LAMONT GEOLOGICAL OBSERVATORY
(COLUMBIA UNIVERSITY)
(Contribution No. 10).

References

British National Committee for Geodesy and Geophysics
1948. "Report of Seismic Work on the North German Explosions, 1946–47." I.G.U. meeting, Oslo, 1948.

Byerly, Perry
1946. "The Seismic Waves from the Port Chicago Explosion," *Bull. Seism. Soc. Am.*, 36:331–348.

Gutenberg, B.
1944. "Reflected and Minor Phases in Records of Near-by Earthquakes in Southern California," *Bull. Seism. Soc. Am.*, 34:137–160; see p. 140.
1946. "Interpretations of Records Obtained from the New Mexico Atomic Bomb Test, July 16, 1945," *Bull. Seism. Soc. Am.*, 36:327–330.

Hodgson, J. H.
1947. "Analysis of Travel Times from Rockbursts at Kirkland Lake, Ontario," *Bull. Seism. Soc. Am.*, 37:5–17.

Jeffreys, Harold
1939. "A Further Study of Near Earthquakes," *Mon. Not. Roy. Astron. Soc., Geophys. Suppl.*, 4:196; also pp. 510 and 571.

Leet, L. D.
1938. "Travel Times for New England," *Bull. Seism. Soc. Am.*, 28:45–48.
1941. "Trial Travel Times for Northeastern America," *Bull. Seism. Soc. Am.*, 31: 325–334.

Reich, H., G. A. Schulze, and O. Förtsch
1948. "Das geophysikalische Ergebnis der Sprengung von Haslach im sudlichen Schwarzwald," *Geologische Rundschau*, 36:85–96.

Rothé, J. P., E. Peterschmitt, and P. Stahl
1948. "*Les Ondes séismiques des explosions.*" I.G.U. meeting, Oslo.

Tuve, M. A., *et. al.*
1948. "Studies of Deep Crustal Layers by Explosive Shots," *Trans. Am. Geophys. Union*, Vol 29, No. 6, p. 772.

Part III

SUBMARINE TOPOGRAPHY AND STRUCTURE

Editor's Comments on Papers 13 Through 16

A recurring theme that runs through the history of global tectonics concepts is the importance of technological advances in methods of gathering data. Such advances were particularly important in marine geology where traditional methods of field study are impossible. Even a simple, fundamental task like determining the topography of the sea floor is complicated by the fact that sea water is, for distances of more than a few hundred feet, opaque to visible light; by the fact that, until the development of the deep-diving research submarine, the sea floor could not be examined firsthand by the investigator; by the enormous size of the unmapped terrain; and by the difficulty of precise position determination.

These factors, and others, combined to retard knowledge of sea-floor topography, so much so that it was not until after World War II that the sea floor ceased being a *terra incognita.* However, the device that would remedy at least part of the situation, the continuous-recording echo sounder, became available in 1922 and was first used in that year to chart a continuous bottom profile across the Atlantic Ocean from Newport, Rhode Island, to Gibraltar (Paper 13).

Although the existence and detailed topography of the Mid-Atlantic Ridge had been documented with discrete soundings by the work of Maury (1855) before the American Civil War, by the H. M. S. *Challenger* in the 1870s (see Thomson, 1885), and by the German research ship *Meteor* in the 1920s (see Maurer and Stocks, 1933), the voyage of the U.S. Coast Guard ship *Stewart* provided the first finely detailed picture of the ridge and its relationship to the bordering abyssal plains. Unfortunately for the development of marine geology, it was not until after World War II that the continuous depth recorder found its way into common use by oceanographic research vessels. For example, such devices were used for the work described in Paper 16.

The work of Wiseman and Sewell in the Indian Ocean in the 1930s (Paper 14) is noteworthy for a variety of reasons but particularly for their discovery of the graben valley that runs along the center of the Carlsberg Ridge. It is interesting that several of the profiles made by the *Meteor* in the 1920s clearly show the graben but, because those profiles were widely spaced, the valley was not recognized as being continuous. Not until almost twenty years later, when M. Ewing and B. C. Heezen (1956) and J. D. Hill (independently, see Hill, 1954) announced the discovery of the Mid-Atlantic Ridge graben did this revealing characteristic of oceanic ridges receive much attention, and within a few years, new theories incorporating the graben began to appear (see Heezen, 1959, 1960).

A similar delay occurred in the case of the enormous submarine fault scarps (later and more generally referred to as fracture zones) that are now known to be characteristically associated with transform faults. The first indication of the occurrence of such fracture zones and their association with oceanic ridges was reported by E. von Drygalski in 1904 when he described an east-west gap across the equatorial Mid-Atlantic Ridge at what is now known as the Romanche fracture zone. Discovery of the first of the scarps was announced in 1939 by H. W. Murray (Paper 15) who described the easternmost part of the Mendocino fracture zone where it terminates abruptly against and offsets the California continental shelf. However, only after many additional years of surveying was the extent of the Mendocino fault (see Shepard and Emery, 1941; Paper 16; Dietz, 1952; Menard and Dietz, 1952; Menard, 1955; Hess, 1955), the existence of other similar faults in the northwestern Pacific (and elsewhere), and their relationship to oceanic ridges and marine magnetism worked out (see Coode, 1965; Wilson, 1965).

After World War II, many lines of investigation in marine geology were transformed by an influx of financial support, particularly from the U.S. Navy. Nowhere is this better illustrated than in the work done during the late 1940s and early 1950s on oceanic ridges as typified by Paper 16, a detailed description of the Mid-Atlantic Ridge that includes, among other things, discovery of a fracture zone that transects the ridge, discovery of the nearly horizontal plains that border the ridge, and recognition of two contrasting major topographic provinces within the ridge—namely, the main range rising above 3000 meters depth and the terraced flanks. These conclusions and others were the result of the first extensive use in deep-ocean exploration of the continuous depth recorder. Once the breakthrough was made, however, scientists from many nations (notably, Great Britain, Sweden, and the USSR) were quick to realize the benefits of the new tool, and knowledge of submarine topography rapidly improved by orders of magnitude.

REFERENCES

Coode, A. M., 1965, A Note on Oceanic Transcurrent Faults, *Canadian Jour. Earth Sci.* **2:**400–401.

Dietz, R. S., 1952, Some Pacific and Antarctic Sea-Floor Features Discovered during the U.S. Navy Antarctic Expedition, 1946–47, *7th Pacific Sci. Cong. Proc.* 335–344.

Drygalski, E. von, 1904, *Zum Kontinent des eisgen sudens, Deutsche sud-polar Expedition fahreten und forschungen des Gauss 1901–1903,* George Reimer, Berlin.

Ewing, M., and B. C. Heezen, 1956, Mid-Atlantic Ridge Seismic Belt: (abs.): *Am. Geophys. Union Trans.* **37:**343.

Heezen, B. C., 1959, Paleomagnetism, Continental Displacements, and the Origin of Submarine Topography, in M. Sears, ed., *International Oceanographic Congress Preprints,* American Associaton for the Advancement of Science, Washington, D.C., pp. 26–28.

Heezen, B. C., 1960, The Rift in the Ocean Floor, *Sci. American* **203**(4):98–110.

Hess, H. H., 1955, The Oceanic Crust, *Jour. Marine Research* **14:**423–439.

Hill, M. N., 1954, The Topography of the Mid-Atlantic Ridge, *Assoc. Oceanog. Phys. Proces-Verbaux,* no. 6, Rome, 269.

Maurer, H. and T. Stocks, 1933, *Die Echolotungen des Meteor: Wissenschaftliche Ergebnisse der Deutschen Atlantischen Expedition auf dem forschungs und vermessungsschiff Meteor 1925–1927, Band II, Part 1 (text), Part 2 (Echolot-profile),* Walter de Gruyter and Co., Berlin and Leipzig.

Maury, M. F., 1855, *The Physical Geography of the Sea,* Harper and Brothers, New York, 287p.

Menard, H. W., 1955, Deformation of the Northeastern Pacific Basin and the West Coast of North America, *Geol. Soc. America Bull.* **66:**1149–1198.

Menard, H. W., and R. S. Dietz, 1952, Mendocino Submarine Escarpment, *Jour. Geology* **60:**266–278.

Shepard, F. P., and K. O. Emery, 1941, Submarine Topography off the California Coast, *Geol. Soc. America Spec. Paper 31,* 171p.

Thomson, C. W., and J. Murray, 1885, *Report of the Scientific Results of the Voyage of H. M. S.* Challenger *During the Years 1873–1876: v. 1, pt. 1, Narrative,* printed and sold for Her Majesty's Stationery Office by publishers in London, Edinburgh, and Dublin, 509p.

Wilson, J. T., 1965, A New Class of Faults and Their bearing on Continental Drift, *Nature* **207:**343–347.

13

Reprinted from *U.S. Naval Inst. Proc.* **69:**216–223 (1943)

SONIC SOUNDING

As Developed by the U. S. Navy

By THOMAS H. WHITCROFT, *Hydrographic Engineer, U.S. Naval Hydrographic Office*

About the time of the discovery of America, Leonardo da Vinci, famous engineer, physicist, and painter of his time, performed many experiments in the physics of water. During these experiments, he discovered that sound traveled through water. It is recorded in his notes that if one end of a tube is placed in the water and the other end in the ear, while on board a stationary ship, any other ships moving in the distance can be heard.

No use appears to have been made of the discovery, and it was not until some 300 years later, or somewhat more than a century ago, that Jean Francois Arago, eminent astronomer, engineer, physicist, and discoverer of the rotary polarization of quartz, proposed that the principles relating to the propagation of sound waves in water be used to measure the depths of the sea.

From that time until the beginning of the present century, there ensued a period of experimentation by many physicists, who discovered new principles and invented devices, which, by process of combination and improvement, have since developed into the present sonic sounding appliances.

Early in the twentieth century, these experiments began to bear fruit. In 1902, the microphone for receiving faint sounds was invented by Grey and Mundy; in 1910–14, the oscillator, for generating submarine sounds, was invented by Dr. R. A. Fessenden; in 1918, the hydrophone, for locating the direction of submarine sounds, was invented by Dr. Hayes; and lastly, in the early 1920's, Doctors Hayes, Fessenden, Behm, and Marti perfected timing devices which proved sufficiently accurate. The combination of these various parts made possible the accurate determination of ocean depths by the timing of sound waves in water.

The first experiments on submarine sound by the U. S. Navy were made by Lieutenant M. F. Maury, U. S. Navy, when he attempted depth measurements in the ocean. In 1854, he records

> . . . By exploding petards or ringing bells in the deep sea, . . . the echo or reverberations from the bottom might be heard, and the depth determined from the rate at which sound travels through the water. But, though the concussion took place many feet below the surface, echo was silent. . . .

This experiment was unsuccessful as it was not generally known that to hear subaqueous sound there had to be a definite connection between the ear and the water.

In 1911, Admiral Winterhalter, U. S. Navy, on the U.S.S. *Washington,* carried out experiments to find the distance between his ship and the Nantucket lightship by the use of sound in water. A radio signal, a signal from a submarine bell, and an air signal using a steam whistle were sent out simultaneously from the lightship. They were received on the *Washington* in the natural order—the radio signal instantaneously, then the submarine signal, and lastly the sound of the whistle. This gave the time of transit of sound waves in both air and water, which doubly determined the distance. This was the first work on finding distance by acoustic methods, and is the forerunner of the science of radio acoustic ranging, now used in hydrographic surveying. Further experiments on this work were carried out in 1917 by Commander Hooper, also of the U. S. Navy.

It was due to the numerous sinkings of ships by submarines on the Atlantic seaboard in the early part of the first World War that Dr. Hayes, at that time Professor of Physics at Swarthmore College, Pennsylvania, began making experiments in subaqueous sound, using the college swimming pool and the sounds of approaching trains on the near-by railroad. He sent his findings to the U. S. Navy Department, which immediately requested that he report to Newport, Rhode Island, where a Naval Research Laboratory and a corps of specialists were working on experiments in submarine detection. This group of men perfected the M-V-hydrophone, a microphone which could be focused on the direction of any sound in water.

It is interesting to note that the research of the U. S. Navy in submarine acoustics was first directed to measurements in a horizontal plane. The development of vertical measurements was made later because it was discovered that the instruments could be applied to that purpose. Conversely, it will be shown as this article progresses that the instrument originally designed for horizontal measurements could be used to measure relatively shallow depths with precision; and again, in seeking a method of measuring distances between ships for long-range firing, that the machine designed for that purpose proved to be an accurate depth-measurer, even in the greatest depths.

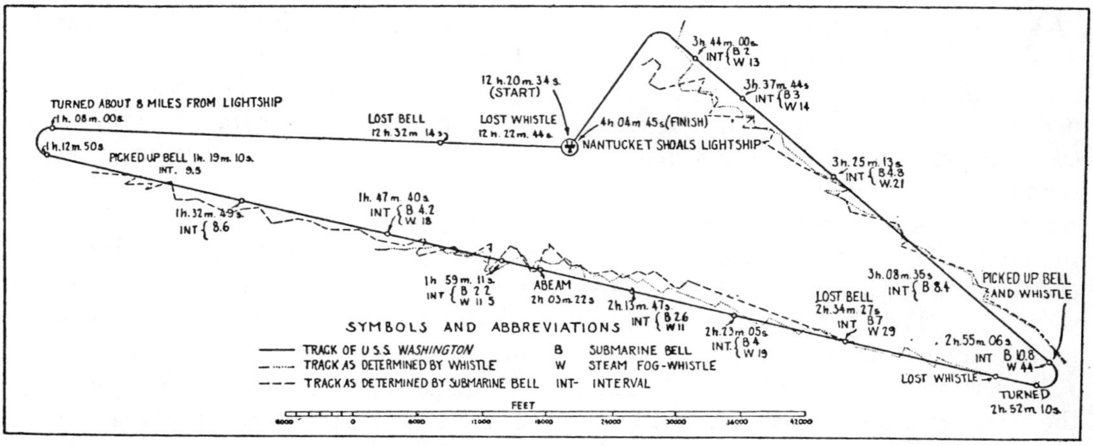

AN ACOUSTIC RANGING EXPERIMENT BY THE U. S. NAVY CARRIED OUT BETWEEN NANTUCKET LIGHT VESSEL AND THE U.S.S. *WASHINGTON* IN 1911

With these instruments, affording the first practical use of sound as an aid to navigation, the first continuous profile of the Atlantic Ocean was obtained, giving data for the first comprehensive bathymetric chart. Also, by means of these instruments, the first application of radio acoustic ranging, as it is carried out today, was made.

In 1918, the U. S. Navy transport *Von Steuben* was equipped with the latest listening device for the detection of submarines. It was decided to test the apparatus as a safeguard against collisions. Dr. Harvey C. Hayes, then Chief Physicist of the U. S. Navy, was aboard to carry out the final tests of the instrument that he had helped to design. Coming out of New York Harbor, the apparatus worked perfectly, all ships near the *Von Steuben* being easily detected and their positions ascertained. The good performance continued until the ship reached the edge of the Continental Shelf. There the listening device failed. The whole apparatus was taken down and checked. Nothing being found out of order, the apparatus was reassembled. Still it would not function. The success thought to have been accomplished had apparently turned out to be a failure.

For the rest of the voyage to Brest, France, a watch was kept on the instrument. At two o'clock one morning, Dr. Hayes was notified that the operator on watch had begun picking up the noises of the *Von Steuben's* propellers, but he reported that they appeared to be located amidships instead of astern. This disclosed the reason for the failure to pick up the propeller noises of other ships while in deep water. The sound received had been a reflection from the bottom which in deep water had been lost. This experience led to the adaptation of her sound-ranging device to depth finding. A table was constructed giving the directions in terms of depths when the listening device was focused on the ship's propellers.

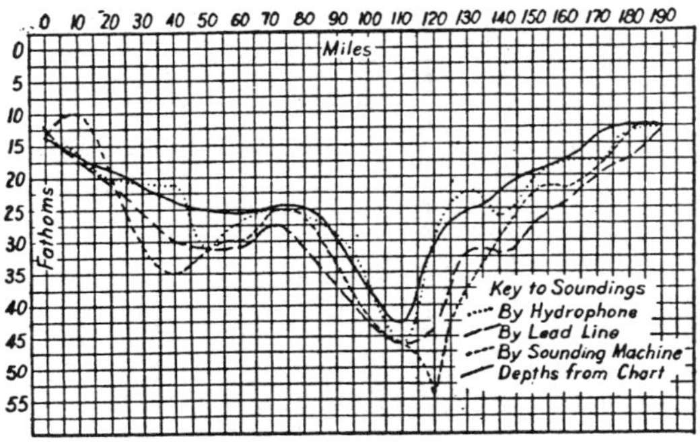

A COMPARISON OF ACOUSTIC SOUNDING (ANGLE METHOD) WITH OTHER METHODS; U.S.S. *BRECKENRIDGE* EN ROUTE CHARLESTON TO KEY WEST, JANUARY 14, 1920

It was on the *Von Steuben*, while approaching Long Island in a thick fog, that the first practical use was made of submarine sound for navigation purposes. The dead-reckoning track of the ship was changed to conform to the soundings received with the new device, and a perfect landfall was made on Ambrose Light. Had the original course been continued, the ship might have grounded on Long Island.

In the next few years, improvements were made on this acoustical ranging apparatus. Many ships of the U. S. Navy were equipped with the M-V-hydrophone, and, at that time, it was used for soundings. On January 14, 1920, the U.S.S. *Breckenridge*, en route from Charleston, South Carolina, to Key West, Florida, ran a line of soundings using three methods: sound, lead line, and sounding machine (wire). The depths were plotted on a graph along with the chart depths. Although conditions were not the most favorable for comparison, it was considered that the method by sound gave the most reliable results. Another profile was run by this method on May 27, 1920, by the U.S.S. *Blakely*, from Newport, Rhode Island, to the 100-fathom curve and return over the same track. Sound and lead line were used, and the results compared closely with the chart.

These runs proved that the sound equipment gave very accurate results in depths less than 100 fathoms. Thus, an automatic sounding machine to aid the navigator in making a landfall was perfected and proved accurate. Its use became more or less universal in the U. S. Navy. Little or no interest was shown in devising an instrument for taking soundings in water deeper than 100 fathoms, because there appeared to be little need for such an instrument at that time.

Research by the U. S. Navy in the use of sound was continued in studies of devices applying to the horizontal plane in sea water. With the submarine sound transmitter and the M-V-hydrophone, which worked very well, all that was necessary to measure distances was an accurate timing device. Early in 1922, such a device was perfected by Dr. H. C. Hayes in the Naval Engineering Experiment Station, and was installed on the U.S.S. *Ohio* for a final test. The purpose of this project at the time was to measure accurately the distance between two ships, in order to obtain an accurate base for long-range firing.

The procedure used was the same as with present-day radio acoustic ranging (R.A.R.). Two ships established the base to be measured. One of the ships would generate underwater sound waves. The other ship would receive these through a hydrophone connected to an open-circuit radio transmitter and thus transmit them back to the first ship by radio. By measuring the interval between the instant of the original sound impulse and the corresponding instant when it was received on the same ship by radio, the distance between the two ships could be calculated. In present practice in R.A.R. work, the "sono" buoy takes the place of the second ship.

The installation for carrying out these tests had been made on the U.S.S. *Ohio*, but at the time of the final trials a second ship was not available. It was accordingly decided that the simplest method for checking the time-measuring device was the use of the *Ohio* as both sending and receiving ship. This was done by reflecting the sound waves off the bottom, which, in reality, amounted to measuring the depth of the ocean. It was reasoned that if the apparatus would measure the distance from the surface to the bottom and return, it would also measure a corresponding horizontal distance.

In February, 1922, the *Ohio* departed from New York for Annapolis Roads to test the apparatus. The result of this cruise was a conclusive proof that a workable deep-sea sonic sounding device had been developed. A line of soundings was run over known depths, and these soundings were checked with those on the chart. On this cruise, the second use of submarine sound waves as an aid to the navigator was recorded. The approach to the entrance to Chesapeake Bay was made in a heavy fog, and the course and successive positions of the ship were ascertained by comparing the profile along the ship's course with the chart. This was the first time that sound waves in deep water proved to be of practical value to the navigator in locating his position.

Now that the U. S. Navy had means to measure any depth of ocean without requiring ships to stop for deep soundings or otherwise interfering with their movements in any way, it was decided to do something of real import with the apparatus. There was to be a transfer of ships from the Atlantic Fleet to the Asiatic Squadron. This offered an opportunity to give the world a continuous profile of the Atlantic Ocean—a feat that had never been done before because of the inability to obtain soundings in sufficient number on a continuous line. It was decided to install the apparatus on one of the destroyers, to run this line of soundings and, if practicable, to continue it around the world.

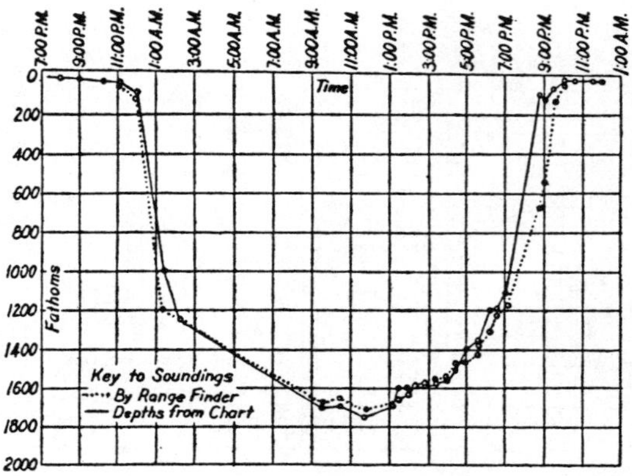

PORTION OF THE SONIC RANGE-FINDER PROFILE; U.S.S. *OHIO* EN ROUTE NEW YORK TO ANNAPOLIS, FEBRUARY 15–16, 1922

It was the U.S.S. *Stewart* that received the apparatus, and in June, 1922, she left Newport, Rhode Island, for Chefoo, China. Dr. Hayes was aboard to direct the use of the sounding machines and to see that they functioned properly. The main project was to run a continuous line of soundings along a great-circle course from Newport, Rhode

Island, to the Azores, and thence to Gibraltar, crossing the Josephine and Gettysburg Banks. Older soundings on the Gettysburg Bank showed a bottom of sand and rounded pebbles, the latter giving rise to the theory of geologists that this Bank had at one time been above water, and so had been exposed to erosion. The sonic sounding machine on the *Stewart* gave soundings which when plotted in profile showed evidence of submerged shore terraces, thus fortifying the theory. This was the first

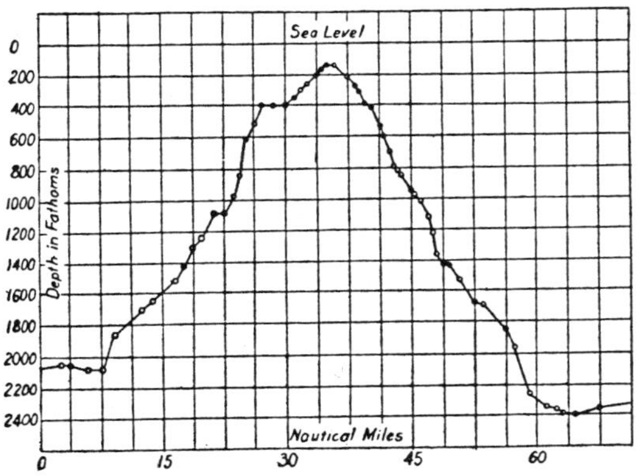

CONTINUOUS PROFILE ACROSS THE GETTYSBURG BANK, U.S.S. *STEWART*, JUNE 28, 1922

contribution to geology in which data on submarine geological formations were supplied by sonic depth equipment. It would have been impracticable to obtain such a dense line of soundings with the wire sounding machine.

Destroyer Division 15 of which the *Stewart* was a part, with Captain T. T. Craven, U. S. Navy, as Division Commander, continued through the Mediterranean Sea toward the China Station. Dr. Hayes had left the *Stewart* at Gibraltar, and the soundings were continued by the Navigator, Lieutenant S. G. Lamb, U. S. Navy. The Commander of the *Stewart* was Lieutenant Commander N. R. Van Der Veer, U. S. Navy. The line of soundings was continued from port to port through the Mediterranean Sea to Port Said. Over 6,500 nautical miles of continuous soundings were run, and this without interfering with the routine duties of the ship. Nothing like that had ever been done before. The complete sounding of the oceans in the future had become a possibility.

The records do not show that soundings were taken in the Red Sea, but at Aden the line was begun again and was continued until the ship entered the port of Singapore, thus adding 2,800 nautical miles to the already long line of soundings ending at Port Said.

This work of the *Stewart* constituted the first real sonic-sounding survey. Today, in accordance with a settled policy, all ships of the U. S. Navy take sonic soundings when cruising in the deep sea. This was the beginning of the great work since accomplished by the U. S. Navy, a contribution to date of over 50 per cent of the data on deep-ocean soundings of the world.

The sounding work of the *Stewart* attracted the attention of scientists in many fields. The first group to take advantage of this means of obtaining information were the seismologists of the west coast of the United States. They knew the land structure well, but that was only half of the story. The other half was buried beneath the sea, and up to that time there had been no economical and practical method to get hold of the data necessary to complete their knowledge. They requested the U. S. Navy Department to furnish this information. Accordingly, the destroyers *Hull* and *Corry* were assigned to survey the coast of California from San Francisco to Pt. Descanso, between the 100- and the 2000-fathom curves.

The work was completed during October and November, 1922. Never before had such a large area of deep water been accurately surveyed on such a large scale. The 34,000 square miles comprising the area were covered with sounding lines 10 miles apart in the northern half and 5 miles apart in the southern half. The project was completed in 38 days, at an average of almost 1,000 square miles per day. The results received were used to construct the first comprehensive bathymetric chart. The latter caused a revival of interest by making possible a more detailed study of the basins of the ocean.

Activities in oceanic sounding continued, and by 1929 the U. S. Hydrographic Office was receiving reports of deep-sea soundings daily. At that time, practically all ships had been equipped with sound depth apparatus of the Fessenden type, developed by the Submarine Signal Corporation. This new type had proved to be more adaptable, both in depth range and in operation, than the original instrument invented by Dr. Hayes.

Each new ship commissioned was equipped with the new sounding device, and on her shakedown cruise many miles of deep-sea soundings were reported from areas never sounded before. All ships were requested to report soundings over areas previously unexplored. Ships of the U. S. Navy about to make extensive cruises were given special orders to develop doubtful dangers and special areas near their courses, and were furnished with instructions and standard blank forms designed to facilitate plotting their results. The survey ships increased the extent of their surveys many miles beyond the 100-fathom curve which had been their former limit. In present practice even survey launches are equipped with sonic depth finders, which have proved to give more uniform depth measurements than the older methods. The Merchant Marine became "sounding conscious," and many valuable reports have been received from that source.

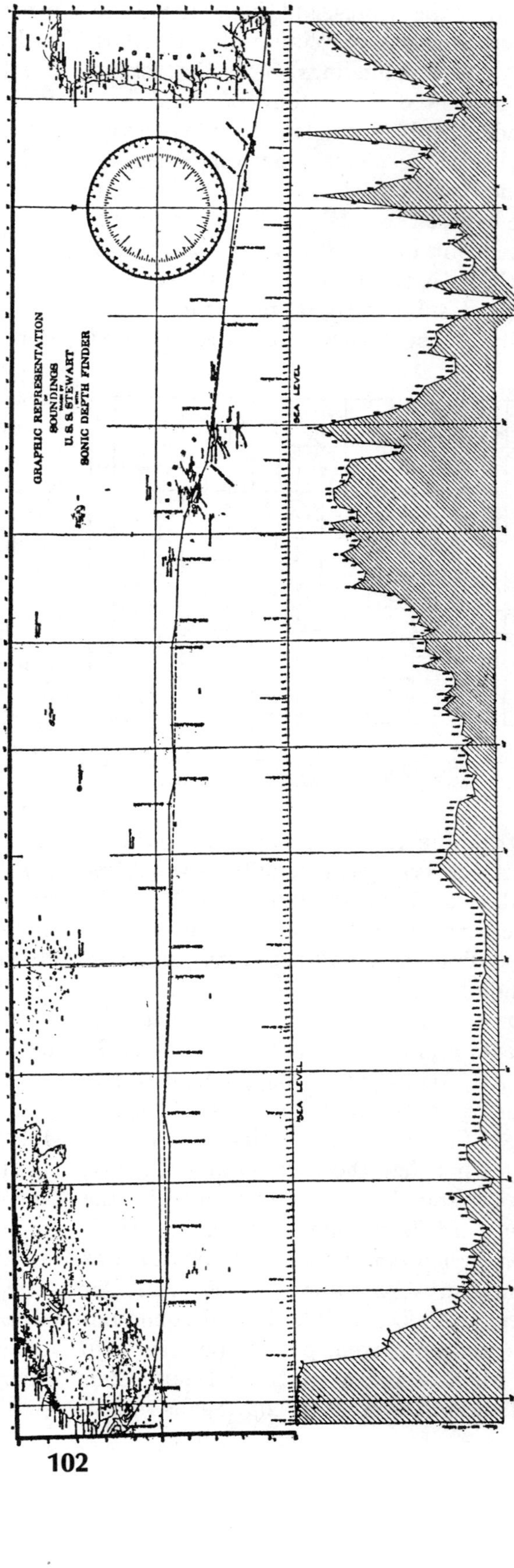

REPRODUCTION OF THE ORIGINAL REPORT OF THE FIRST CONTINUOUS PROFILE OF ANY MAJOR OCEAN; U.S.S. *STEWART*, JUNE, 22–29, 1922

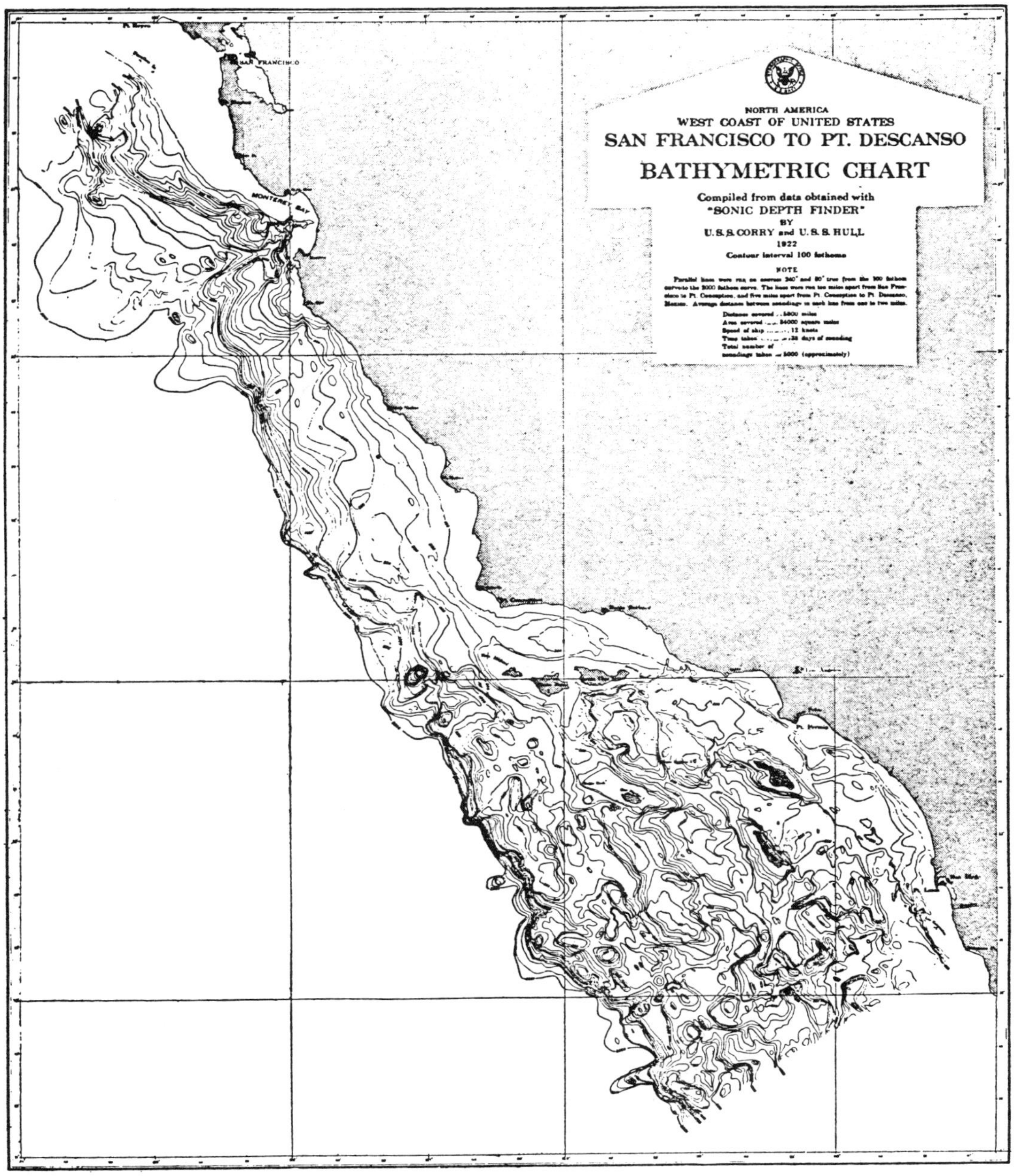

FIRST COMPREHENSIVE BATHYMETRIC CHART BASED ON SONIC SOUNDINGS; U.S.S. *HULL* AND *CORRY*, 1922

The miles of deep-sea sounding lines received in the U. S. Hydrographic Office increased annually from 10,000 nautical miles in 1929 to 100,000 nautical miles in 1939. This mass of information exceeded the total of all other agencies in the world.

It is difficult to give credit to any specific person or organization for the origination of the acoustic depth-measuring instrument. This difficulty is due to several factors—the secrecy surrounding the early attempts, the several separate devices necessary to make up a depth finder, and the many attempts that were successful but not really practical. However, credit must be given to the U. S. Navy for its investigations of submarine acoustics, and to Dr. H. C. Hayes for developing the first practical sonic sounding instrument, showing the world that sound waves in the sea could be used practically for ocean depths. The Hydrographic Office of the U. S. Navy should also be given credit for the greatest contribution to the bathymetric information of the world today.

In 1911, the U. S. Navy had proved that the principle, used in radio acoustic ranging today, was practicable, and around 1920 was well in the forefront of the pioneers of sonic sounding methods. The achievements of the *Stewart*, the *Hull*, and the *Corry* revived an interest in oceanic soundings that had been nearly dormant since the time of Lieutenant M. F. Maury, U. S. Navy, and Mr. Cyrus W. Field. The U. S. Navy today can have the same feeling of pride in its achievements in submarine acoustics that was felt in Maury's time in his work with the deep-sea lead. The Secretary of the Navy of the period, James C. Dobbins, in his annual report of 1856, referring to Maury's work, said:

> I confess I felt some pride in having the science and Naval genius of our country to continue foremost in these great ocean surveys. . . . these deep-sea soundings, these studies of winds, currents, and temperatures of the oceans, these gradual approaches to a greater familiarity with the wonders of the great deep, are pregnant with incalculable usefulness to those who conduct mighty navies, as well as to all who go down to the sea in ships.

The U. S. Navy has not reached its ultimate goal, which is to have complete information of all navigable waters of the globe; however, it can be proud of its achievements that tend to bring this goal within the realm of possibility.

Author's Note.—Grateful acknowledgment is given to Mr. H. Abbey of the Bureau of Ships and to Dr. Hayes of the Naval Research Laboratory for their assistance in collecting data and to Mr. J. G. Kellar of the Hydrographic Office for his helpful criticism in the preparation and careful editing of this article.

14

Reprinted from *Geol. Mag.* **74**:219–230 (1937)

The Floor of the Arabian Sea.

By JOHN D. H. WISEMAN, Mineral Department, British Museum of Natural History, and R. B. SEYMOUR SEWELL, Leader of the "John Murray" Expedition to the Indian Ocean.

(PLATE XI.)

DURING the course of the John Murray Expedition some 22,000 miles of the floor of the Arabian Sea and the neighbouring areas of the Indian Ocean were mapped by means of the Echo-sounding apparatus. The region between India and Africa was crossed four times, namely :

(i) from Aden to Karachi,
(ii) from Bombay to Mombasa,
(iii) from Zanzibar to Colombo, and
(iv) from Colombo to Aden,

while more detailed surveys were carried out in the Gulf of Aden, the Gulf of Oman, and the region off Zanzibar.

The data thus acquired have been charted by Lieut.-Commander Farquharson, R.N., of the Hydrographic Department of the Royal Navy, who accompanied the expedition as Navigator and Surveyor ;

and we now know that this region is traversed by several submarine mountain chains of varying length and height (Fig. 1), namely :

(*a*) Along the southern coast of Baluchistan and Persia is a line of interrupted ridges, that forms a barrier some 60 miles off the present coast-line, rising from the general level of the floor of the Gulf of Oman to within 732–914 metres of the surface, and separated from the mainland by basins that contain water of a depth of some 366–549 metres deeper than the barrier itself. This range is almost certainly a part of the Zagros System of Persia.

(*b*) Commencing from the neighbourhood of Cape Monze, near Karachi, a double ridge, with an enclosed deep gully, runs towards the south-west across the entrance to the Gulf of Oman. On the north-west side of the gully lies the " Murray " Ridge, which appears to be a continuation of the Khirthar range of Sind and rises to within 1,127–1,185 metres of the surface, while on the south-east side lies a second ridge that at one place has a depth of only 755 metres of water. The floor of the Gulf of Oman has an almost uniform depth of 3,383 metres, but in the gully the depth is 3,658 metres. As we trace these ridges and the enclosed gully to the south-west we find that the deep water of the gully seems to be continued for some distance, while the ridge on its south-east side also appears to be continued, for there is a series of shallow soundings that occur in line with it.

(*c*) These soundings, taken together, indicate the possibility that this " Murray " system may be connected with the great Carlsberg Ridge, which was so named by the late Dr. Johannes Schmidt,[1] who crossed the ridge in the *Dana* in about lat. 1° N., and who postulated its extension to join Socotra. As we trace this great mountain chain from the neighbourhood of Socotra, it is found to run at first towards the south-east and then after coming into close relationship with the submarine base on which the Chagos Archipelago is situated, bends first south and then west of south, and may be continued as far as the island of Roderigues, though the petrographical evidence does not support such a continuation. Throughout the greater part of its length from Socotra to the Chagos Archipelago this ridge appears to be double. Along the two crests numerous soundings between 2,291 and 3,059 metres have been made and the least water detected over them is 836 metres near the northern end, and 1,569 and 1,752 metres on the two ridges respectively in about lat. 1° 30′ N. where there is a depth of over 3,383 metres in the enclosed gully. On either side of this great ridge are deep areas in which the depth is about 5,121–5,303 metres. The Carlsberg Ridge seems to cease just to the south of Roderigues, but it is continued on as a low ridge, with some 3,650 metres of water over it, far to the south, and in all probability is connected

[1] Schmidt, J., *Dana's Togt Omkring Jorden, 1928–1930,* Copenhagen, 1932, 255, fig. 198.

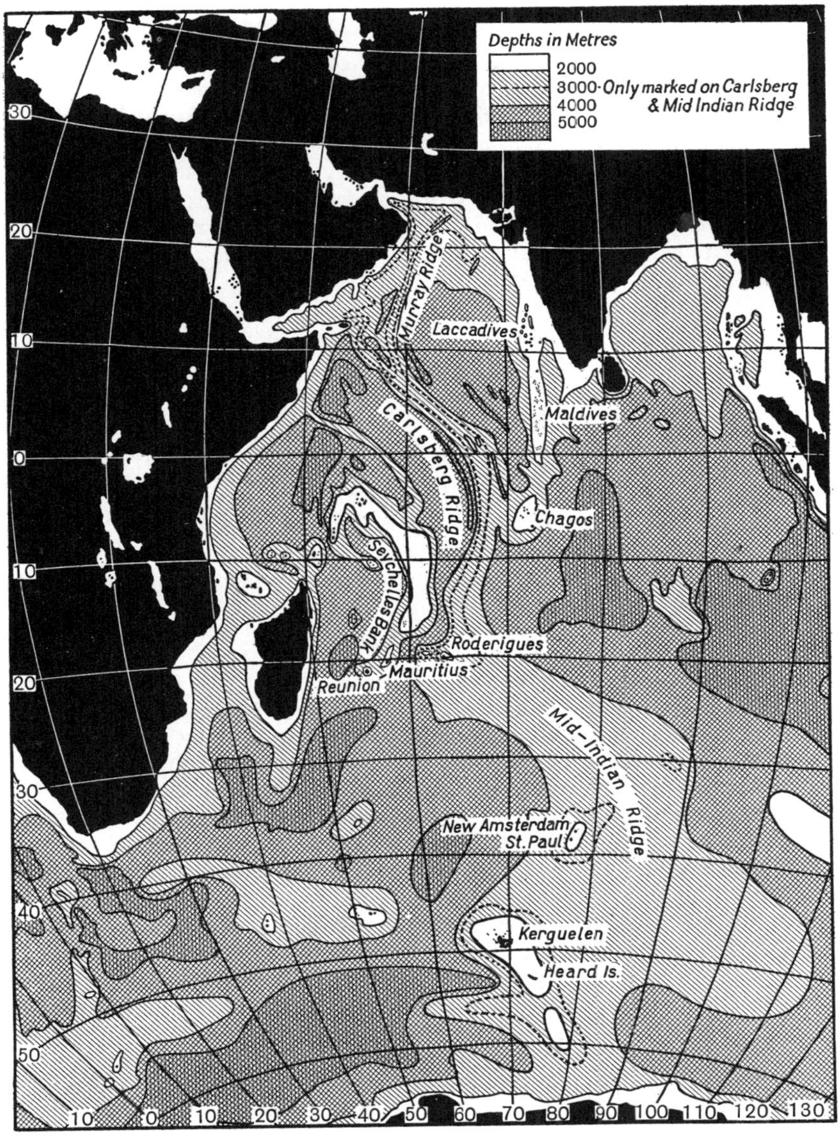

FIG. 1.—Bathymetric chart of the Indian Ocean.

with the Kerguelen Plateau,[1] from which rise the islands of St. Paul, New Amsterdam, Kerguelen, and Heard, and finally blends with the Antarctic continent.

(*d*) Commencing on the African coast in about lat. 4° N. and running a course concentric with the Carlsberg Ridge is another ridge of lesser height, but rising some 1,830 metres above the sea-floor.

(*e*) At its southern end this ridge may be continuous with the curved Seychelles–Mauritius Bank, on which lies near its northern end the Seychelles group, and at its southern end Mauritius. Between these are a series of banks, such as Saya da Malha, and Nazareth, that rise up to or very nearly to the surface.

(*f*) Finally between the Seychelles and Madagascar lies the base on which are situated Providence Island, Farquhar Island, and the Cosmoledo group, while in about lat. 16° S. we have the isolated island of Tromelin.

The great Carlsberg Ridge and its southern continuation thus form a Mid-Indian Ridge, running from north to south down the Indian Ocean, that reminds one of the similar Mid-Atlantic Ridge in the western ocean. This ridge divides the Indian Ocean into two, east and west, areas, while at the northern end there are several subsidiary basins, namely :—

(i) The north-east or Arabian Basin, bounded on its south-west by the Carlsberg Ridge and on the east by the coast of India and the Laccadive-Maldive Ridge, while at the northern end it is cut off by the " Murray " Ridge.

(ii) The basin of the Gulf of Oman.

(iii) The Somali Basin, lying to the south-west of the Carlsberg Ridge ; this latter basin, which is divided into two by the above-mentioned ridge (*d*) and the Seychelles–Mauritius Bank, has a maximum depth of 5,303 metres and each part is connected with the main western area of the Indian Ocean by a deep gully, having depths of some 4,570 metres. These gullies run respectively between Madagascar and the Seychelles, and the Seychelles and the Carlsberg Ridge.

In these three basins the nature of the sea-floor is quite different :—

(i) In the Gulf of Oman the basin is in its shallower regions floored with green mud, which in the neighbourhood of Cape Ras al Hadd is strongly impregnated with H_2S, while in the deeper areas grey clay occurs. This area is practically devoid of animal life between some 200 and 1,550 metres.[2]

(ii) Throughout the Somali Basin the sea-floor is composed of globigerina ooze, and supports, apparently, a normal animal population.

[1] Wüst, G., " Anzeichen von Beziehungen zwischen Bodenstrom und Relief in der Tiefsee des Indischen Ozeans," *Die Naturwissenschaften*, 1934, 241.

[2] Sewell, R. B. Seymour, " The John Murray Expedition to the Arabian Sea," *Nature*, cxxxiii, 1934, 88.

(iii) The Arabian Basin in its deeper part is floored with red clay, with, at any rate in certain areas, a plentiful formation of manganese nodules. It is an interesting fact that this basin is also devoid of animal life in its deeper parts.[1]

The Character of the Coastal Regions.

Turning now to the character of the coastal regions, it has long been recognized that the west coast of India, in all probability, has been formed by extensive scarp faulting. Similarly Blanford [2] has put forward the view that there has been in the past a great fault running along the Makran and Baluchistan coasts, though Pilgrim [3] is of the opinion that the character of this coast is to be attributed to a widespread submergence, which " buried fathoms deep the steep mountain valleys, river systems and sea cliffs, which had been carved out during so many ages previously ".

The south-east coast of Arabia between long. 55° E. and Masira Island runs in a S.W.–N.E. direction, in a line that is an almost exact continuation of the African coast from Cape Guardafui southwards. Along this stretch of coast the geological formation consists of horizontally stratified sandstone alternating with layers of a harder material. This formation, which is clearly a sub-aqueous deposit, ends abruptly at the coast in cliffs that reach 400–600 feet in height. The continental shelf here is narrow, and at its seaward edge drops steeply and irregularly into deep water. As Farquharson [4] points out, " the slopes are extremely irregular, consisting of alternate slopes and valleys running out to seaward, in appearance similar to that of a mountain side." The same narrowness of the continental shelf and abrupt descent of the continental slope can be traced along the African coast as far as Point Malindi, just north of Mombasa. On the edge of the continental shelf opposite the Arabian coast lie the Kuria Muria Islands and Masira Island; in the former the horizontal sandstone formation can be well seen in the eastern part, but at the western end of the main island, Jezirat Hallaniya, the strata are tilted upwards (Plate XI), and the extreme west end of this island and the adjoining Soda Island are composed of granite. The granite is intrusive into the sandstone formation, and its slopes can be traced down the continental slope to a depth of at least 1,370 metres. Clearly the stratified sandstone must have extended originally further to the south and south-east, and this, taken in conjunction with the general character of the coast, indicates that there has, in all probability, been an extensive scarp faulting along this and the adjacent African coast.

[1] Sewell, R. B. Seymour, " The John Murray Expedition to the Arabian Sea," *Nature*, cxxxiv, 1934, 686.

[2] Medlicott, H. B., and Blanford, W. T., *A Manual of the Geology of India*, Calcutta, 1879, lxxi.

[3] Pilgrim, G. E., *Mem. Geol. Sur. India*, xxxiv, pt. 4, 1908, 60.

[4] Farquharson, W. I., *The John Murray Expedition, Scientific Reports*, i, 1936, No. 2, 49.

Geological Evidence.

The Expedition was fortunate enough to obtain fragments of rock from the sea-bottom, both on the Carlsberg ridge and from the deep basin (4,800 metres) on its north-eastern side. All of these proved to be of basaltic type. Analyses by one of us (J. D. H. W.) have shown that these sub-marine basalts, as well as a third from the Andaman Sea Basin (depth, 2,286 metres), differ in a striking manner from the average composition of the Deccan Trap in India, for they contain considerably less total iron and in addition the Carlsberg Ridge rocks are poorer in potash. It thus seems clear that these basalts from the Indian Ocean cannot be regarded as a westward continuation of the Deccan Trap. Since geophysicists regard the earth as showing a gradual increase in iron from the periphery to the core, the question may be asked whether it is possible for continental basalts, which presumably have been thrust to the surface through a great thickness of sial, to originate at greater depths and consequently be richer in iron than those erupted on the ocean floor. Although at present it is impossible to give a complete answer to this question, it is interesting to note that Daly's average world plateau basalt (column 5 of the accompanying table) is richer in total iron than the rocks from the Indian Ocean.

Table of Analyses of Basaltic Rocks

	(1)	(2)	(3)	(4)	(5)
SiO_2	51·12	46·55	50·65	50·61	48·80
TiO_2	1·72	1·17	1·19	1·91	2·19
Al_2O_3	14·81	18·13	17·07	13·58	13·98
Fe_2O_3	2·35 } 9·14	5·89 } 10·59	2·16 } 8·17	3·19 } 13·11	3·59 } 13·37
FeO	6·79	4·70	6·01	9·92	9·78
MnO	0·17	0·08	0·17	0·16	0·17
MgO	7·56	3·82	7·74	5·46	6·70
CaO	8·44	13·69	10·09	9·45	9·38
Na_2O	3·93	2·34	3·15	2·60	2·59
K_2O	0·13	0·57	0·36	0·72	0·69
H_2O+	2·36	1·33	0·80	1·70	1·80
H_2O-	0·54	1·03	0·31	0·43	
P_2O_5	0·14	0·92	0·18	0·39	0·33
CO_2	*Nil*	*Nil*	*Nil*	*Nil*	—
Incl.	—	0·04	—	—	—
	100·06	100·33	99·98	100·12	100·00

(1) Average of three unoxidized basalts from the Carlsberg Ridge. Analyst: J. D. H. Wiseman.

(2) Oxidized Variolitic Basalt from the North-East Basin, Indian Ocean. Analyst: J. D. H. Wiseman. As a certain amount of FeO has been oxidized by sea-water to Fe_2O_3, the total FeO + Fe_2O_3 is consequently slightly too high.

(3) Basalt from the Andaman Sea. Analyst: J. D. H. Wiseman.

(4) Average of eleven Deccan Trap analyses. *Bull. Geol. Soc. Amer.*, xxxiii, 1922, 774.

(5) Average World Plateau Basalt. *Igneous Rocks and the Depths of the Earth*, New York and London, 1933, 201.

A further very interesting difference between these sub-marine basalts and those of sub-aerial origin is to be found in the radium content. Dr. J. H. J. Poole has very kindly examined our specimens and has sent us the following data :—

	Radium-content in 10^{-12} gram per gram unit.
Basalt from the Andaman Sea . .	0·43
Station 133, Augite-basalt . . .	0·46
,, ,, Variolitic basalt . .	0·49
,, ,, Hornblende-augite-dolerite	0·49
,, 166, Variolitic basalt , .	0·46
Average . , , .	0·466
Mean for Deccan basalts[1] . . .	0·77
Basalt from Indian Ocean Islands[1] . (Three specimens)	0·86
Average	0·815

Thus the sub-marine basalts contain approximately half the amount of radium as compared with land basalts. At present we are unable to put forward any explanation of this, though it may be possible that continental basalts acquire some of their radio-activity through contamination with richer layers during their upward passage.

At the southern end of the Carlsberg Ridge, the island of Roderigues is volcanic and so also are the islands of St. Paul, New Amsterdam, Kerguelen, and Heard, that rise from the southern part of the continuation of the ridge, namely the Kerguelen Plateau. The Seychelles–Mauritius Bank is also very largely volcanic, as is shown by the geological character of the islands Réunion and Mauritius; and, though the Seychelles group is largely composed of granite, a certain amount of basalt is also present. Between the Seychelles and Madagascar the bank on which Providence Island is situated is also now known to be volcanic.[2] The island of Socotra, like the Seychelles and the western part of the Kuria Muria Islands, is partly composed of granite.

We would here call attention to the apparent similarity which exists between the topography of the floor of the Arabian Sea and the region to the west of it that is characterized by the presence of the Great Rift Valley (Fig. 2). The Rift Valley commences at its northern end in the deep tectonic rift of the Red Sea, and its northern continuation the Jordan valley of Palestine is bounded on the west by the hills of Judaea. In the Arabian Sea we have, corresponding to this, the deep Murray Gorge, bounded on the north-west by the Murray Ridge and its probable northern continuation, the Khirthar Range of Sind. At its southern end the Red Sea opens into the Gulf of Aden which runs from west to east. At its eastern end the Gulf comes into close relations with the south-west extension of the Murray Gorge, thus forming, as it were, a connecting link between the Rift

[1] Poole, J. H. J., *Phil. Mag.*, xlviii, 1924, 827.
[2] Wiseman, J. D. H., *Trans. Linn. Soc., Zool.*, Ser. 2, xix, 1936, 437.

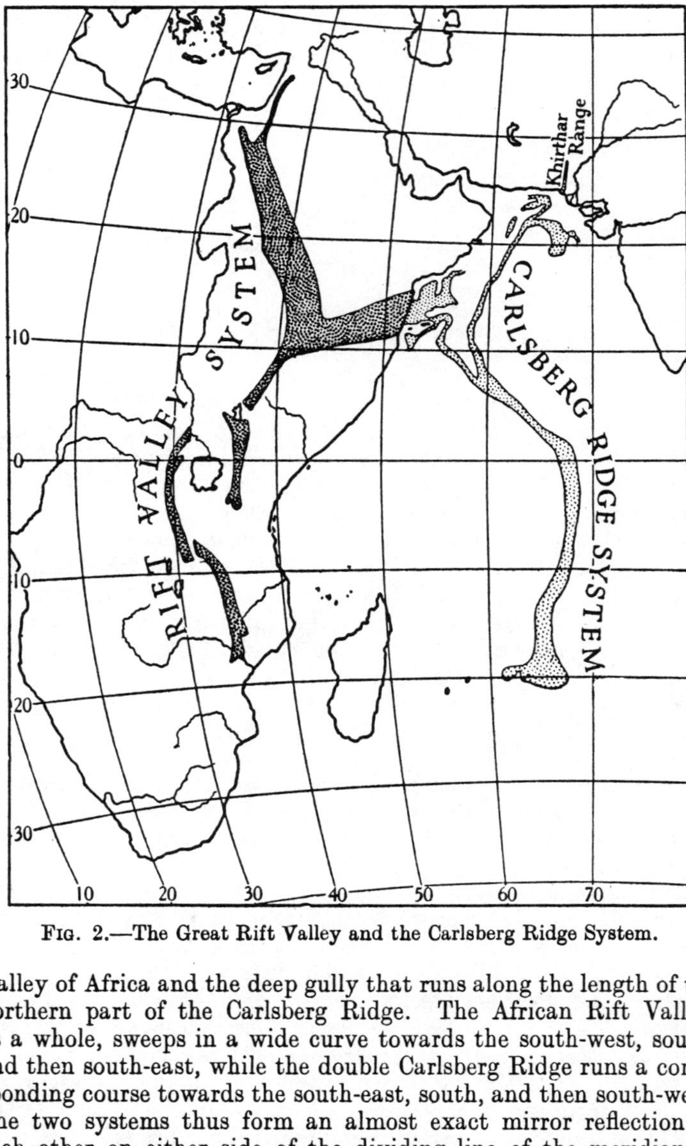

FIG. 2.—The Great Rift Valley and the Carlsberg Ridge System.

Valley of Africa and the deep gully that runs along the length of the northern part of the Carlsberg Ridge. The African Rift Valley, as a whole, sweeps in a wide curve towards the south-west, south, and then south-east, while the double Carlsberg Ridge runs a corresponding course towards the south-east, south, and then south-west. The two systems thus form an almost exact mirror reflection of each other on either side of the dividing line of the meridian of 50° E. (Fig. 2), and each appears to end in the neighbourhood of the parallel of 20° S. In our chart of the sea-floor the Carlsberg Ridge is shown as a continuous double fold, running approximately in the form of an arc. We know that the Rift Valley has in places been interrupted and the main line of its trend is shifted to east or west; but it must be borne in mind that as yet we are acquainted only with the broad lines of the Carlsberg Ridge and that there are stretches of several hundred miles in which we have only very few soundings. It is only the general trend of this great mountain chain that is known, and it is possible, indeed one might say probable, that there are interruptions along its length, as there are along the concentric ridge of the Seychelles–Mauritius Bank.

Seismological Evidence.

Heck [1] has recently published a paper in which he calls attention to the existence of a great earthquake belt, running round the whole earth. The major belt in the eastern hemisphere runs along the line of the Alpine–Himalayan mountain system; but secondary belts have been traced running to north and south. Of these secondary belts one is shown to extend southward along the Red Sea and then down the Rift Valley, and another can be traced across the Gulf of Oman and then down the middle of the Arabian Sea (Fig. 3), where we now know that the Carlsberg Ridge runs its course. A similar secondary belt can be traced down the length of the Mid-Atlantic Ridge. While providing us with an additional point of similarity between the Rift Valley and the Carlsberg Ridge, the presence of these earthquake belts indicates the lines along which the earth's crust is in a condition of instability, and this seems to be borne out by the variations in gravity that have been investigated by Meinesz and others.

Geodetic Observations.

During our voyage through the Maldive Archipelago, we were fortunate enough to have on board Lieut.-Col. E. A. Glennie, R.E., D.S.O., of the Geodetic Branch of the Survey of India, and gravity observations were carried out at nine stations in this group of islands. As a result Glennie [2] reached the conclusion that throughout the length of the Laccadive Archipelago there is a positive anomaly, which he connects with a similar positive variation extending along the length of the Aravalli Mountains and the subsidiary Satpara Range in India. He suggests that the line of upwarp, which has formed this oldest range in India, is continued south and forms the base on which the coral islands of the Laccadives have been built. In the Maldive area the gravity anomaly was found to be negative, and Glennie interprets this as a result of a downwarp in the sima layer which he correlates with the downwarp present over the southern end of the Indian peninsula. If this be so, then it seems reasonable to conclude that there has been faulting between the Laccadive and Maldive Archipelagoes along the line of the Eight Degree Channel.

[1] Heck, N. H., *The Geographical Review*, xxv, 1935, 125.
[2] Glennie, E. A., *The John Murray Expedition, Scientific Reports*, i, 1936, No. 4, 96.

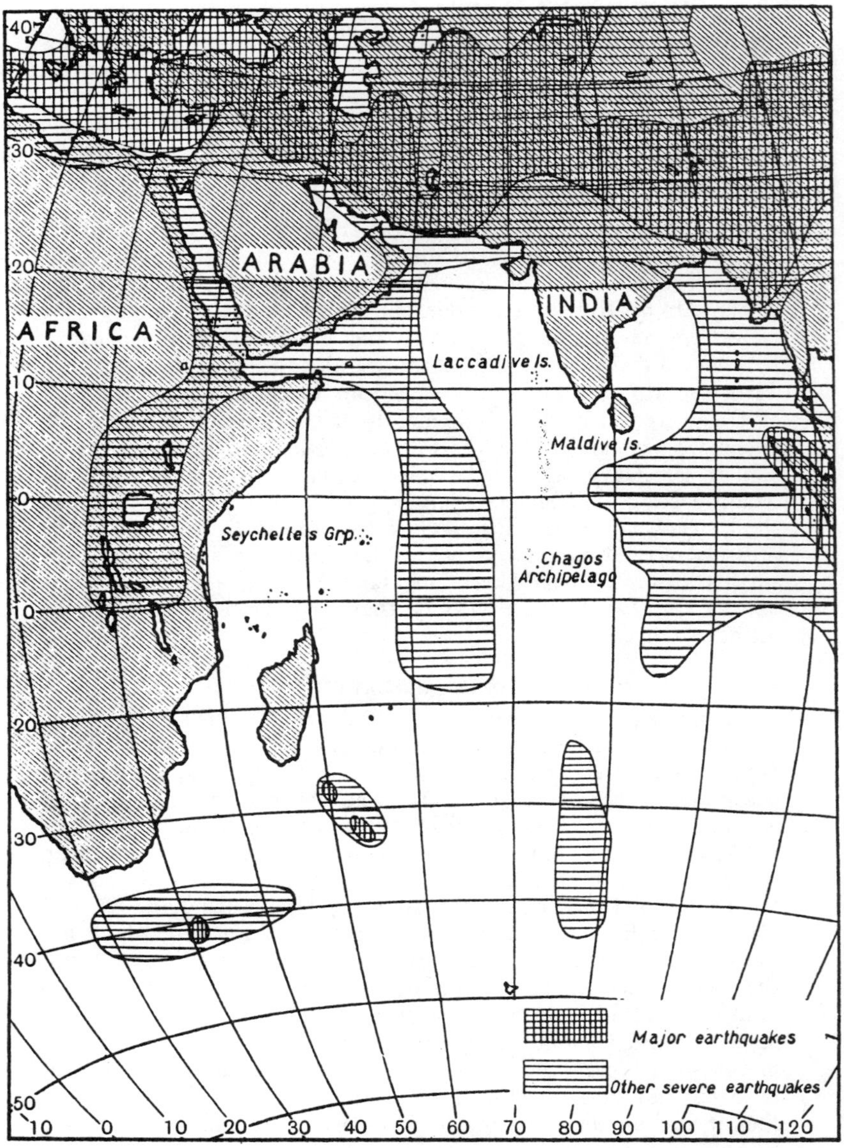

FIG. 3.—The Earthquake Zones of the Indian Ocean and neighouring areas (*after Heck*).

Meinesz [1] in his voyage across the Indian Ocean crossed the northern end of the Carlsberg Ridge and obtained evidence of a rise in the gravity anomaly in the region of the ridge, for his regional anomalies at the following stations were :—

Lat.	*Long.*	*Regional anomaly.*
10° 02′	55° 25′	− 9 milligal.
7° 57′	61° 54′	+ 1 ,,
7° 53′	65° 58′	− 27 ,,
7° 56′	68° 46′	− 31 ,,
8° 06′	72° 48′	− 61 ,,

Unfortunately the existence of the Ridge was not known at the time of his voyage, and Meinesz's stations just miss the major elevation ; but the variations in this area strongly remind one of the similar variation that has been shown to be present over the similar Mid-Atlantic Ridge. In a more recent paper Bullard [2] has shown that in East Africa gravity is on the average slightly deficient. A similar result was found by Meinesz for five stations in the Indian Ocean. Bullard also considers that " the isostatic anomalies are generally small over the plateau but that large negative values occur over most parts of the Rift Valleys, particularly over the western branch ". In addition the same author has shown that at some stations a little distant from the Rift, the anomalies are positive (+ 16 to + 21), and this would agree with the positive variation found by Meinesz to the east of the Carlsberg Ridge in lat. 7° 57′ N., long. 61° 54′ E. Along the African coast the gravity results seem to support the view that there is an off-shore rift of which the deep Pemba Channel is part. He concludes that " the distribution of matter under the Rifts can be accounted for if they have been formed by faulting and folding under horizontal compression, but not if they have been formed by tension ".

The Age of the Carlsberg Ridge System.

We now come to the consideration of the age of the Carlsberg Ridge and its southern continuation, the Mid-Indian Ridge. We have already pointed out that the whole of the northern part of the Arabian Sea is surrounded by a series of faults. Blanford and other geologists have placed the age of the great fault down the west coast of India as Pliocene or Post-Pliocene. If, as we think, the Murray Ridge and its associated gorge is a continuation of the Khirthar Range of India, then its formation must have taken place in Eocene–Miocene times, and the fracture across it must have been subsequent to this. The faulting along the south-east coast of Arabia cuts across stratified sandstones and limestones that are almost certainly of the same age as the Makran Group of sandstones, namely Miocene or a little later, and hence this faulting must be subsequent to this, namely Pliocene or Post-Pliocene.

[1] Meinesz, F. A. V., *Gravity Expeditions at Sea*, Delft, 1934, ii, 89.
[2] Bullard, E. C., *Phil. Trans. Roy. Soc.*, Series **A**, ccxxxv, 1936, 445.

The topographical similarity between the Carlsberg Ridge and the great Rift Valley suggests the possibility that they are contemporaneous, and, if so, then the Carlsberg Ridge is Tertiary in age. It is significant to note that the islands at the southern end of the Mid-Indian Ridge are in all probability Tertiary, a fact which seems to have been definitely established for Kerguelen and is probable for the others, namely St. Paul, New Amsterdam, and Heard. Finally the volcanic agglomerate at the base of Providence Reef has been shown by one of us (J. D. H. W.) to be Eocene–Oligocene in age and probably contemporaneous with " the volcanic activity in the Ampasindara petrographical province, Madagascar ".

It is impossible to estimate the age of the basalts of the Carlsberg Ridge, but it seems highly probable that the floor of the north-west part of the Indian Ocean, as we know it to-day, assumed its present form as a result of compression in Tertiary times, probably contemporaneously with the upheaval of the Alpine-Himalayan mountain system and the arcs of the Malay Archipelago and the formation of the Rift Valley. Subsequently in Pliocene or Post-Pliocene times the area of land that once filled the triangle now bounded by the northern part of the East African coast and its continuation, the south-east coast of Arabia, the Baluchistan coast, and west coast of India, became separated off by a series of faults and was submerged to its present depth. There is little or no indication that any older continental mass or land isthmus such as the hypothetical continent of Gondwanaland or the isthmus of Lemuria, ever existed except in the granitic mass of the Seychelles and perhaps the corresponding granites of Socotra and the Kuria Muria Islands, and it is possible that these are all intrusives, similar to the intrusive granite masses that have been found in the Himalayas.[1]

EXPLANATION OF PLATE XI.

The northern coast of Jezirat Hallaniya, Kuria Muria Islands, showing upward tilting of the limestone strata.

[1] Wadia, D. N., *Current Science*, iv, 1936, 691.

THE NORTHERN COAST OF JEZIRAT HALLANIYA, KURIA MURIA ISLANDS. NOTE THE UPWARD TILTING OF THE LIMESTONE STRATA.

[*Face p.* 230.

15

Reprinted from *Field Engineers Bull.* **13:**27-33 (1939), by permission of the National Ocean Survey, NOAA, U.S. Department of Commerce.

SUBMARINE SCARP OFF CAPE MENDOCINO, CALIFORNIA

Harold W. Murray, Associate Cartographic Engineer
U. S. Coast and Geodetic Survey

The significance of Cape Mendocino and the nearby area to navigators using these waters is a classic example of the use of submarine and shore configuration in position determination. The Pacific Coast Pilot describes the cape as a "mountainous headland" and "famous landmark of the old Spanish navigators". The cape is a turning point for nearly all vessels bound north or south and, in view of the dangers in this vicinity, must be approached with great caution in thick weather because the currents and irregular bottom tend to make the ordinary methods of navigation uncertain. It is also in a region of great climatic change, and the meteorological conditions northward of the cape are quite different from those to the southward. Fog, for example, is more prevalent to the southward, whereas rainfall is heavier and the northwesterly winds of summer are more violent to the northward. The combined influence of these factors on navigation is emphasized by the number of wrecks which occurred during the earlier years of the present century when the then existing hydrographic surveys were so woefully inadequate.

A vessel running the courses commonly used in this locality would approach Cape Mendocino from the north on a course of 176° and would expect to pass about six miles off the cape to clear safely the known dangers. When the cape is abeam and bearing about East, the course would be changed to 160°, a 16° change to the East. That is, to exaggerate, one might say that a vessel has to run out and around the cape, keeping six miles off. It is evident that if one did not run far enough or, in other words, changed course too soon, the vessel would run onto the shoals or onto the cape itself. The currents in this locality are strong and variable, some wind effects are little known, and fog is more prevalent than in most places on the coasts of the United States, there usually being approximately 1,300 hours of fog per year. A vessel passing the cape may well have been running for many hours in fog without sight of any fixed object and be entirely dependent on dead reckoning and soundings for an approximate position.

It was during a period of foggy weather in the year 1916 that the master of the Steamer BEAR, "when his reckoning put him about 15 miles northward of Cape Mendocino, began to take soundings to locate his position" and feel his way to the lightship anchored off the cape. The soundings immediately indicated that the vessel was proceeding over deep water of 100 fathoms or more. When the depths began to shoal from 80 fathoms to 34 fathoms and subsequent soundings showed substantially deeper water, it appeared from the published chart then in use that the vessel had safely passed the cape and the course was changed as usual. About an hour later the vessel stranded two miles north of the cape with a loss of six lives. The contributing factor in the disaster was that the misleading soundings had been obtained in a reported but unsurveyed and, consequently, inadequately charted submarine depression (Eel Canyon) several miles northward of Cape Mendocino.*

* See Coast and Geodetic Survey Special Publication No. 48 (1918), "The Neglected Waters of the Pacific Coast", for details of this and other accidents.

Between the years 1899 and 1917, 15 wrecks or strandings occurred in this area and an additional 50 occurred at other points along the California coast. Accidents off the coasts of Oregon and Washington in this period totaled 26 and 15 respectively. In each case the lack of hydrographic surveys, insufficient knowledge of currents, and inadequate charts were the contributing though not necessarily the sole factors involved.

Hydrography previously executed by the Coast and Geodetic Survey in the vicinity of Cape Mendocino, aside from earlier reconnaissance surveys, consisted principally of 1:10,000 and 1:20,000 scale surveys accomplished between the years 1872 and 1886. These surveys, generally speaking, extended from six to eight miles offshore and included the heads of all submarine depressions terminating within the above mentioned limits. All soundings were vertical casts and probably did not exceed 20,000 in number.

It was not until the years 1919 to 1921 that the Coast and Geodetic Survey Ships WENONAH and LYDONIA, R. R. Lukens and E. H. Pagenhart commanding, surveyed the offshore area of Cape Mendocino and exposed the rather complex submarine topography existing offshore for a distance of at least 67 statute miles. The deeper offshore soundings, totaling about 6,500, were obtained by the laborious and time-consuming vertical cast method, horizontally controlled in part by dead reckoning and by three-point fixes on shore objects. The intensity of hydrography was naturally limited by the methods used, by the length of time available, and by the funds allotted to the vessels, but was accepted as adequate for the needs of navigation and the heretofore reported but mysterious Eel Canyon was now definitely surveyed and firmly secured within the confines of a geographic projection.

In 1935 the Coast and Geodetic Survey's plan of making modern and more intensely developed surveys, a project begun in 1932 at the southern limit of the State of California, had progressed northward to the vicinity of Cape Mendocino. The submarine topography revealed in the waters contiguous to the cape is shown by the submarine contours in the accompanying illustration.

The hydrography represented in the illustration was obtained at selected intervals during the period from 1935 to 1938 by the Coast and Geodetic Survey Ship GUIDE, F. H. Hardy, O. W. Swainson, and E. W. Eickelberg commanding. These surveys consisted of three series. One series of nine 1:10,000 and one 1:20,000 scale surveys embracing the area between the shore line and the 20-fathom curve consisted of 48,000 soundings. The second series was composed of one 1:20,000 and three 1:40,000 scale surveys extending from the 20-fathom curve to distances of three to seven statute miles offshore and consisting of about 25,000 soundings. The last series consisted of one 1:120,000 scale survey with about 7,000 soundings extending from the last mentioned limits to more than 66 miles offshore. In all, a total of more than 80,000 soundings have been taken within the area of the illustration. The soundings obtained are principally echo soundings supplemented by vertical cast and hand lead soundings in the waters adjacent to the shore line. Horizontal control consisted of three-point fix angles on shore objects in the inshore area and radio acoustic ranging in the offshore area.

The contour interval shown on the illustration is 100 fathoms (600 feet) for depths of 100 to 1,800 fathoms. In depths less than 100 fathoms the 10-, 20-, 30-, 40-, 50-, and 75-fathom contours are

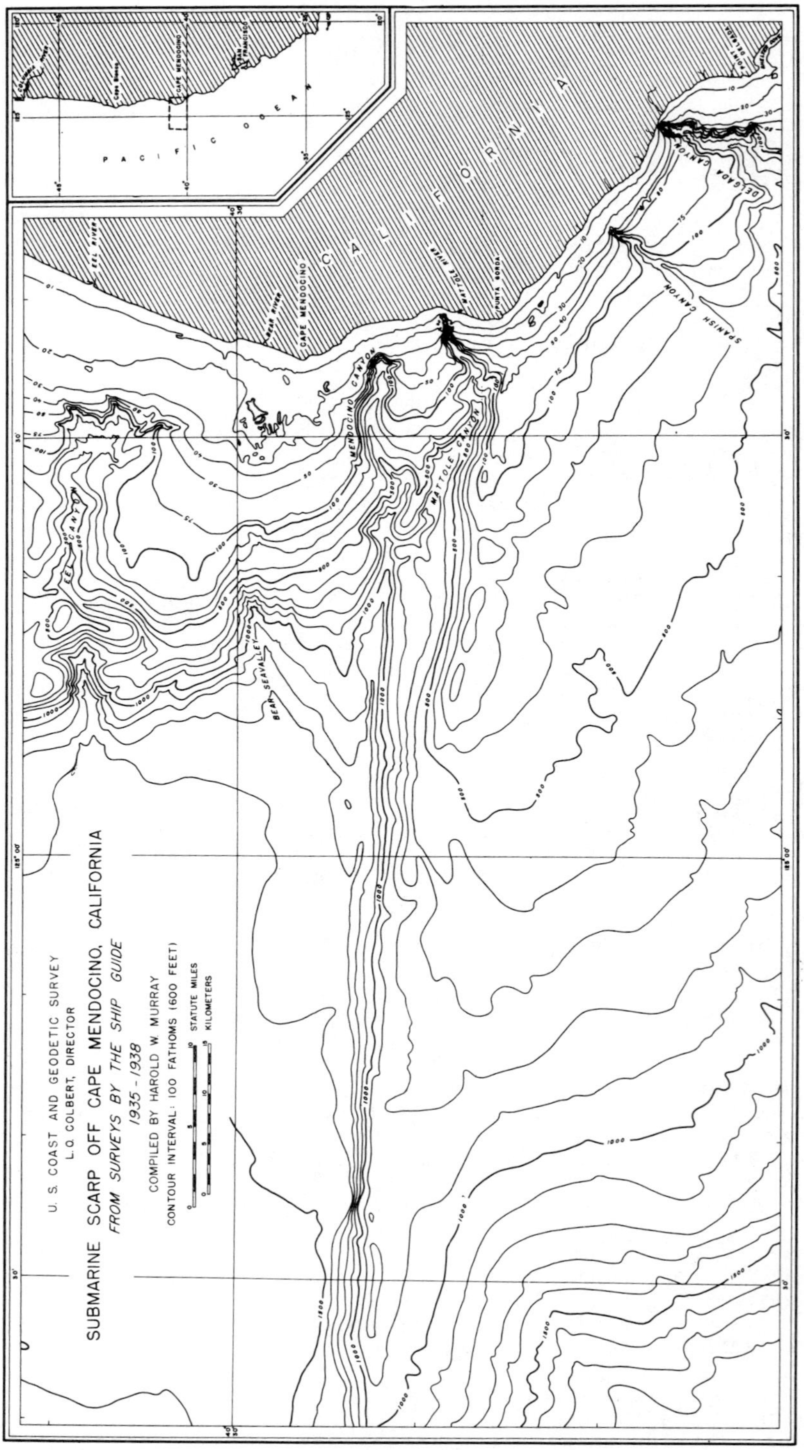
U. S. COAST AND GEODETIC SURVEY
L. O. COLBERT, DIRECTOR
SUBMARINE SCARP OFF CAPE MENDOCINO, CALIFORNIA
FROM SURVEYS BY THE SHIP GUIDE
1935 - 1938
COMPILED BY HAROLD W. MURRAY
CONTOUR INTERVAL: 100 FATHOMS (600 FEET)
STATUTE MILES
KILOMETERS
CALIFORNIA
PACIFIC OCEAN
CAPE MENDOCINO
MENDOCINO CANYON
MATTOLE CANYON
SPANISH CANYON
DELGADA CANYON
EEL CANYON
BEAR SEAVALLEY
PUNTA GORDA
EEL RIVER
BEAR RIVER
SAN FRANCISCO

shown. The contour interval used in the 1:120,000 scale survey previously mentioned was 25 fathoms, or four times greater. This insured a more accurate contour delineation in areas where echo soundings were practically continuous on rather widely-spaced sounding lines.

The diversity of submarine topography expressed in the illustration is self-evident. Heading the list is the long submarine scarp one-half to one mile in height extending more than 66 statute miles from shore. The western extremity of this feature has not as yet been ascertained. Portions of the face of the scarp plunge downward to the north with a steepness of from 24 to as much as 100 per cent. The downward slope of the top of this scarp, measured from the closed 200-fathom contour to the closed 800-fathom contour, is 1.7 per cent. However, from the western extremity of the 800-fathom contour to the western limit of the 1,000-fathom contour (outside the limit of the illustration), the rate of descent has increased to 3.9 per cent. The ocean bottom to the north and northeast of the scarp is quite flat and about one and three-fourths miles below the surface of the ocean, whereas the bottom to the south of the scarp slopes gently southwestward at a rate of 3.8 to 7.6 per cent accompanied by a depth change of from one-fifth to two miles. This scarp with the three types of contrasting topography constitutes a submarine feature too unusual to possess a known rival on the entire West Coast.

The 100-fathom contour closely approximates the limit of the continental shelf which is broader on the north than on the south. The continental slope beginning at the 100-fathom contour slopes away in the broader areas at a rate of about 4.3 per cent just above Spanish Canyon and is as great as 11 to 19 per cent on either side of Bear Seavalley where it is considerably shorter in length. Here again contrasting topography is presented in that the rate of slope on the north is about three to four times greater than that to the southward.

Submarine canyons are well entrenched on either side of the scarp and protrude several miles into the continental shelf. Mattole and Delgada canyons to the southward of the cape are remarkable in that they extend so close to shore. Eel Canyon on the north has a broad head about five miles wide with five pronounced tributaries. It traverses a distance of 32 statute miles between the 30- and 1,400-fathom contours. The bottom gradient between the 75- and 900-fathom contours, a distance of about 25 miles, is from 5.0 to 3.0 per cent. At the 900-fathom contour the bottom slopes steeply to a depth of 1,300 fathoms with gradients as great as 30.3 per cent after which it lessens to about 2.4 per cent. The submarine knoll existing near the mouth of the canyon, around which the stream channel has had to travel 11 miles, is a phenomenon in deflection of submarine canyon courses. This knoll will evoke an interesting discussion as to whether it is younger or older than the canyon, or contemporary with a portion of the canyon's history. The fact that the mouth of the canyon including that portion of the canyon just eastward of the knoll approaches a straight line would imply, for example, that the canyon was fault-controlled and that the knoll was a subsequent intrusion occurring at some time after the canyon was well established.

Bear Seavalley is about 18 miles long between the 75- and 1,400-fathom contours. Its gradient is about 14 per cent down to the 500-fathom contour, thence 30 per cent to a depth of 1,000 fathoms after which it levels out from 11 to as low as 2.9 per cent. The name of this feature was supplied by the writer, all other names shown on the illustration being in use on the later editions of the Coast and Geodetic Survey charts of this area. The term "seavalley", however, is a recent decision of the United States Board on Geographical Names and is applied to submarine depressions that are of valley form but unaccompanied by steep adjacent parallel walls such as are found in canyons.

Although outside the scope of this article, it is nevertheless of practical interest to note that another recent decision is "seamount". This new term is being applied more frequently off the West Coast and is used to denote a submarine elevation of mountain form. As a specific example, the first feature to receive this designation was a submarine mountain discovered by the Coast and Geodetic Survey Ship GUIDE in 1933 about 75 miles west of Point Piedras Blancas, California. This feature rises from a depth of 1,900 fathoms to 729 fathoms and has a net elevation above the ocean floor of 1,171 fathoms or 7,026 feet. It was named "Davidson Seamount" in honor of George Davidson (1825-1911) of the U. S. Coast and Geodetic Survey.

Mendocino and Mattole canyons join at a depth of around 900 fathoms. Their lengths inshoreward from this point are 14 miles to the 40-fathom contour and 18 miles to the 10-fathom contour respectively. Two alternate outlets into the broad region of the 1,400-fathom contour are possible: one where Bear Seavalley enters, and the other about nine miles farther westward. The total lengths, in the case of the longer Mattole Canyon, to the two outlets are about 39 and 48 miles. Mendocino Canyon is more direct and has a gradient of about 6.4 per cent from a depth of 200 to 1,100 fathoms after which it levels out to about 2.2 per cent. The gradient along the major portion of Mattole Canyon is about 5.4 per cent or slightly less. Portions of the side walls of these two canyons are similar and yet contrasting. Near the apex of the 400-fathom contour, Mendocino Canyon has a steep wall slope of about 49 per cent on the north side, whereas Mattole Canyon has its steeper side slope of about 52 per cent of the south side where the face of the scarp serves as a side wall. In the same vicinities the opposing walls of each canyon are also similar in that they have lesser slopes of 21 and 16 per cent respectively.

Spanish and Delgada canyons are only partially shown on the illustration. Spanish Canyon is fairly straight and has a gradient of about 7.2 per cent from the 30- to the 300-fathom contour after which it lessens to about 2.4 per cent. Delgada's gradient is about 15 per cent from the 10- to the 100-fathom contour, thence 11 per cent to the 200-fathom contour after which it changes to about 2.7 per cent.

The contouring of several features represented in the illustration has revealed the desirability of additional development for further geological and seismological researches. Such additional development will necessarily be more comprehensive than that needed for purposes of navigation and will be accomplished by the Coast and Geodetic Survey Ship GUIDE, E. W. Eickelberg commanding.

The new development will consist briefly of:

1. A zigzag echo sounding line extending along the fault scarp to ascertain its present unknown western extremity.

2. A system of short, closely-spaced, vertical cast sounding lines crossing the submarine scarp approximately normal to the depth curves between the junction of Mattole and Mendocino canyons and the western limit of the illustration. These lines will be spaced about five nautical miles apart. The sounding interval will be as close as 100 meters on the steepest portion of the scarp. Bottom specimens will also be obtained at each vertical cast sounding. This development will permit a more exact determination of the face of the scarp and will, of course, be intensified if any unusual features, such as cliffs or terraces, are discovered.

3. Spanish Canyon will be further developed by a system of lines running approximately at an angle of 45° with the depth curves. Other areas to the northeastward, including the longest tributary of Eel Canyon will also receive further development.

While the careful development of these features and the intimate knowledge obtained from them are of considerable scientific value, probably the greatest present value is to the navigator. Previously in this article a short description was given of the stranding of the Steamer BEAR near Cape Mendocino. If the master of that vessel had been provided with a modern chart of this locality, he would not have made the assumption that the cape had been safely passed and the disaster probably would have been averted.

At that period, of course, soundings were obtained by vertical casts or pressure tubes. Nowadays, sounding is much simplified by the use of an echo sounding device with which more and more commercial vessels are being equipped. With such a device and a modern chart, a ship rounding Cape Mendocino in fog can determine its position almost as accurately as though the weather were clear, simply by adjusting the profile of this unique bottom as given by a continuous line of soundings, to the chart.

It is pleasant to think that Nature, while making Cape Mendocino one of the most dangerous points on our coasts, with its reefs, fogs, and generally unfavorable weather, has so patterned the submarine topography that the very peculiarities of its features increase the safety of navigation in this locality.

Following are some remarks by Captain N. H. Heck concerning the seismological aspects of the submarine scarp.

COMMENT

N. H. Heck, Hydrographic and Geodetic Engineer
Chief, Division of Terrestrial Magnetism and Seismology
U. S. Coast and Geodetic Survey

Recent confirmation of early surveys which indicated a vertical scarp running west from Cape Mendocino in about latitude 40°18' raises the interesting question as to whether or not this is an extension of the San Andreas Fault. Since this means a sharp change in direction of the fault as it leaves the coast, it is important to introduce the seismic evidence. It is to be understood that the San Andreas Fault is a major earth feature extending from the vicinity of Cape Mendocino through California into Mexico and possibly even farther. The only similar formation on land, comparable in extent, is the great fault which begins in northern Syria, extends through Palestine and across the Red Sea into Africa. It is possible, however, that the great submarine troughs are similar phenomena.

It is necessary first to appraise the accuracy of the locations of the known earthquakes of the region. Several elements are essential. The instrument should be suited for recording the earthquakes that occur, there should be a proper distribution of stations, the time scale on the records should be open so that the time of arrival of an earthquake phase is accurately known, and the velocities of the earth waves of the region should be known. If we were discussing the problem in the vicinity of San Francisco or southern California, these conditions would be met, but they are far from being met in the Cape Mendocino region.

During recent years there have been seismographs at Ferndale and Ukiah, and it is probable that the earthquake epicenters are reasonably well recorded off Cape Mendocino, but these are not enough stations for best results; and, since we are dealing with a geological phenomena, the history should be as long as possible. Prior to the installation of the two stations mentioned, only approximations of the epicenter could be made. A number of vessels reported feeling shocks in the vicinity, but the difficulty is that no matter what the relation of the vessel to the epicenter may be, if it feels the shock, it appears to come directly from beneath the vessel. It is probable that the recent determinations of epicenters are 10 or 15 miles in error and the earlier ones in still greater error, so that the direct association of earthquakes with the well-defined scarp cannot be made. However, the definition of the scarp is very important in connection with future more precise determinations. While the evidence is incomplete there is reason to think that there is not the same degree of earthquake activity to the northward of Cape Mendocino, either on land or sea, as there is just south of that vicinity.

Professor Perry Byerly of the University of California, in a paper presented at the Richmond meeting in 1938 of the American Association for the Advancement of Science, made the interesting suggestion that the fault continues only until it meets the region where there is a transfer from continental to oceanic crustal conditions (i.e., from 40 km. thickness to practically zero). This is a very interesting speculation, and it is in part supported by the westerly trend of the scarp.

In any case this is an excellent example of the need for accurate depiction of submarine features in connection with geological and seismological problems.

16

Reprinted from *Geol. Soc. America Bull.* **60:**1527–1540 (1949), courtesy of the Geological Society of America

NORTH ATLANTIC HYDROGRAPHY AND THE MID-ATLANTIC RIDGE

By Ivan Tolstoy and Maurice Ewing

CONTENTS

ILLUSTRATIONS

Abstract

This paper presents primarily the soundings obtained by the Research Vessel Atlantis during the summer field season of 1947. It also utilizes some of the data obtained by later expeditions of the Atlantis and by several other vessels. The Atlantis is equipped with a continuously recording fathometer with a range up to 4000 fathoms.

The outstanding features of the area investigated in 1947 (Atlantis cruise #150) are:

(1) A conspicuously flat plain at a depth of 2900 fathoms occupies the deeper sections of the North Atlantic basin between Bermuda and the Azores. More recent data indicate that this horizontal stretch of sea floor extends at least as far south as Lat. 29°N. and possibly as far north as Lat. 40°N., between Long. 50°W. and 56°W., and also that at the same depth a similar plain is found east of the Ridge in the Northern Canary basin. They are dotted with small sea mounts of variable height whose exact shapes have not yet been determined. The size and number of these appear to increase toward the south, and the southern extremities of the plains consist of smooth horizontal stretches interrupted by pronounced elevations and depressions.

(2) The Mid-Atlantic Ridge is characterized by two strongly contrasting types of topography: (a) The central backbone of the Ridge, or Main Range, which is shoaler than 1600 fathoms, consists of a series of parallel ranges trending NE.–SW. Several of these rise to less than 800 fathoms. Their trend follows roughly that of the Main Range. The portion of the Main Range to which the most attention was devoted is between Lat. 30° and 34°N. In this area its width is of the order of 150 miles, and it is of the order of several hundred miles in the area of the Azores.

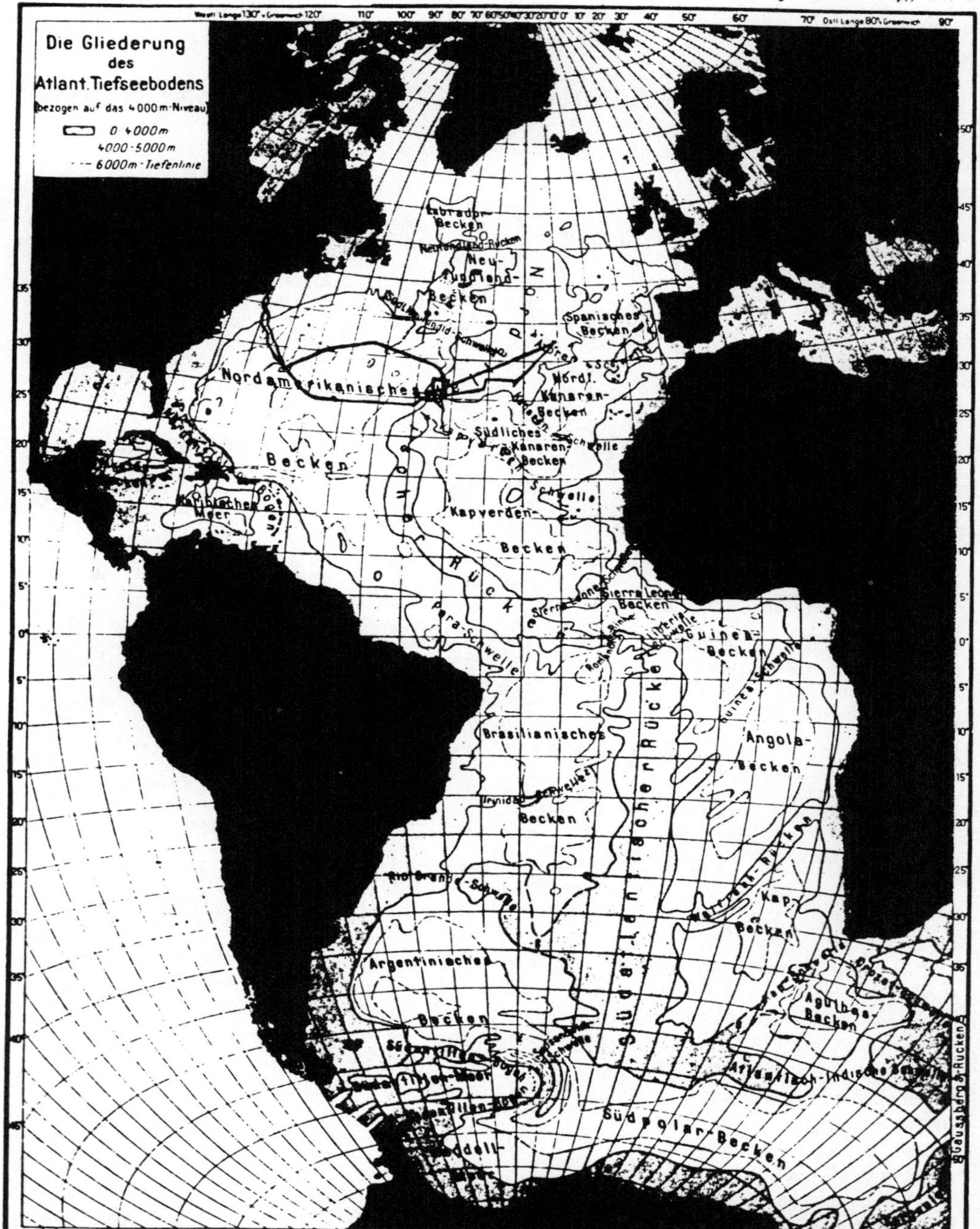

THE ATLANTIC OCEAN, AFTER MAURER AND STOCKS

The path of the ATLANTIS during the summer of 1947 (Cruise 150) is shown by the black lines

(b) The flanks, between the 1600- and 2500-fathom isobaths, consist of a succession of smooth shelves, each from 1 to 50 nautical miles or more in width. This zone is 200–300 miles in width.

(c) The region between the limit of the 2900-fathom plain and the foot of the first of the series of shelves characterizing the flanks of the Ridge, at 2500 fathoms, in some localities stands out as a distinct physiographic province.

(3) The following local features are noteworthy:

(a) Close to 31°N.Lat. a deep east-west trench extends from about 41° to 43°W.Long. and cuts deep into the Main Range. Its deepest point is at 2800 fathoms. Crushed and metamorphosed ultrabasics were brought to the surface by dredging its flanks.

(b) At 30°15′W., 34°N., a flat-topped sea mount rises to 180 fathoms, about 100 miles southeast of the Main Range. The dredge brought up calcareous discs of probably Cenozoic age off its top. They were about 15 cm in diameter and about 4 cm thick.

(c) Another sea mount was found at 33°43′N., 62°30′W., 150 miles NE. of Bermuda, shoaling to 780 fathoms, having a NW.-SE. elongation, an unknown total length, and a width of 4 miles across its flat top.

(d) Another flat suboceanic plain occupies the bottom of the basin separating Bermuda from the Northeastern United States, at a depth of 2650 fathoms.

Introduction

The authors' intention is to bring as rapidly as possible to the public's attention the existence of certain very important features of ocean-bottom topography. It is hoped that they will attract the attention of investigators to certain crucial points which can to a large extent be elucidated by a more widespread use of recording fathometers with a range of 0 to 4000 fathoms.

The present paper is the first of a series which will describe the data obtained by the ATLANTIS cruises in the North Atlantic since the summer of 1947. It was judged sufficient to illustrate it by means of maps and fathograms. Of course, the latter provide only somewhat distorted profiles of the sea floor. However, this distortion is always small enough so that all the striking features stand out very satisfactorily. Part II of this paper will give a series of corrected profile drawings.

Since the 1947 summer cruise (cruise #150) the ATLANTIS has made three cruises across the North Atlantic (cruises 151, 152, 153). Also, three almost uninterrupted fathograms of other ships crossing the Atlantic, as well as several shorter ones and numerous discrete sounding lines, were made available for examination, thanks to the kindness and co-operation of the officers of the U. S. Navy Hydrographic Office. Some of the main features of these data will be mentioned here in passing. Their detailed presentation is reserved for a later paper.

Acknowledgments

This work was made possible through the support of the National Geographic Society, The Woods Hole Oceanographic Institution, The Geological Society of America, and the Bureau of Ships, Navy Department. Most of the data were obtained by the Research Vessel ATLANTIS, belonging to the Woods Hole Oceanographic Institution. The authors are very grateful to Dr. Columbus O.'D. Iselin for facilitating in many ways the task of obtaining data from cruises other than those with which the authors were directly connected. The authors are also much indebted to Dr. J. B. Hersey for the loan of his data obtained on ATLANTIS cruises prior to cruise 150. Through the courtesy and co-operation of the U. S. Navy Hydrographic Office many additional data have been made available to the authors. It is expected that in co-operation with this institution the authors will soon obtain a much more complete and satisfactory picture of the North Atlantic Ocean bottom. Finally the co-operation and work of Adrian Lane, Master of the ATLANTIS, and of his crew were greatly appreciated at all stages of the field work.

The Mid-Atlantic Ridge: Generalities

The Mid-Atlantic Ridge (or Swell, or Rise) has attracted much attention since the end of the nineteenth century. It is the most conspicuous feature of the Atlantic Ocean basin and is essentially a long narrow submarine mountain range which extends from about Lat. 65°N. (Iceland) to Lat. 50°S. It parallels the American, European, and African coasts (Pl. 1). At the equator this parallelism is strikingly illustrated by a pronounced E.-W. flexure.

Plate 1 shows the axis of the Ridge to be almost entirely above 2240 fathoms (4000 m), and bordered by deeper regions which are themselves split up into a series of basins, whose depths are of the order of 3000 fathoms, by ridges branching out from the main body of the Mid-Atlantic Ridge. These diverging mountain ridges give a herringbone-like aspect to the relief of the Atlantic floor, particularly in the southern hemisphere. The high parts of the Ridge shoal to less than 1000 fathoms in places, and the highest peaks form groups of islands such as the Azores and St. Paul's Rocks. North of Lat. 45°N. the Mid-Atlantic Ridge broadens and rises gradually to Iceland. South of Lat. 50°S. the Ridge bends eastward, parallel to the South African coast, and is believed by some authors (Wiseman and Seymour Sewell, 1937) to extend to the Carlsberg Ridge in the Indian Ocean.

Previous Hydrographic Work

The map shown in Plate 1 was published in the account of the Meteor oceanographic expedition of 1925–1927 (Maurer and Stocks, 1933), during which 14 echo-sounding profiles, with soundings spaced every 20 minutes of time, were run from coast to coast between Lat. 20°N. and 50°S. The Meteor (Dietrich, 1939) and Altair (Wüst, 1939) did more surveying in the North Atlantic in 1934–1939, on the ridge north of the Azores and on the Azores plateau itself. The U.S.S. Muir made two crossings of the Atlantic: from Philadelphia to Dakar and back in 1945, taking soundings every half hour.

All these soundings have contributed greatly to our knowledge of the bathymetry of the Atlantic Ocean basin and have in particular helped to emphasize the distribution of submarine mountain ranges with respect to the deeper basins. However, these surveys consisted of a series of discrete soundings. Until now no continuous fathograms of any portion of the Mid-Atlantic Ridge have been available. As will be shown, important features of ocean-bottom structure may escape notice if continuous fathograms are not available.

Gravity Measurements

Vening-Meinesz (1934) has shown that the Mid-Atlantic Ridge does not seem to be associated with any conspicuous gravity anomalies. However, the isostatic anomalies tend to be positive, particularly in the area of the Azores, where the regional isostatic anomalies are the smallest, showing that the compensation of the Azores plateau, like that of many other island blocks, is regional.

Seismicity

Though this is one of the principal seismic belts of our globe, its activity is in no way comparable with that of the Circum-Pacific Belt. All the shocks known to have originated in the area of the Mid-Atlantic Ridge are of the shallow type and are mostly of the c and d classes (Gutenberg and Richter, 1949). There seem to be three regions on the Ridge in which seismic activity is concentrated: the eastern extremity of the equatorial flexure, in the vicinity of St. Paul's Rocks; the western extremity of the flexure, around 10°N. 40°W.; and in the vicinity of the Azores. The portion of the Ridge between Lat. 25° and 35°N. also seems to be somewhat more active than the rest of the Ridge, though this may be due to the favorable location of these epicenters with respect to the existing seismological stations. The first three localities mentioned have produced much larger shocks than have been noted elsewhere on the Ridge—class b and even a (Azores, Nov. 25, 1941). The Mid-Atlantic Ridge is thus the locus of a certain amount of unreleased stress, which seems to be mostly concentrated at the two ends of the equatorial flexure and also in the vicinity of the Azores plateau.

Theories on the Origin of the Ridge

There has been much speculation as to the origin of the Mid-Atlantic Ridge. Haug (1900) believed it to be the median anticline of what he called the Mid-Atlantic "geosyncline".

Taylor (1910) considered it as a horst. Wegener (1924) believed it to be the bottom of the rift that was opened up prior to the separation and drifting apart of the bordering continents. Molengraaff (1928) was of essentially the same opinion. H. S. Washington (1930) stresses the geological significance of the ultrabasics of St. Paul's Rocks. These rocks are peridotites that indicate crushing and metamorphism by pressure. He regards them as evidence of lateral pressure in the building of the Ridge. He also remarks that Prinz's theory of differential torsion of the two hemispheres might explain the prevalent trends of the Ridge, and in particular the equatorial flexure.

Bucher (1940) believes in the essential similarity of the pattern of continental and oceanic structures in the North Atlantic and surroundings areas. He says the pattern of the ocean floor seems to be "comparable to that of the basins and swells of the continental areas outside the orogenic belts, although the scale is both larger horizontally and vertically on the oceanic surfaces."

Umbgrove (1947) emphasizes the symmetry of the distribution of basins and ridges in the South Atlantic and in the African continent and suggests that the prevalent trends are here undoubtedly those of the two major linear patterns of the earth's crust (the N.–S., E.–W., and bisecting systems). Hans Cloos (1939) on the basis of an experiment on clay has shown that the type of topography characterizing the Azores plateau can be explained in terms of doming of a section of the earth's crust with formation of fractures parallel to the longest dimension of the domed section (NW.–SE.). Subsequent extrusion of magma through fissures would have formed the NW.–SE. ridges surveyed by the ALTAIR.

Agostinho (1936) also believes the major topographic features of the Azores plateau are due to the extrusion of volcanics through predominantly northwest-southeast fractures. In addition he thinks the origin of this big bulge in the Ridge is somehow related to the fact that it is situated at what he considers the intersection of two major submarine trends: the almost north-south trend of the Mid-Atlantic Ridge and the almost east-west trend formed by the succession of banks beginning with the Great Banks, through Milne Bank, the Azores, and the banks west of Gibraltar.

Thus the status of the problem of the origin of the Mid-Atlantic Ridge is somewhat confused. The present authors are not prepared to take any definite stand on this aspect of the problem. However, the possible implications of the results obtained so far are indicated, since our investigation has revealed a series of hitherto unsuspected features of the ocean floor, both on the Mid-Atlantic Ridge itself and in other parts of the North Atlantic.

ACCURACY OF DEPTHS AND POSITION

Contours of the ocean bottom were drawn where such a procedure was judged practicable —*i.e.*, when intersections between various parts of the ship's track occurred, or when the distance between two tracks was sufficiently small. No absolute criteria for the justification of such a procedure can be established, since the determining factors, such as the type of topography and accuracy of navigation, are subject to considerable variations from place to place and from day to day. Accuracy of the topographic maps varied and depended upon numerous factors, the relative importance of which cannot always be determined. The following causes of error should be noted, and kept in mind when considering the results here reproduced.

(1) The ship's position was obtained by astronomical methods and dead reckoning. An accuracy of ±1 mile is a reasonable estimate for the ship's position as obtained in the evening and morning from star fixes, since especial care was taken. The noon fix should not be relied upon to better than ±2 miles. The largest possible errors would occur for intermediate times.

The effect of currents and wind on a small ship, cruising at 8 knots, can be very great and had to be considered several times in the drafting of the hydrographic maps. Such corrections are made possible by the closing, at one or several points, of the network of ships' tracks. Accuracy is proportional to the number of ships' tracks available, particularly when associated with star fixes. Such corrections were made by mutual adjustment of the ship's supposed courses drawn on tracing paper, which process was continued until all intersections were made consistent with the soundings. In many cases the necessary adjustments were small, and in several cases unnecessary. Several times when some interesting feature was encountered, such as an unusually high ridge, box surveys were conducted. A zigzagging course was adopted, each change in course being determined from the readings on the fathogram, in order to try to enclose the feature within a network of shortest possible tracks. Accuracy of positioning of the soundings increases with the number of intersecting courses and with more rugged topography. However, absolute geographical positioning will remain open to errors of a larger order of magnitude, except when there are one or several star fixes in the area. These are always fixed points in the transformations of the adjustment.

(2) The errors of sounding may be due to three causes:

(a) Departure from 60 cycles of the frequency of the generator supply to the synchronous motor running the stylus arm of the recording apparatus. The vertical divisions of the recording tape are calibrated to give exact depth readings for a veolocity of the stylus arm corresponding to 60 cycles. A departure from this figure of ± 1 cycle will give a reading which will be in error by $\pm \frac{1}{60}$ of the true depth, which at a depth of 2000 fathoms would mean an absolute error of 33 fathoms. Most of the recording is free from errors of ± 1 cycle or more—or, at least, these will usually be readily detectable on the fathogram. However, since there was no automatic frequency-control device, changes of the order of half a cycle may have been fairly common and may have been overlooked. Therefore, in oceanic depths the soundings should not be relied upon to better than ± 20 fathoms.

(b) The angle at the apex of the outgoing cone of sound, which leads sometimes to the so-called slope correction (Adams, 1942). This error is not sufficiently large to be taken into account in the plotting of the maps. Thus if a vertical wall were supposed to rise from the ocean floor to sufficient height from a depth of 2000 fathoms, and if we suppose the half angle at the apex of our cone of sound to be of the order of 25°, the error in the positioning of the foot of the slope would be of 12000 sin 25° = 5000 feet approximately. If this wall were 1000 fathoms high, it would therefore record as a slope of $\tan^{-1}\frac{6}{5}$ or 50° approximately. Now since 5000 feet is always an error of lesser magnitude than those involved in dead-reckoning navigation, and since the steepest continuous average slopes of any length that were recorded were of the order of 25°–30°, there is little reason for applying this correction to the drafting of our topographic maps. We must just keep in mind that the slopes shown on the fathograms are a little less steep than the true slopes, and that vertical cliffs such as might be expected to occur in connection with long continuous rises of many thousands of feet over a few miles will not show up as such on fathograms.

(c) Variations in the velocity stratification of the ocean. The errors due to this cause are not very large (of the order of 1 per cent of the total depth or less). We here follow the general policy of such centers of hydrographic study as the U. S. Hydrographic Office and the U. S. Coast and Geodetic Survey and disregard them.

General Description of the Areas Investigated

Plate 2 shows the various sounding lines now available. They include Atlantis cruises Nos. 143, 144, 147, 150, 151, 152, and 153, and most of the lines made available to us through the courtesy of the U. S. Navy Hydrographic Office. Since new data are coming in continuously, it was deemed inadvisable to make a detailed presentation of the material examined so far. Therefore, except for a few detailed local surveys, we will limit ourselves to a very generalized description of the various types of submarine topography in the North Atlantic.

Plate 2 was drawn to illustrate (1) the general trend of the Ridge between Lat. 40°N. and 29°N.; (2) the distribution of the known conspicuous sea mounts; (3) the extent of the data upon which we have determined the generality of certain types of submarine topography; (4) the control upon which to a large extent our 1600-fathom contour was based. Where no lines are shown, we either omitted this contour or used the few scattered soundings available on the H. O. 0955 and 0956 charts.

The fathograms shown in Plates 3, 4, 5, and 6 were obtained during the Atlantis cruise No. 150 and illustrate the major types of topography in the North Atlantic. We will confine ourselves mostly to a description of the area covered on this cruise.

A study of the lines in Plate 2 has shown that many of the types of topography encountered are characteristic of broad areas of the ocean floor.

On cruise 150 the Atlantis went from Woods Hole to Bermuda, from Bermuda to the Azores and back, thus crossing the northwestern basin of the Atlantic (the "North American Basin") and the Mid-Atlantic Ridge. These areas were characterized by several distinct types of bottom topography. Thus the slightly shoaler part of the North American basin, which was crossed on the trip from Woods Hole to Bermuda, has somewhat different characteristics from those of the deeper part, between Bermuda and the Ridge. The central and deepest part of the latter is occupied by a great horizontal plain at the depth of 2900 fathoms. The areas bordering this basin, on the other hand, were characterized by more rugged topography, and the eastern border was quite different from the western. The rugged and high central part of the Mid-Atlantic Ridge is quite different from its shelved flanks. The reader should realize that this subdivision of the ocean into separate zones is by no means arbitrary, but corresponds to what appear to be distinct physiographic provinces of the ocean bottom.

Ocean Bottom Between Cape Cod and Bermuda

On cruise 150 one continuous fathogram from Woods Hole to Bermuda was obtained. This has since been matched by numerous other fathograms taken by the Atlantis and other ships. Portions of the fathogram of cruise 150 are reproduced in Plate 3. This shows a succession of steplike horizontal or imperceptibly sloping levels at 1500, 1950, 2100, and 2500 fathoms (Pl. 3a). The last three are more pronounced than on a typical crossing of the continental slope. However, a 1500 ± 100 level is a fairly general feature of the continental slope between Lat. 40° and 35°N. The deepest section of ocean floor along this line occurs about half way between Woods Hole and Bermuda, at 2650–2700 fathoms. It is about 250 miles wide and is remarkably flat and featureless, except for one fair-size sea mount at 36°38′N., 67°55′W. Unfortunately this sea mount was encountered on the homeward bound course, when the fathometer was not recording. Subsequent cruises by the Atlantis (151 and 153) and the R/V Caryn (winter of 1947–1948) have shown that it rises to 1590 fathoms.[1] During cruise 150 the Atlantis obtained a core 20 miles southeast of this mount. It showed coarse quartz sand stratifications in the upper section of the core. This sand is now being studied by D. Ericson. Recent (Trask, Phleger, and Stetson, 1947) investigation has shown the presence of sand in the deep basin of the Gulf of Mexico. More recent cruises of the Atlantis have found sand at depths of many feet in various parts of the North American basin. The presence of these sands poses one of the most baffling problems of submarine geology.

At Lat. 34°N. on the track of A150, at the end of the 2700-fathom stretch, the bottom starts a gentle overall rise toward Bermuda. The ocean floor on this slope is quite different from that noted off the continental shelf. The succession of levels is not so apparent as it was on the continental rise. Numerous steep sea mounts (10- to 30-degree slopes, rising between 100 and 500 fathoms above the ocean floor on which they are situated) dot the gentle overall rise (Pl. 3). Only in between these peaks is there any indication of the existence of levels. Thus Plate 3 shows a shelf at 2500 fathoms, before the steeper rise toward Bermuda starts. This shelf is even more pronounced to the east (A151). It is preceded on all available fathograms by a sharp peak.

The Vicinity of Bermuda

Figure 1 shows a bathymetric map of the approaches to Bermuda, on a plotting-sheet scale, and the eight Atlantis tracks upon which these contours are based. The base of the island has a NE.-SW. trend and is about 100 miles wide in a NW.-SE. direction. Approaching Bermuda from the north (Pl. 3d), at the depth of 2450–2500 fathoms the first change in slope occurs in the slow upward grade toward the south. After this the bottom rises much more rapidly. These fathograms show the large sea mount 25 miles north of Bermuda, rising to less than 500 fathoms. The steepest part of the slope is between 850 and 100 fathoms; its average inclination is of the order of 20°–25°.

[1] Recent data show that the whole section of the North American basin northeast of the Atlantis 150 track is characterized by a flat plain at 2650 fathoms, dotted by a number of large flat-topped sea mounts some of which rise to less than 800 fathoms (Pl. 2). A number of these mounts appear to string out in a southeast direction, away from Cape Cod. A detailed report on this area is now in preparation.

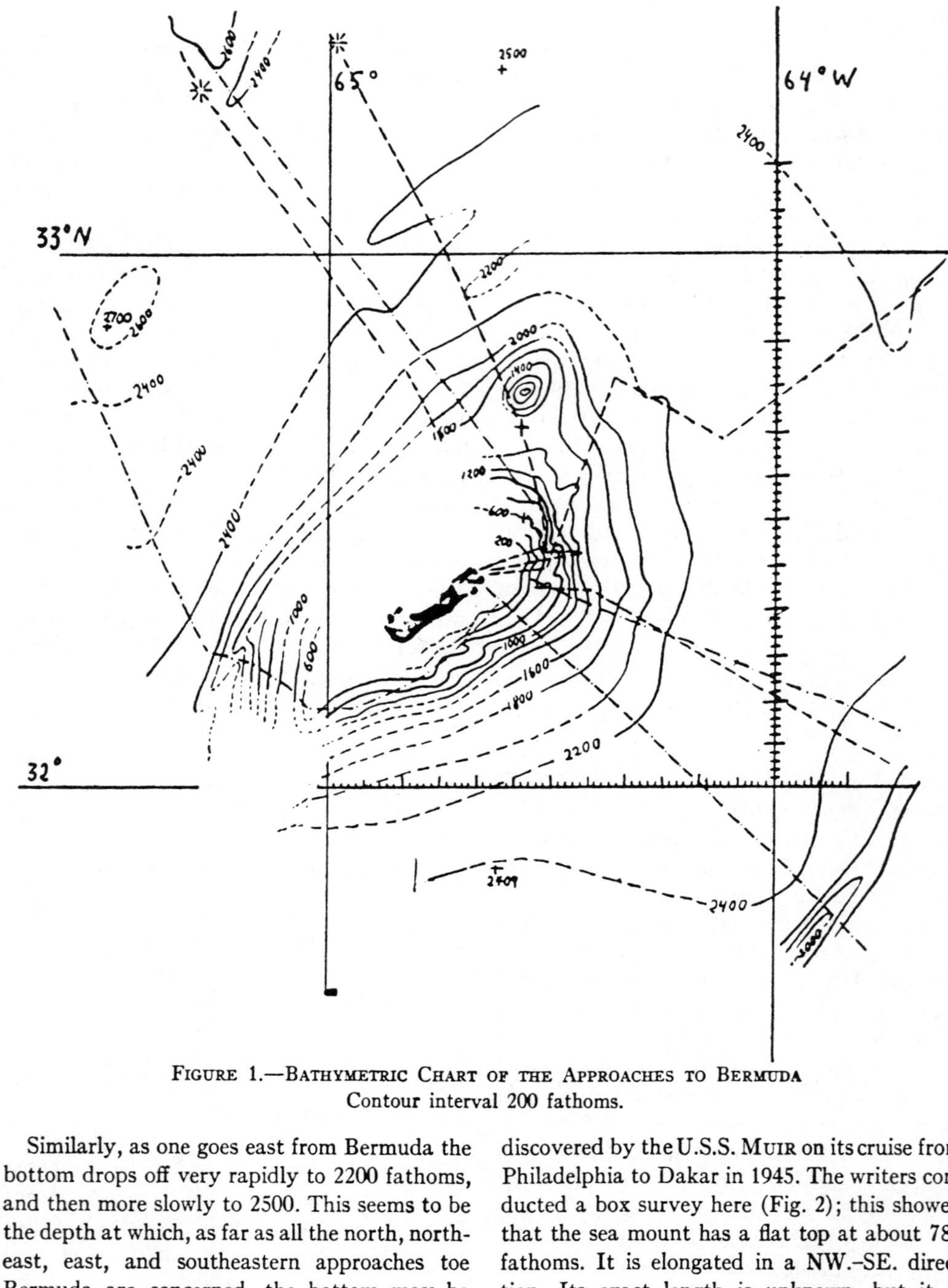

FIGURE 1.—BATHYMETRIC CHART OF THE APPROACHES TO BERMUDA
Contour interval 200 fathoms.

Similarly, as one goes east from Bermuda the bottom drops off very rapidly to 2200 fathoms, and then more slowly to 2500. This seems to be the depth at which, as far as all the north, northeast, east, and southeastern approaches toe Bermuda are concerned, the bottom may be said to start feeling the influence of the island. On the northeastern course from Bermuda the 2500-2550 fathom stretch of sea floor extends for about 32 miles. It is then followed by a rise, gentle at first, then steeper and steeper until the depth of 780 fathoms is reached 130 miles from Bermuda. This sea mount was originally discovered by the U.S.S. MUIR on its cruise from Philadelphia to Dakar in 1945. The writers conducted a box survey here (Fig. 2); this showed that the sea mount has a flat top at about 780 fathoms. It is elongated in a NW.–SE. direction. Its exact length is unknown, but it is probably less than 30 miles.

The fathogram is reproduced in Plate 3d. Its appearance recalls somewhat the description given by Hess (1946) of the Pacific guyots, and the sea mounts of the Gulf of Alaska described by Murray (1945). A short core obtained from the summit shows a discontinuity between a

13-cm layer of Recent sediment and an underlying chalky mud containing Eocene Foraminifera, suggesting that in recent times some scouring agent has been active close to the surface miles wide. Thence a generally level plateau at 2400 fathoms extends for about 60 miles, followed by a steplike succession of flat stretches at 2550–2600 fathoms and 2750–2800 fathoms,

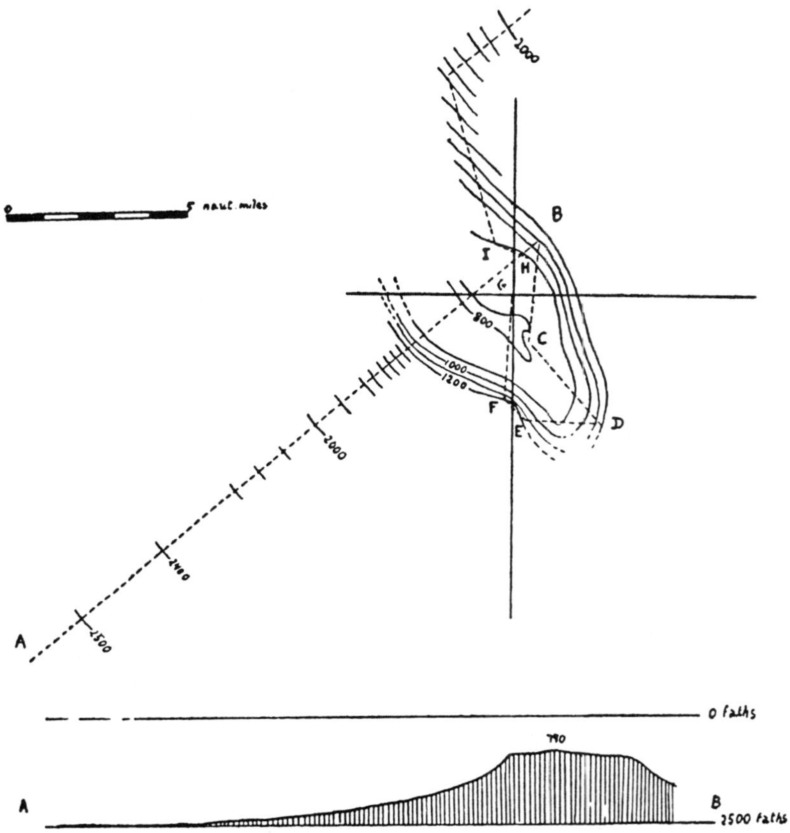

FIGURE 2.—SEA MOUNT NORTHEAST OF BERMUDA

The letters corresponding to changes of course are shown at the corresponding places on Plate 3d. Contour interval 100 fathoms. The vertical and horizontal scales on the profile are the same.

of this sea mount. Northeast of this mount, the bottom drops off rapidly to 2700 fathoms.

THE OCEAN FLOOR BETWEEN BERMUDA AND THE 2900-FATHOM PLAIN

The section of ocean floor here considered extends from points A to B of Plate 2 on the northern A150 track.

The fathogram reproduced in Plate 4a shows gentle topography in this section of the ocean floor. After a short (3 miles) 2700-fathom stretch northeast of the sea mount just described, it is characterized essentially by a long gentle swell which rises to 2300 fathoms and is about 60 with perhaps an intermediate flat step at 2650 fathoms having a width of 20 miles (Pl. 4a).

The end of the 2750 level seems to be characterized by a narrow (2 miles) flat-bottomed trough (2840 fathoms) followed by a sharp peak (2350 fathoms) and a broad hump rising to 2560 fathoms and 20 miles wide (Pl. 4a).

To the south the rise from the 2900-fathom plain to Bermuda is more rugged. Thus all soundings pertaining to this area show that there are numerous peaks, separated by deep valleys.

The fathograms obtained between Bermuda and the 2900-fathom plain confirm this. They show that the bottom consists of a succession of much sharper peaks, separated by deep valleys and narrow flat-bottomed trenches reaching much greater depths than any valley on the northern track (2900 to 3100 fathoms). Judging from existing data, their trend appears to be northeast-southwest.

The 2900–fathom Plain

At point B on the northern route (Pl. 2) a depth of 2900–2880 fathoms is reached. At this point and depth begins what is perhaps the most remarkably flat stretch of ocean floor in the Atlantic. It extends, with a few knolls here and there, for over 300 miles of track (B to C on Pl. 2).

A typical fathogram obtained in this area is reproduced in Plate 4b. Both the western (Pl. 4b) and the eastern extremities (Pl. 4c) of this plain are characterized by a multiplication of the otherwise small number of knolls. These look like small pimples interrupting the otherwise horizontal ocean floor. This 2900-fathom plain is now known to occupy a very large section of the North American basin. Though we cannot yet delineate its boundaries precisely, owing to the insufficient amount of continuous fathograms, it may extend as far north as Lat. 40°N. In these latitudes it might be 50 or 100 fathoms shoaler. Several fathograms show it to be several hundred miles wide in the vicinity of Lat. 31°N., though it appears to be marked by more numerous and higher knolls. Many of these rise to over 100 fathoms above the level plain. Also several higher peaks rise 1000 fathoms or more above this level. Some of these appear to be ridges having either due north or east-northeast trends. The level plain is somewhat deeper (2950 to 3000 fathoms). Its beginning, at 31°N., 58°W. on the fathograms of A151 is characterized by a high peak (about 1000 fathoms in height), having slopes of the order of 20° to 30° on the fathogram. (The width of its base is about 9 miles.) In other words, when the slope correction is considered, its walls would perhaps be more than 40° steep (Fig. 3). It is preceded by a small trough, 2–3 miles wide, the bottom of which is horizontal at 3100 fathoms.

The fathograms of Atlantis cruise 151 show a similar plain occupying the floor of the basin east of the Ridge, between the Ridge and the northwestern coast of Africa (the northern Canary basin). It appears therefore that this 2900-fathom plain is of great regional significance. Its aspect on the fathograms suggests that vast portions of the ocean floor have remained tectonically undisturbed since their formation. The

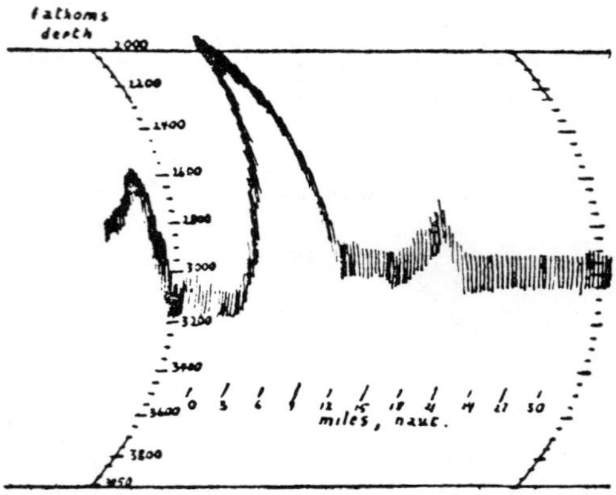

Figure 3.—Sea Mount at the Western Edge of the 2900-Fathom Plain at Lat. 31°N.

This is a tracing of a fathogram showing a sea mount located at the point marked X on Plate 2. (31°N., 58°W., approximately).

small knolls may be due to one of several causes: block faulting, lava escaping through fissures, or submarine volcanoes. Their true shapes and distribution pattern over the ocean floor remain to be ascertained.

The Ocean Floor between the Eastern Edge of the 2900-fathom Plain and the Foot of the 2500-fathom Terrace

The 2900-fathom plain ends rather aburptly at point E on Plate 2 against a series of high peaks, one or two of which have slopes exceeding 30°, and heights of as much as 500 fathoms (Pl. 4c). All available fathograms between Lat. 36° and 30°N. show the presence of this mountainous region between the 2900–3000 fathom plain and the "Terraced Zone" of the Ridge. Seismic work by Ewing and Press aboard the Atlantis on cruise 150 has shown that this area gives a different type of reflection, indicating the presence of sizable amounts of mud (several hundred feet), as against practically none on the 2900-fathom plain and much larger quantities in the Terraced Zone of the Ridge.

A sea mount of exceptional dimensions is located at 35°12′N., 45°33′W. Its western foot is marked by a small trough (2600 fathoms). It rises from this depth to 1300 fathoms (Pl. 4d). It marks the end of the mountainous area and the beginning of the Terraced Zone of the Ridge.

The Terraced Zone

On the eastern side of the peak we have just described the bottom drops off to 2500 fathoms, at which depth it remains approximately flat for about 35 miles (Pl. 4d). On the track of Atlantis 150, along Lat. 31°N., a 2500-fathom level stretch of the same width was found, preceded also by a high peak at approximately 31°N., 45°30′W. shoaling to 1800 fathoms. One cannot say what relation these two remarkably similar structures bear to one another.

This marks the beginning of the succession of flat stretches characterizing the lower flanks of the Ridge, which, for lack of a better term, we shall call terraces. The area characterized by this type of topography we shall henceforth call the Terraced Zone. The authors do not wish to imply anything concerning the origin of these flat stretches by the use of the name "terrace." It was merely deemed more advisable to use some familiar term instead of coining a new name.

Along the Atlantis 150 track the 2500-fathom terrace ends against a 2000-fathom peak bordering a 2300–2350-fathom terrace, approximately 35 miles wide (Pl. 4d). The inner eastern edge of this terrace is marked by a broad hump of small elevation, itself broken up into a series of narrow ridges shoaling to 2100 fathoms. Next comes a 2250–2270-fathom terrace 25 miles wide and characterized by its remarkable flatness. Similar terraces appear on all fathograms approaching the Ridge between Lat. 35° and 30°N. at approximately the same depths. Conceivably the differences in depth for terraces found along parallel lines would not always be real, but would be due in part to instrumental causes.

The 2250-fathom terrace marks the beginning of that portion of the Ridge and its flanks which was closely investigated during the summer of 1947. The zone between the 2250 and 1600 isobaths is 60–100 miles wide on both sides of the Ridge. It is characterized by a discrete series of flat terraces that occur at the same depths on both sides of the Ridge, at points hundreds of miles apart. The reflection shooting studies done during the same cruise by Ewing and Press indicate great thicknesses of sediments on these terraces (up to 1000 feet and more). The fathograms reveal that the most conspicuous terraces, remarkable both in numbers, in different localities, and in total length, are at depths of 2250, 2150, 2100, 2050, 1950, 1850, 1800, 1750, 1700, and 1600 fathoms. Of these the most numerous were at 2250, 2050, 1950, 1850, and 1800 fathoms. Plate 5 reproduces a few typical records from the terraced zone. We have given no detailed charts of the exact distribution of all the individual terraces enumerated above because the number of tracks available over most of the terraced zone is insufficient as yet to enable us to establish any definite correlations between most of them. However, in a few cases it has been possible to establish without much doubt the continuity of some of these terraces over great distances.

In Plate 7 the 1900-fathom terrace might be continuous for more than 90 miles and appears to follow closely the general NE.–SW. trend of the Ridge. The 2050-fathom terrace might be continuous for about 80 miles. The corresponding fathograms are reproduced in Plate 5a. The fathograms of the crossing of the Atlantis 153 cruise over this area show that the 2100-fathom level (Pl. 7) extends for perhaps another 100 miles farther west. Possibly, therefore, these terraces form a series of horizontal fringes contouring the outer edge of the higher portions of the Mid-Atlantic Ridge.

Plate 5b shows the fathogram obtained in the region of point D on Plate 2. The summit of the peak on the extreme right corresponds to the change in course made at that point, indicating that the change occurred before reaching the true summit of the sea mount. It emphasizes the possibility of a correlation between the 1700-, 1750-, and 1850-fathom terraces on each side of the sea mount. The portions of these visible on the fathogram would be sections of what may well be a continuous series of terraces circling the flanks of the sea mount.

Many of the terraces described are horizontal, but in some there is evidence of tilting. This is particularly well illustrated by the fathogram shown in Plate 5c in which a break is suggested in the center of a 1750-fathom terrace with subsequent tilting of the left (southwestern) portion. Many of these terraces are bordered on their outer edge by peaks from 50 to several hundred fathoms in height. Others are enclosed on both sides by steep walls having slopes of as much as 30° to 35°. Examination of the fathograms obtained by the ATLANTIS cruise 151 reveals that the same terraces, at exactly the same depths and characterized by the same horizontality over many miles, are to be found between 1600 and 2500 fathoms off the coast of North West Africa. Plate 5d shows the aspect of the terraced zone immediately south of the area represented by Plate 7.

The Main Range of the Mid-Atlantic Ridge

This part of the Ridge is notable for the absence of terracing. It consists essentially of a series of parallel ridges trending NE.–SW., and shoaling in places to less than 800 fathoms. Plates 6a and b show a few samples of the records obtained over the Main Range.

Only at the outward edge of the Main Range is any evidence of terracing found; it is in the form of flat-bottomed valleys (Pl. 6a, b; Pl. 5a) that appear mostly in the transition zone between the Main Range and the Terraced Zone The depths of their level floors are those of the shoaler series of terraces recognized elsewhere in the terraced zone: 1600, 1700, 1750, 1800, and 1850 fathoms.

Wherever box surveys were conducted over the Main Range or where the intersection of two of the ship's courses made the drawing of isobaths possible, the ridges show a remarkably constant NE.–SW. trend. This is also the trend of the Main Range as a whole in the area surveyed, as may be seen from Plate 2. Plate 7 emphasizes the NE.–SW. trends of the individual ridges of the Main Range.

Dredging the eastern flank of the steep ridge running along the northwest edge of the 2100-fathom terrace at point G of Plate 7 brought up a large block of manganese-encrusted limestone of probably Cenozoic age.

A major exception to the NE.–SW. trends of this area is a deep trough running E.–W. along Lat. 30°N. from 41°W. to 43°W., which can be seen in the southwestern section of Plate 7. A detailed survey was performed in this area (Pl. 8; Pl. 6c, d). The trough is bounded by steep walls having in places average slopes of the order of 30°. The deepest point of the trough is at 2800 fathoms in the western part. The bottom appears to be irregular; two deeper sections (more than 2600 fathoms) are separated by a narrow pass at 2200 fathoms. Several dredging stations were made here, on both the northern and southern flanks of the trough, in the general area of the end points of the tracks. In each case numerous rocks were brought to the surface. Shand (1949) identified them as being strongly or partly metamorphosed ultrabasics: peridotites, olivine gabbro, and olivine basalt. Many of them show evidence of crushing and slickensiding. One or two specimens are practically fault breccia. Some of the basalt fragments show a glassy surface, indicating rapid chilling. Both the topography of this trough and the nature of the rocks obtained therefrom suggest that it may be a graben. It seems significant that this feature is in the center of one of the seismically more active regions of the Mid-Atlantic Ridge (between Lat. 25°N. and 35°N.). This and the absence of traces of the terraced type of topography along the walls of this trough would indicate that it is a fairly recent feature, superimposed upon the typical Mid-Atlantic Ridge type of topography.

Finally we call attention to the remarkably large sea mount at point E on Plate 2. (*See also* Figure 4 and Plate 6c, left end.) Here is another sea mount of the flat-topped variety, which rises to less than 200 fathoms from the surrounding bottom at 1900 fathoms. Dredging the top yielded a large number of calcareous discs characterized by peculiar cavities in their centers. Whereas the outside of the pebbles is relatively smooth, the interior of these cavities is rough. Their dimensions are variable; they average about 15 cm long and about 4 cm thick. Thin sections of these discs reveal the presence of numerous pteropods. According to D. Ericson, all are species that could have inhabited the ocean from the early Tertiary to the present. None are sufficiently characteristic of a specific period to date this limestone.

The fathograms taken in the vicinity of the

Azores show that the plateau on which these islands are situated is limited on its southwest side by a steep slope about 600 fathoms high (Pl. 6c). The topographic map which resulted

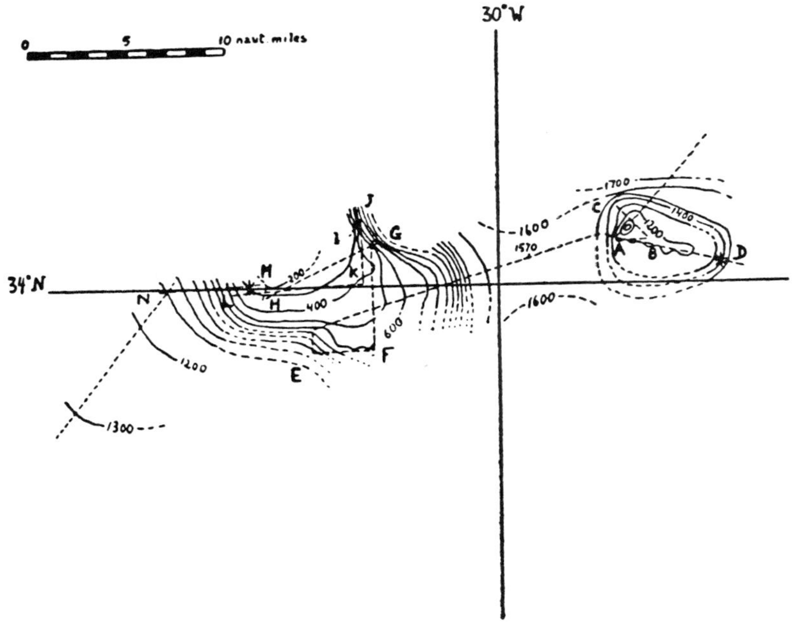

FIGURE 4.—SEA MOUNT AT POINT E OF PLATE 2

This is a sketch-map interpretation of the fathograms shown in Plate 6c of the flat-topped sea mount southwest of the Azores. The letters correspond to those of Plate 6c. Contour interval 100 fathoms.

from the investigations of the ALTAIR shows that the Azores plateau is limited on its sides by such a slope and forms an easterly bulge of the Mid-Atlantic Ridge.

The type of topography shown on the fathograms on the Azores plateau obtained by the ATLANTIS was quite different from any noted on the rest of the Ridge. The surface of the plateau is characterized by a series of ridges, which the ALTAIR expedition plotted as trending NW.-SE. (*i.e.*, approximately at right angles to the trends noted in the Main Range farther south), and whose spacing is entirely different from that of the ridges forming the Main Range of the Mid-Atlantic Ridge. Instead of being close together with only a narrow steep-walled valley separating each two ridges, the ridges on the Azores plateau are separated by broad valleys. It seems improbable that two such different types of structure could have been built by the same process.

SEISMIC EVIDENCE

Reflection-shooting studies of the ocean bottom by Ewing and Press on board the ATLANTIS during the summer of 1947 indicate a greater thickness of sediment in the terraced zone than on the 2900-fathom plain west of the Ridge or on the Main Range. The latter seems to be devoid of any sediment covering the mountain slopes. Each time a mountain was approached, the sediment covering the terraces thinned conspicuously.

INADEQUACY OF DATA OBTAINED BY SOUNDINGS WITHOUT CONTINUOUS FATHOGRAMS

Until now the horizontal plains which, judging by the results of the most recent data, seem to be characteristic of the deeper section of the floor of the North Atlantic Ocean, had gone unnoticed. These sections of the ocean floor were known to be more or less horizontal, without any conspicuous sea mounts. But their unusual flatness, interrupted here and there by a few knolls, which makes them so remarkable, had escaped the notice of all previous observers. Until quite recently continuously recording fathometers had not been used in the deep ocean. All soundings pertaining to this and other parts of the ocean were made by a visual dial arrangement, and the result was a series of discrete soundings several miles apart; each sounding was probably accurate only to within ± 30 to 50 fathoms. The profile of the ocean bottom obtained from these was like an interpolated curve, with a possible cumulative error between two successive soundings of the order of 50 to 100 fathoms. It is therefore not surprising that these flat stretches of ocean floor, great portions of which are horizontal within the limits of detection by the human eye on a fathometer recording—*i.e.*, to within ±5 fathoms—, remained unsuspected for so long. With the ever more extensive use of continuously recording fathometers, such flat plains are being discovered in the Pacific Ocean also. The late H. W. Murray of the U. S. Coast and Geodetic Survey had given us the following verbal communication:

"In the Gulf of Alaska there is a flat plain lying 140 miles offshore from Yaquina head. It has a length of 200 miles in a northwest-southwest direction and lies at a depth of 1461 fathoms at its northwest end and at 1668 fathoms at its southeast end. It seems to slope or tilt toward the continental slope, where it lies in water 200 fathoms deeper."

Conclusions

The Mid-Atlantic Ridge is characterized by two very distinct types of topography:

(1) The Terraced Zone, a succession of horizontal flat stretches occurring at depths ranging from 2500 fathoms to 1600 fathoms, is found along both flanks of the Ridge, at the same depths. From the few locations that could be correlated, it appears probable that they form fringes along the foot of the Main Range. They are covered by thick sediments. These terraces, at the same depths, recur off the coast of northwest Africa. In a few cases they exhibit evidence of tilting.

(2) The Main Range of the Mid-Atlantic Ridge, above 1600 fathoms, is a succession of parallel mountain ridges, trending NE.-SW. There are no terraces in this section of the Ridge, although the higher terraces are found in valleys, the floors of which are below 1600 fathoms.

Of all the hypotheses on the origin of sections of the Mid-Atlantic Ridge perhaps the least speculative are those proposed by Hans Cloos and J. Agostinho for the origin of the type of topography on the Azores plateau. Though these hypotheses appear reasonable enough for the specific case of the Azores, it does not appear to us to apply necessarily to the rest of the Ridge, for, as we have already noted, the Azores plateau appears to be a feature distinct from the rest of the Ridge, characterized by its own particular type of submarine topography.

Apart from a few highly interesting but scanty geological facts, the evidence about the nature of the Ridge obtained thus far by us is mostly topographic, and even this leaves much to be desired. We therefore cannot conclude much about the origin of the Mid-Atlantic Ridge.

One of the big surprises of the 1947 Atlantis expedition (cruise 150) was the proof of the existence of consolidated Cenozoic limestone on the Mid-Atlantic Ridge. Possibly considerable areas of the Ridge consist of sedimentary rocks. If so, the hypothesis of folded ranges in certain parts of the Ridge should not be rejected arbitrarily, on the sole basis that it is difficult in the light of our scanty knowledge of the history of the earth's crust to explain the why and how of such a process. It must be disproved by experimental and observational data; until then it remains a possibility.

Obviously many more data in the form of hydrographic surveys and fathograms are needed. An accurate bathymetric chart of the Atlantic Ocean would contribute greatly to our knowledge of the structural history of the Earth. Such a chart would also be useful for navigational purposes, since the Atlantic sea floor is marked by numerous features which would provide excellent "landmarks".

Such a project would be greatly facilitated by a more extensive use of continuously recording fathometers. Formation surveying—*i.e.*, surveying by means of a flotilla of ships keeping radar formation and equipped with continuously recording fathometers—would enable this work to be carried out speedily and efficiently.

References Cited

Adams, K. T. (1942) *Hydrographic manual*, U. S. Dept. Comm., Coast Geod. Survey, Spec. Pub. no. 140.

Agostinho, J. (1936) *Volcanoes of the Azores*, Bull. Volcanologique, Organe l'Assoc. Volcanologie de l'Un. Geodes. Geophys. Inter., p. 123–138.

Bucher, W. H. (1940) *Submarine valleys and related geological problems of the North Atlantic*, Geol. Soc. Am., Bull., vol. 51, p. 489–512.

Cloos, H. (1939) *Hebung-Spaltung-Vulcanismus*, Geol. Rundschau, vol. 30, Zwischenheft 4A, p. 506–510.

Dietrich, G. (1939) *Einige morphologische Ergebnisse des Meteor, fahrt Januar bis Mai 1938*, Ber zweite Teilfahrt Meteor, p. 20–23.

Gutenberg, B., and Richter, C. F. (1949) *Seismicity of the Earth*, Princeton Univ. Press.

Haug, E. (1900) *Les geosynclinaux et les aires continentales*, Soc. Géol. France, 3e ser., Bull. 28.

Hess, H. H. (1946) *Drowned ancient islands of the Pacific basin*, Am. Jour. Sci., vol. 244, p. 772–791.

Maurer, M. and Stocks, Th. (1933) *Die Echolotungen des . . . Meteor . . .*, Wiss. Ergebn. Deutsch Atlant. Esped. . . . Meteor . . . , 1925–1927, Bd. II, Beil. 4–32.

Molengraaff, G. A. F. (1928) *Theory of Continental Drift* (Symposium), Am. Assoc. Petrol. Ge p. 91.

Murray, H. W. (1945) *Profiles of the Aleutian tre* Geol. Soc. Am., Bull., vol. 56, p. 757–782.

Shand, S. J. (1949) *Rocks of the Mid-Atlantic Rid* Jour. Geol., vol. 57, p. 89–92.

Taylor, F. B. (1910) *Bearing of the Tertiary mounta belt on the origin of the earth's plan*, Geol. Sc Am., Bull., vol. 21, p. 179–226.

Trask, P. D., Phleger, F. B., and Stetson, H. C (1947) *Recent changes in sedimentation in th Gulf of Mexico*, Science, vol. 106, p. 460–461.

Umbgrove, J. H. F. (1947) *The pulse of the Earth*, The Hague.

Vening Meinesz, H. F. (1934) *Gravity expeditions a sea*, vol. 2, Delft.

Washington, H. S. (1930) *The origin of the Mid-Atlantic Ridge*, Md. Acad. Sci., Jour., vol. 1, p. 20–29.

Wegener, A. (1924) *The origin of continents and oceans*, Translated from the third German editions of 1922 by J. G. A. Skerl.

Wiseman, S. D. H., and Seymour Sewell, R. B. (1907) *The floor of the Arabian Sea*, Geol. Mag., vol. 74.

Wust, G. (1939) *Das submarine Relief bei den Azoren*, Abh. Preuss. Akad. Wiss., Phys. Math, Kl., Jahrg. 1939, no. 5, p. 46–58; map 1:3,000,000, Fig. 4, p. 52.

Columbia University, New York, N. Y.
Manuscript Received by the Secretary of the Society, September 7, 1948.
Contribution No. 498, The Woods Hole Oceanographic Institution

[*Editor's Note:* The original article contained seven larger-than-page-size plates that are not reproduced here. Four of those plates consist of reproduced fathograms; one is an index of fathogram locations and ship's tracks; one is a sketch map of the main area surveyed on the Mid-Atlantic Ridge; and one is a sketch map of an east-west trough discovered during the study.]

Part IV

THEORIES

Editor's Comments on Papers 17, 18, and 19

Since (as has frequently been pointed out) it is theory that determines to a considerable extent what and how we observe, no scientific field of endeavor can progress very far without appropriate theories to guide investigation. In this regard, the field of global tectonics has been fortunate in the great diversity of theories that have been proposed and been available for testing and for suggesting fruitful lines of investigation.

Among the theories that strongly influenced the development of geological thought along lines of continental drift were the ideas presented in Osmond Fisher's provocative short article (Paper 17) built on the work of Charles Darwin's son George (1879) who suggested that the moon formed by splitting away from the earth due to rotation and solar-tidal effects (see also Pickering, 1924). Fisher hypothesized that the deep scar left by this fission formed the Pacific Ocean basin and that, as the earth adjusted itself to this catastrophic loss of material by flowing to fill in the opening, parts of the existing continents were carried along with the moving mantle material. According to Fisher, the Atlantic basin is "a great rent" formed as North and South America drifted westward by "flow in the upper fluid layers toward the cavity." Fisher's theory is thus one of a number of catastrophic drift theories that envisioned drift as a one-time, short-term phenomenon. Some authors have claimed, in fact, that the supposed catastrophic, anti-uniformitarian aspects of continental drift strongly retarded its acceptance by the geological community (see, for example, Westoll, 1965; Rupke, 1970; Marvin, 1973, p. 38; Vine, 1977).

Wegener's role in the development of continental drift theory needs no elaboration here. He is widely regarded as being the father of continental drift, despite the fact that he was far from being the first to propose such a theory. What Wegener did do in Paper 18 and in Wegener, 1966 was to propose a detailed, comprehensive theory that included (1) the details of continental movement through time, (2) a force to drive the motion, (3) a detailed explanation of the mechanics of what was moving with respect to what, (4) an explanation of the world's great modern fold mountain ranges by drift, (5) a description of an enormous diversity of evidence supporting the theory, and (6) the suggestion that drift is still occuring and a proposal as to how it could be measured. Of course, we now know that Wegener was wrong in many of his ideas, but the amazing thing is how often he was correct, even with respect to subtle details of theory and evidence.

The last of the theoretical papers reprinted here is a remarkable effort by a remarkable geologist. Arthur Holmes was a world leader in developing the enormously extended geological time scale developed through use of radiometric dating during the early part of the twentieth century. Extending his work on radiometric dating, Holmes considered the implications of heat release by radioactive isotopes in the interior of the earth (Paper 19) and came very close to developing a theory of sea-floor spreading almost 30 years before the publication of Hess's classic work in 1962. Unfortunately, Holmes stopped short of the critical step, remarking only that "on the convection current hypothesis a certain amount of drift is inevitable mechanically," and left us with two tantalizing diagrams (his figures 12 and 13), the second of which comes very close to being a definitive modern picture of sea-floor spreading.

There were, of course, many other continental drift theories proposed in the latter part of the nineteenth and in the first half of the twentieth century (see Reclus, 1872; Warring, 1887; Mantovani, 1889–1890; Taylor, 1910; Baker, 1911; Daly, 1923; Argand, 1924; Joly, 1925; Gutenberg, 1927; Holmes, 1928; Choubert, 1935; Du Toit, 1937). It is not surprising for a variety of reasons, including the fact that the ocean basins were insufficiently studied, that all of these theories were deficient in one way or another. The three papers reprinted here were some of the more successful, but as can be seen, the evidence available before the era of the post–World War II exploration of the ocean basins simply did not provide an adequate observational basis for the development of a clear theoretical model of global tectonics.

REFERENCES

Argand, E., 1924, La tectonique de l'Asie, *Cong. Geol. Internat. C. R., XIII* **1:**171–372.

Baker, H. B., 1911, Origin of the Moon; Interesting Theory Advanced by Dr. Howard B. Baker, *Detroit Free Press,* 23 April, Feature Magazine section, p. 7.

Choubert, B., 1935, Recherches sur la genese des chaines Paleozoiques et Ante-cambriennes, *Rev. Geographie Phys. et Geologie Dynam.* **8**(fasc. 1):5-50.

Daly, R. A., 1923, The Earth's Crust and its Stability, *Am. Jour. Sci.* **205:**349-371.

Darwin, G. H., 1879, On the Precession of a Viscous Spheroid, and on the Remote History of the Earth, *Royal Soc. London Philos. Trans.* **170**(pt. 2):447-538.

Du Toit, A. L., 1937, *Our Wandering Continents, An Hypothesis of Continental Drifting,* Hafner Publishing Co., New York, 366p.

Gutenberg, B., 1927, Die Veranderung der Erdkrusts durch Fliessbewegungen der Kontinentalscholle, *Gerlands Beitr. Geophysik* **16:**239-247 and **18:**281-291.

Hess, H. H., 1962, History of Ocean Basins, in *Petrological Studies,* A. E. J. Engel et al., eds., Geological Society of America, New York, pp. 599-620.

Holmes, A., 1928, Radioactivity and Earth Movements, *Geol. Soc. Glasgow Trans.* **18:**559-606.

Joly, J., 1925, *The Surface History of the Earth,* 2nd ed., Clarendon Press, Oxford, 211p.

Mantovani, R., 1889-1890, Les fractures de l'ecorse terrestre et la theorie de Laplace, *Soc. Arts et Lettres de la Reunion (St. Denis) Bull.,* pp. 41-53.

Marvin, U. B., 1973, *Continental Drift—The Evolution of a Concept,* Smithsonian Inst. Press, Washington, D.C., 239p.

Pickering, W. H., 1924, The Separation of the Continents by Fission, *Geol. Mag.* **61:**31-35.

Reclus, E., 1872, *The Earth,* Harper and Brothers, New York, 567p.

Rupke, N. A., 1970, Continental Drift before 1900, *Nature* **227:**349-350.

Taylor, F. B., 1910, Bearing of the Tertiary Mountain Belt on the Origin of the Earth's Plan, *Geol. Soc. America Bull.* **21:**179-226.

Vine, F. J., 1977, The Continental Drift Debate, *Nature* **266:**19-22.

Warring, C. B., 1887, The Evolution of Continents, *Vasser Brothers Inst. Trans.* **4:**256-274.

Wegener, A., 1966, *The Origin of Continents and Oceans,* Dover, New York, 246p.

Westoll, T. S., 1965, Geological Evidence Bearing upon Continental Drift, in P. M. S. Blackett et al., eds., *A Symposium on Continental Drift,* Royal Society, London, pp. 12-26.

17

Reprinted from *Nature* **25**:243-244 (1882)

ON THE PHYSICAL CAUSE OF THE OCEAN BASINS

O. Fisher

GEOLOGISTS have reason to thank Prof. Ball for directing their attention to the remarkable investigations of Mr. G. H. Darwin upon "The Precession of a Viscous Spheroid, and the Remote History of the Earth," (*Phil. Trans*. Roy. Soc., Part ii., 1879). Prof. Hull has already been led to point out one result which appeared to him to flow from them, in showing how the ancient tides may have produced the planes of marine denudation, though Mr. Darwin has since expressed doubts as to the legitimacy of this conclusion. I wish to offer another speculation arising from Mr. Darwin's work, which I think may account for the hitherto unexplained distribution of land and water upon the surface of the globe.

Herschel remarked long ago, in his "Physical Geography," that the prevalence of land and water over two opposite hemispheres "proves that the force by which the continents are sustained is one of *tumefaction*, inasmuch as it indicates a situation of the centre of gravity of the total mass of the earth somewhat eccentric relatively to that of the general figure of the external surface —the eccentricity lying in the direction of our antipodes: and is therefore a proof of the comparative *lightness* of the materials of the terrestrial hemisphere." In my "Physics of the Earth's Crust," just published, I have shown reasons for thinking that the distribution of the materials of the earth, which gives rise to this condition, is of the following kind. I accept on the whole the theory that the earth is a hot globe, of which the superficial crust is rendered solid by having become cool, and that the central part is solid, either from great pressure, or from whatever other cause may be assigned; an intervening layer beneath the cooled crust still remaining liquid. The layers of which the whole is composed are arranged in order of their density. Now I have given reasons for believing that Herschel's "comparative lightness of the materials of the terrestrial hemisphere" arises from the fact that the cooled crust beneath the continents is intrinsically less dense than that beneath the great oceans. I think that the crust beneath the continents consists of the cooled acid, or granitic, and therefore lighter magma, which ought naturally to have formed originally the entire superficial portion of the globe. But I conclude that the bottoms of the great oceans consist nevertheless of a crust formed out of the cooled basic layer. Beneath the cooled crust the laws of hydrostatic equilibrium would require that, if the substratum is truly liquid, it should be of the same density under both these areas. I also conclude that the upper surface of the basic crust which forms the floor of the oceans is really depressed below the mean surface of figure.

To these conclusions I arrived without being able to suggest any satisfactory explanation of the facts. I saw that they agreed with, and were supported by, the view of those geologists who assert that the great oceanic and continental areas have never changed places; but neither could I any better see the reason for this.

Let us now inquire whether Mr. Darwin's researches throw any light upon the subject. I shall refer chiefly to the summary and discussion of results appended to his paper, for it is small blame to a sexagenarian, not a professed mathematician, to admit that to follow the *calculations* is beyond the scope of his powers. As I understand Mr. Darwin, he thinks it probable that the moon and the earth were once a single mass, and that at the time when this mass was rotating at the rate of about one revolution in five hours the whole separated into two portions, the smaller of which went to form the moon; and that the moon then began to recede from the earth, until now, after the lapse of fifty-four millions of years *or more*, it is at its present distance. The ellipticity of the mass when rotating at the above-named speed would be about 1-12th. [This would make the mass very much less compressed than an ordinary orange.] He does not think it probable that this amount of ellipticity would cause the spheroid to break up simply from the centrifugal effect of the rotation; but he suggests, judging from the calculated period of a gravitational oscillation of a fluid spheroid, of uniform density equal to the mean of the earth, viz. 1 hour 34 minutes, that the period of the free oscillation of a spheroid "consisting of a denser nucleus and a rarer surface," but of the same mean density as the earth, might coincide with the period of the bodily solar tide at that time. "It seems to be quite possible that two complete gravitational oscillations of the earth in its primitive state might occupy four or five hours." "Accordingly the solar tides would be of enormous height." He then adds: "Does it not then seem possible that, if the rotation were fast enough to bring the spheroid into anything near the unstable condition, then the large solar tides might rupture the body into two or more parts? In this case one would conjecture that it would not be a ring that would detach itself."

I now proceed to build my speculation upon his. It is obvious that, according to the above theory, the act of fissiparturition by which the moon was born must have been sudden. One of the two solar tidal protuberances broke away from the earth to inchoate a separate existence. A great but shallow hole must consequently have been formed, whose centre would have been on or near the equator. Prof. Ball says: "Not for long would that fragment retain an irregular form ; the mutual attraction of the particles would draw the mass together. By the same gentle ministrations the wound on the earth would soon be healed. In the lapse of time the earth would become as whole as ever, and at last it would not retain even a scar to testify to the mighty catastrophe."

I form a less hopeful prognostication. I think the ocean basins are the scar, which still testify to the place of separation.

The density of the moon is 0·56654 times that of the earth. Putting the mean density of the earth at 5·5, this makes the density of the moon 3·1. The density of granite is about 2·68, and that of basalt 2·96. Consequently the density of the moon is a little greater than that of the basic layer of the earth's surface, which I think we may expect to occur at the sea-board at a depth of about 25 miles. The entire mass of the moon is 0·011364 of the mass of the earth.

Accordingly, it would require a layer of about 31 miles thick, of the density of granite, to be taken off the surface of the primitive mass to make a body of the mass of the moon ; and if the mean density of the matter removed was the same as that of the moon, a somewhat thinner layer would suffice. But if we reduce the area of the skin removed to the area of the oceans, it would require to be $\frac{197}{146} \times 31$, or about 41 miles deep. Hence a uniform layer rather less than 41 miles thick taken off the oceanic areas would be sufficient to make the moon.

Of course the layer removed would not, in fact, have been of uniform thickness. But the above estimate gives an idea of the size of the cavity which would be produced. What then would happen? This would depend upon whether the surface had already become at all solid. I conceive this would be the case at a very early stage, judging from the manner in which a solid layer forms on the liquid lava of Kilauea. The hole would therefore fill up by the rise of the liquid from below, rather than by the lateral approach of the edges of the wound. When the raw surface again solidified we should have a crust of greater density over the area in question, because formed from a lower and denser layer, which would have risen not quite to the level of the lighter crust. There would, however, have necessarily been a certain amount of flow in the upper fluid layers towards the cavity, and this would have carried the cooled granitic crust which, floating on it, still remained upon the earth along with it. What was left of the granitic crust would therefore be broken up into fragmentary areas, now represented by the continents. This would make the Atlantic a great rent, and explain the rude parallelism which exists between the contours of America and the Old World.

The sudden rupture of so considerable a fragment from the rotating spheroid, would alter its mass, form, and moment of momentum. It appears then that its axis of rotation would be altered, which might account for the fact, that the approximate pole of the oceanic area is not in the equator.

The volcanic surface of the moon, if volcanic it be, would lend considerable support to the view which I maintain, that the water substance emitted by volcanoes is an integral constituent of the fluid substratum. For when the moon broke away from the earth it would carry with it the aqueous constituent of the magma. Owing to the much smaller force of gravity in the moon, the pressure under which this would there be placed would be much less than in the earth. Consequently it would more easily escape, and the signs of volcanic action would be more pronounced. But the difficulties surrounding terrestrial vulcanism are so great, that one is hardly tempted to add the lunar to them. O. FISHER

18

THE ORIGIN OF CONTINENTS

Alfred Wegener

These excerpts were translated expressly for this Benchmark volume by Rhodes W. Fairbridge from Petermanns Geog. Mitteilungen **58:**185–195, 253–256, and 305–309 (1912). *The first and second parts (to page 188 of the original) are translated verbatim. The conclusion (pp. 305 through 309) is presented in summary.*

In the following article a first rough attempt will be made to explain genetically the major features of the earth's surface, that is to say, the continental platforms and ocean basins, by means of an all-embracing concept, the Principle of the Horizontal Mobility of Continental Blocks [*Translator's note*: usually given as "Continental Drift" in English]. Ancient land connections between continents, formerly explained by an assumed subsidence of the intervening ocean basins, will now be interpreted by an assumed rifting and drifting apart of the continental blocks. The picture obtained is new and in many ways paradoxical, but, as will be shown, it does not lack some basis in physics. And on the other hand, using only the main points, the geology and geophysics contain so many surprising generalizations and correlations that it seems justified at this time to present a new working hypothesis in place of the old idea of sunken continents. The need for this new approach has evidently been growing, along with the arguments against the commonly accepted Principle of Permanence of the Oceans (1). In spite of these broad-based foundations I am only offering this new principle as a working hypothesis until it has at least been possible to demonstrate by means of astronomically controlled geodetic measurements the continuity of these horizontal movements in the past with an exactitude excluding all doubt.

In the first chapter, on the basis of geophysical and general geological experiences, the question will be posed whether the essential idea of horizontal drifting of plates could be considered within an apparently rigid earth's crust, and in what manner they might take place. The next chapter makes a first tentative attempt to follow out the main fractures and displacements of the pole. In the third chapter there will be a brief consideration of the type of geodetic astronomical measurements that seem to be most suitable for demonstrating a measurable and continuing drift at the present time. Also an attempt will be made to explain polar migrations.

(1) E. Böse: *Die Erdbeben* ("Die Natur" Series, n.d., p. 16) recognizes the present state of affairs with the words "the Contraction Theory is no longer fully accepted, and in the meantime no other theory has been proposed which can either replace it entirely or explain all the facts."

Before we begin the presentation, a few brief historical notes should be introduced. In England at the present time, thanks especially to the book by O. Fisher *Physics of the Earth's Crust,* there is a widespread hypothesis that postulates the Pacific Ocean as a last remnant of a once much larger depression whence the Moon is supposed to have separated from the Earth. [*Editor's Note:* See Paper 17 in this Benchmark volume.] This theory is based partly on G. H. Darwin's study of the evolution of the Moon and on the celebrated work of Poincaré on the isostatic figures of rotating globes. We will not go into the matter further at this time except to indicate that the theory should be rejected following the objections by Schwarzschild and by Liapunow (ref. Rudzki: *Physik der Erde,* Leipzig, 1911, 210p.) It should be noted that W. H. Pickering has also developed the idea that America has been rifted away from Europe and Africa and is now separated from them by the width of the Atlantic (*Journal of Geology,* v. 15, no. 1; see also *Gaea,* v. 43, 1907, p. 385). This hypothesis rests mainly on the parallelism of the coasts and, correctly in my opinion, rifts the American continents away from the Old World, but it is based on obviously incorrect hypotheses that put the movement in some unspecified and untestable period of the remote past, so that the idea is only of historical interest.

On the other hand, F. B. Taylor's paper recently published ("Bearing of the Tertiary mountain belt on the origin of the earth's plan", *Bull. Geol. Soc. Amer.,* v. 21, 1910, pp. 179–226.) can be regarded as a precursor to our present work, although both were developed quite independently of one another. Taylor, just as we do, assumes that horizontal drifting of continents has taken place at specific geologic times, especially in the Tertiary period, and thus has led to the development of the great Tertiary orogenic belts. Considerable stress is laid on the rifting away of Greenland from North America, specifically on the evidence of coastal parallelism. Thus, for a second part of the Earth's crust, he has already expressed our own opinion. In the case of the Atlantic, however, he assumed that an appreciable part of its width was originally an extension of the North American block, whereas in our opinion the original marginal fractures would be almost at the present coastlines. We also do not agree with Taylor in his interpretation of the Mid-Atlantic Ridge, in spite of its great depth, as a remnant of a former intermediate land connection. Taylor failed to recognize the tremendous consequences involved in announcing the drift of continents and offered no sort of test or proof for their existence in the face of contrary views, so that his ideas were received mostly with skepticism.

We will proceed now with the presentation.

I. GEOPHYSICAL ARGUMENTS [p. 186]

1. The Problem of Continental Platforms.

This is an old question. It concerns the platform-like disposition of the continents with respect to the deep-sea floors. As long ago as 1878 Heim wrote:

> "Thus the continents appear to be thick, broad platforms ('Sockel'), in which the great orogenic belts rise like low ribs, occupying only one five-hundredth part of the total volume. . . . The crustal movements that occur along the boundaries between continents and oceans are probably of a different type from the orogenic motions within the continental crust, although the forces are probably the same. . . .To me it appears that until more precise observations are made on the continental oscillations of the past . . . and until we have more complete measurements of the degree of relative compression of the major orogenic belts, we can scarcely expect any important or reliable advance in the knowledge of the original nature of the orogenic belts and continents and of the internal form of the latter" (Heim: *Untersuchungen über der Mechanismus der Gegirgsbilding* usw., Basel, part 2, 1878, p. 237).

From the next figure (from Krümmel: *Handbuch der Ozeanographie,* Stuttgart, v. 1, 1907, p. 87, i.e., the hypsometric curve) it may be seen that there are two distinctive levels corresponding to the floor of the deep sea and the surface of the continents. Respectively they lie at around 4,300 m below and 700 m above mean sea level. It can be seen that the lower part of the continental platform, the shelf, is overlapped by the sea and is separated by a steep fall-off to the deep-sea floor. Since we will be dealing here with the whole continental unit (i.e., continent plus shelf), it will be necessary to disregard the usual shape of the coastlines that result from marine transgressions over the shelf. In Fig. 2 on Plate 36 a map of the continental platforms has been sketched in this manner (Based mainly on Kegel: *Die Grösse der Schelfflächen auf Grund einer neuen Berechnung,* Göttingen, dissertation, 1910). For the most part, the edges of the platforms coincide with the 200 m contour, but following Krümmel (op. cit., p. 113) it should be 200–300 m for the Norwegian, Barents, and Iceland-Faroe shelves and may even reach 500 m. There are differences in detail also around Britain, Siberia, Southeast Asia and the Bering Strait. It seems also that New Guinea is connected with Australia.

There are probably few hypotheses in natural science that have been marked by more contradiction or confusion than the opinions that have hitherto been expressed on these most remarkable platform-like features of the Earth's crust. Although in this work polemic will in general be avoided, we can hardly avoid a short critical look at some of the earlier views, which we can resolve if we replace them with our own hypothesis. In the first place, European geologists in general hold to the concept of the drying-apple concept whereby a cooling earth would shrink internally, causing its crustal skin to wrinkle. In the latter there would be a constant horizontal stress ["Gewölbedruck," i.e., folding pressure], that leads to the wrinkling, that is, orogenic fold belts. People tend to assume that subsidence of the earth's crust goes on irregularly during the shrinking, so that locally a horst will remain behind, while neighboring blocks sink down. Thus the continents would be horsts. With further shrinking, parts of the horsts sink into the oceans so that an active oscillation between continents and oceans can take place. The main advocate of this theory, Edward

Suess, summarized the idea in a sentence: "The broken-up remnants of the earth's crust are where we live." (*Das Antlitz der Erde,* v. 1, 1885, p. 778). This was one unexpected consequence of the concept worked out by Suess himself, which for many people showed up its weaknesses and thus indirectly opened the way for a more correct view. Since the time that Heim had showed field evidence in favor of the contraction hypothesis, geophysicists have pointed out objection after objection. One basic assumption, that "the earth must be cooling," never seemed to be doubted and has remained untested until the discovery of radioactivity put the question the other way round, that is, that the temperature of the earth's interior might be regarded as rising (Rudzki: *Physik der Erde,* Leipzig, 1911, p. 122). Since then it can be said with greater probability that the earth's core does not consist of easily compressed gases but of essentially incompressible nickel-iron. Thus the originally quoted causes of contraction are no longer valid for explaining the great folds of the earth's orogenic belts, especially now that their size has become known thanks to the development of the nappe theory. The tremendous compressional stress involved in the shrinking along a great circle belt to cause one hemisphere to override the other is found to be physically impossible inasmuch as the molecular forces (compressibility) are not enough to constrain a 100 km wide belt that has to be overthrust. "The crustal slice would not survive but would be broken up into small pieces" (Rudzki, op. cit., p. 244). Thus the shrinking cannot be concentrated along great circles but must be evenly distributed over the whole of the earth's surface, as shown clearly by Ampferer (Über das Bewegungsbild von Faltengebirgen" *Jb. k. k. geol. Reichsanst., Wien,* v. 56, 1906, 539–622; on page 549 he referred to the compressional stress as "enormous") and by Reyer (*Geologische Prinzipienfragen, Leipzig,* 1907, p. 140ff). Further it is difficult to see how such a contraction process in the earth can at one time lead to wrinkling or folding, but at another time to the subsidence of enormous blocks and to the uplift of horsts. The idea of these oscillating, up-and-down motions of continental blocks and oceanic depressions is in contradiction to the well-known fact that almost all sediments that we find on the continental platform originated in shelf seas, being derived from shallow-sea transgressions and were never formed on the floor of the deep oceans. Deposits formerly regarded as abyssal have been reinterpreted in recent times as neritic. This is notably true in the case of the Cretaceous Chalk, as Cayeux demonstrated. The so-called radiolarites of the Alps of Upper Jurassic and Cretaceous age, actually reddish shales containing chert and jasper, are certainly regarded by Steinmann and others as abyssal, but they could also probably develop in a transgressive situation with depths of as little as 1,000 m (There is also other evidence for assuming these seas to have been formerly deeper in places). Unquestionably the red Tertiary shales of Barbados (West Indies) are correctly interpreted by Harrison and Jukes-Brown as representing abyssal red clay. (From the viewpoint of our theory there is no reason why abyssal sediments should not be found here and there in elevated situations.) If, on the other hand, we compare the vast majority of sediments (i.e., of shallow-water type) with the extreme rarity of abyssal facies, one may reasonably generalize that continents have never been part of the deep-sea floor.

If none of these arguments for continental breakup were valid, our hypothesis would have to be rejected. But gravity measurements point the other way. If the ocean floors were merely subsided continental blocks, they would have to consist of the same material. Gravity data show, with unassailable logic, that oceanic crust is appreciably denser than continental crust, and indeed it is so different that the present elevation contrast represents an isostatically compensated state. We will write further about gravity and isostasy in the next section. Anyone who disregards these data will find it hard to grasp the Drift Theory. This contrast between continental and oceanic crust has frequently been emphasized, notably by the specialists in America. But they have either not considered, or have denied the drift hypothesis and the idea of the permanence of continental plates. There is the well-known axiom about the "Permanence of Oceans," which is associated mainly with the names of Dana and of Wallace. Bailey Willis has recently developed the concept, although only in a rough way. (*"Principles of Paleogeography"*, Science, v. 31, N.S., 1910, no. 790., p. 241–260: "The great ocean basins are permanent features of the earth's surface and they have existed, where they are now, with moderate changes of outline, since the first waters gathered." Somewhat harsh is his closing remark: "This conclusion appears to place the permanence of ocean basins outside the category of debatable questions.") The opposition objects to this doctrine, and for good reasons. If it were true, we would be forced to accept the idea of former land bridges across deep ocean basins to join former continents together. Hundreds upon hundreds of paleontological discoveries, ever-increasing in volume, show conclusively that such formerly connected continents experienced unhindered exchange of fauna and flora across what are now deep-sea basins. Perhaps both hypotheses permissable, both permanence of oceans *and* global break-up. Any attempt to retain the permanence-of-oceans concept, by using island links instead of true land bridges or by assuming migratory routes across presently unexplored areas, is so unsatisfactory that we need go no further in this direction.

Thus the two basic assumptions seem to be mutually opposed, and no compromise is possible. Both sides seem to have good incontestable arguments, but they lead to contradictory conclusions. I will attempt to show that the essential requirements of both are quite simply met if one accepts the hypothesis of rifting and continental drift.

2. Gravity Measurements [p. 187]

Since ocean water has a density less than half that of crustal rocks, it is to be expected that gravity over the oceans would be considerably less than over the continents. But it has long been known that this is not the case, and even more surprisingly that the value is almost constant from place to place. The first pendulum observations on oceanic islands, followed by those of Scott-Hansen on the "Fram" Expedition (1893–1896) in the Arctic Ocean, were marred either by the disturbing effect of the island or by the imprecision of the instruments. But on a ship this method of measuring the length of the second pendulum swing is not feasible because a steady base is needed. In order to get satisfactory results Hecker employed a method proposed by Mohn, which, although only one tenth as precise, served its purpose adequately. It employed a mercury barometer together with a "hypsometer" (boiling point thermometer). From a precise measurement of the boiling point of water the atmospheric pressure was obtained while the barometer gave the weight of the column of mercury, that weight varying with gravity. The difference between the measurements (in different places) corresponds to the regional anomalies in gravity. With this equipment Hecker made a voyage from Lisbon to Rio de Janeiro in 1901, and another in 1904–1905 via the Mediterranean and Australia to San Francisco and back to Yokohama (Japan). (Hecker: Bestimmung der Schwerkraft auf dem Atlantischen Ozean, *Veröff d. Kgl. Preuss. Geod. Inst.*, Berlin, 1903, no. 11; Further: Bestimmung der Schwerkraft auf dem Indischen und Grossen Ozean, *ebenda,* 1908, no. 16.) With these results he was able to meet the earlier criticisms and establish definitively that the expected gravity deficit over the oceans does not exist.

The only explanation that seems possible for this result, as was recognized even before Hecker's journeys, by T. Fischer, Faye, Helmert, and others, is that the oceanic crust consists of a distinctly denser material than the continental crust. At the bottom of the deep ocean there is an atmospheric pressure of around 400 Atm, thus the rocks are under constant compression. Working the same way is the mean temperature differential; at the ocean floor it is about 0°C, but at the land surface it averages 15°C. But these differences sould only explain a few percent of the observed density contrast, judging from the physical constants of the rocks themselves as observed under laboratory conditions. We have no alternative but to assume that the rocks of the oceanic crust are specifically denser and chemically different from those of the continents. The main question is how the boundary runs between the two.

The most widespread view is that worked out by Airy already in 1855, later taken up by Stokes and most recently by J. Loukaschewitsch ("Sur la mécanisme de l'écorce terrestre et l'origine des continents," St. Petersburg, 1911). According to the Airy hypothesis, the lithosphere floats buoyantly on a dense magmatic layer, being thick under the continents and thin under the oceans. The magma [i.e., the mantle] rises higher under the oceanic crust, and it is this compensated dense material that is reflected by the higher gravity values there. Although this can explain the gravity values, there are perhaps other alternative possibilities. We cannot go further into this theory of a continuous lithosphere surrounding the whole earth, except to remark that we can visualize another model that could also explain the gravity observations and would likewise introduce some simplification. It is illustrated in the next figure [the wood-block floating in water]. In this model the continents are visualized as fragments of a discontinuous lithosphere, now floating in a denser material.

[*Translator's note:* Sections 3–9 review familiar geophysical data so are only summarized here.]

3. Isostasy [p. 188]

It has been noted that gravity over the oceans is not only generally greater than would be expected from the mass deficit found there, but is typically about the same strength as on the continents (though with some exceptions). This fact is difficult to explain except by assuming a gravitational adjustment, that is, "isostasy" (coined by Dutton in 1892). In this way the lighter continents float in the denser medium, and are so thoroughly in adjustment that the static pressure is balanced, just as with an iceberg floating in the ocean. If we consider two vertical columns in equilibrium, one within a continental plate, the other within an oceanic sector, the weight must be the same. The confining pressure within each will vary with depth, but the contrast will decrease with depth, becoming zero at the base of each column. Here both pressure and material composition are identical. This is known as the compensation surface ["Ausgleichs-fläche"].

One may mention the effect of loading on a buoyant lithosphere. Just as a load added to a floating iceberg depresses it, the lithosphere responds likewise. In the coral atolls, as explained by Charles Darwin, a slow subsidence accompanies the accumulation (confirmed by deep drilling on Funafuti). The isobase maps of Scandinavia prepared by Baron de Geer show a more than 250 m depression of the interior, as reconstructed from the tilting of shorelines. Similar tilting is observed in North America and by the writer himself [Wegener] in Greenland. Rudzki, assuming isostatic adjustment, calculated the former ice thickness for Scandinavia at 933 m, for North America at 1,667 m. The alternative explanation, employing pure elastic deformation, would give the quite unrealistic value of about 7 km for the ice thickness (the volume of which is constrained by knowledge of the range of eustatic oscillations).

Sedimentary loading would play a comparable role in isostatic behavior as recognized by Osmond Fisher, the new surface adjusting steadily to almost the same elevation as the old one. But since sediments are less dense than basement rock, a sedimentary basin would gradually tend to fill up without disturbing the gravitational adjustment.

4. Thickness of Continental Blocks [p. 189]

Hayford estimated the depth to the base of the lithosphere below continents, that is, the compensation level, to average about 114 km. On different grounds, Helmert, using the pendulum observations, found a characteristic pattern at the edge of continents (observed earlier by

Schiötz) that shows a rise in gravity as one approaches the coast from the land, followed by an abrupt fall in values to the foot of the continental slope, beyond which in the open ocean it returns to the normal level. From this Helmert calculated a 120 km thickness for the lithosphere, a result almost identical to Hayford's. Recent confirmation was obtained by Kohlschütter in the East African rift zone. A variation of about 50–200 km around the world is to be expected.

Earthquake studies also give evidence of the thickness of the lithosphere and indicate the existence of a relatively thin elastic layer separated from the earth's interior (lower mantle) by a magma layer [referred to nowadays as low velocity layer], which must lie as depths somewhat greater than 100 km according to Wiechert. At around 50 km a discontinuity is recognized by Mohorovičić (1909), but this is apparently not the same as the base of the lithosphere. The hypocentrum depths for the deeper earthquakes, instrumentally observed and analyzed, indicate 50–170 km depth, which confirms the idea of an average thickness for the elastic lithosphere to be around 100 km.

5. Material [p. 190]

The sedimentary component in the continental crust according to figures published by Haug in 1907 from American data reaches 40,000 feet in the Appalachians. Suess felt, however, that such figures might give an exaggerated idea of basin subsidence since the sedimentary wedge tends to thicken basinward through time and should be measured not just vertically. In places, on the continental platforms, however, the sedimentary cover is much thinner. If we remove the average sedimentary cover the continental block thickness would only be reduced by 100 m, showing once again that the continental/ocean crust contrast is a first-order difference, in which the erosion-sedimentation process only plays a secondary role. The ubiquitous material in the ancient continental crust ("Urgestein") is gneiss. Suess (in *Das Antlitz der Erde,* p. 626) called the latter SAL (better known now as SIAL), while the oceanic lithosphere is SIMA, and core NIFE, each named for their principal elements. The specific gravity of SIAL is 2.5–2.7 and SIMA 3.0–3.3. This difference would be appropriate to the mean surface level contrast between continents and ocean floors.

6. Plasticity [p. 191]

In Wegener's opinion the sialic blocks would be able to move horizontally within the "sea" of SIMA. He was under the impression that the mean melting point of SIAL was 200–300° C higher than that of SIMA, but he pointed out that the greater the depth, the more plastic would be the behavior of the materials. Thus a rigid SIAL could glide over a superficially more mobile SIMA.

Wegener cited heat-flow measurements from Silesia that suggested a thermal gradient of 31.8 m per degree C, which would result in melting temperatures being reached at depths of at around 100 km depth. At least a 1,500 km layer of SIMA would have to behave in a plastic manner. In view of the rising pressure with depth it seemed likely to Wegener that the base of the SIAL would also behave in a plastic manner. In the earth's crust the extremely slow application of the stresses achieves eventually major effects. Black pitch displays the same behavior pattern: a hammer blow will fracture it like glass, whereas a small load applied over a long period will cause plastic deformation. Isostatic reactions give an idea of rates of flow; central Scandinavia was known to be rising at 1m/100 yr from the release of ice loading. This gives a measure of viscosity (in the mantle). Kelvin had already calculated, from the degree of flattening of the geoid, that the whole earth must have a viscosity comparable to that of steel (7.6×10^{11} dynes per cm^3). This would also be appropriate for the 430 day period of the polar wobble, where a perfectly rigid earth would, following Euler, give a 305 day period. But allowing for Weichert's calculation of the earth's core as having a viscosity of 20.2×10^{11} dynes per cm^3, the remaining mantle would only have 0.9×10^{11} dynes. This seems to correspond to that determined for SIMA.

Earthquake seismology also confirms that there is no question of a completely fluid layer at the base of the lithosphere, merely a plastic one.

7. Orogeny and the Evolution of Continental Platforms [p. 193]

The sedimentary layer is always marked by higher plasticity and lower heat-flow than is the underlying crystalline material (gneissic SIAL). We see many large-scale folds in the sediments, in contrast to the complex foliations seen in gneiss. The latter shows flowage, whereas the overlying sediments, according to the nappe theory developed by Bertrand, Schardt, and Lugeon, behave more as superficial sheets. Mountain building results in a thickening of the continental plates.

These observations confirm earlier interpretations of the compression evidence in orogenic belts, but go further. The early idea, proposed by Heim, for example, postulated a four-fifths shortening for the Swiss Jura and a one-half for the Alps. However, the nappe theory suggests that the width of the Alps is only one-fourth to one-eighth of what it was originally. Thus what is today 150 km wide represents 600–1200 km of compressed crust. Before the collision the continental plates must have had completely different outlines, a problem that Wegener felt had not yet been adequately studied. Furthermore, if the Himalayas are a zone of horizontal compression, where would the southern tip of India have originated? Would there still be room for the postulated "sunken continent" of "Lemuria"? [*Translator's note:* This had been proposed as a sort of greater Madagascar].

Wegener draws attention once more to "the geosynclinal principle," that the thickest sedimentary troughs eventually become the orogenic belts as shown long ago by James Hall. In the words of Haug ("Traité", p. 160): "mountain chains form at the sites of geosynclines." As was pointed out already by Mellard Reade, the sediments at the base of a thick geosynclinal column would naturally become heated during the slow subsidence and would be rendered plastic. One should not forget that (in intercontinental collision) some deep water facies would also become involved, and likewise some of the more plastic simatic oceanic crust.

As repeated orogenies thicken the continental plates, the total area of continental crust decreases with time. It is a one-way, irreversible process. Thus Wegener proposed a basic model for the evolution of continental platforms. The sialic basement rocks are never found horizontal or undisturbed over large areas and normally are intensely folded. Wegener suggests the sialic crust was once world-encircling, with perhaps a thickness of 30 km, but has become progressively dismembered and thickened by this alternation of rifting and collision. A 3-km-deep "Panthalassa" universal ocean may once have covered the globe, but with progressive evolution of continental crust has become dismembered into smaller oceans, marginal seas and shelf waters. The shallow seas were formerly much more extensive than today because the continental blocks have gradually become thicker (and thus higher).

8. Volcanism [p. 194]

A serious problem has been so far ignored. With every drift of continental plates (at the trailing edge), the underlying SIMA with its elevated temperature would be exposed on the sea floor. Would not this lead to catastrophic events? Actual experiences by Bergeat (in 1888, 1889, and 1892) with submarine eruptions (at 700–1000 m depth) in the Mediterranean (Vulkano, Lipari Islands) proceeded without producing any cataclysms and were relatively quiet. There seems to be more chance of volcanism at the frontal margin of a drifting continent. The Mediterranean does not seem to be a simple suture zone, but is a belt of fragmented blocks, marked by rotation and grinding motions to and fro.

Although the opening rifts have relatively few volcanoes, the historical record shows certain times to have been marked by very extensive volcanism and to be separated by long periods of quiescence. The former seem to be times of active drifting; the latter of relative inactivity.

9. Remarks on the Causes of Drifting [p. 194]

The question of the forces involved is so critical that I cannot avoid the subject, although I feel it is premature. It is first necessary to establish the facts before one can have any hope of working out the causes.

Obviously, the polar shifts need to be considered. Each such migration must have initiated a train of new centrifugal forces and would bring about major displacements of mass. At the same time it should be remembered that the mass displacements [of drift] also themselves lead to adjustments of the pole.

I feel it is quite probable that the separation of the moon from the earth might be the ultimate cause. The meridional fracture systems around the globe seem to speak for this model. These [lineaments] control the well-known characteristic geometry of the southern continents, being pointed to the south. The former North Pole, in the vicinity of the Bering Strait, also coincides with the sharp projections of North America and eastern Siberia, though younger tectonics in the region have somewhat complicated the picture. At some time probably it will be possible to relate continental drift to convection currents within the mantle. Also for future consideration is the chance of identifying exogenetic causes that lead to stresses during rotation. To this class may belong the old idea of "Flight from the Poles" [Pohlflucht] which has been most recently revived by Taylor. The time is not yet ripe, however, for further discussion of these questions.

II. GEOLOGICAL ARGUMENTS [p. 253]

Before we continue the discussion of the rifting and collision of continents in the course of earth's history that was begun in the first part of this paper, it should once again be emphasized that this is only a first tentative model, in many ways incomplete and, in part, incorrect. Certainly it will need to be very carefully examined. Once the main points are established, the investigation of the details will not be so difficult, and the errors can be weeded out. We shall begin with a discussion of recent fracturing.

1. Graben Faults

In the picture we have been given on graben tectonics, little attention has been paid so far to gravity measurements. Mostly, people have been satisfied if it could be demonstrated that along a given zone of the earth's crust, the superficial layers have subsided, as has been shown for example by Lepsius through borings along the plain of the upper Rhine, whereas sediments of the same age have been found at higher elevations along the sides of the graben. From the gravity measurements, however, it is now evident that beneath the graben deposits there is material of higher specific gravity than on either side. Thus it must be concluded that we have here a rift in the continental crust that has been intruded by dense SIMA, rising to a level of isostatic equilibrium. This would be the case if the SIMA reached up to a level of about 3.5 km below the surface of the continental crust and if assuming that the sides of the graben had been filled in by the collapse of the sedimentary margins (with step faulting, etc.), so that quite naturally the floor of the graben would be completely covered by youthful deposits. In my opinion, all graben structures should be interpreted in this way, that is to say, at the beginning of continental rifting there is either a new rupture or else a revival of an earlier, aborted rift that died out with its driving forces. The latter seems to be the case in the Upper Rhine valley because the fractures developed in the early Oligocene at the same time as North America was pulling away from Europe (see below).

The most interesting example of graben development, however, is found in the East African rift valleys and in their continuation through the Red Sea to the Jordan valley. E. Suess (in: *Beiträge zur geologischen Kenntnis des östlichen Afrika. Die Brüche des östlichen Afrika,* Wien, 1891) has already recognized these faults as part of a gigantic fracture zone on purely geological grounds. He writes:"In a general way, this region has all the appearances of a very long fracture in the earth's crust, divided up into a series of criss-cross splinters, and locally subsided in blocks and wedges to various depths." If the continuation traceable through the Ethiopian (Somali) region had not developed as the Red Sea before the Noachian Flood, for which there is naturally no proof, it would seem that we would also have a very youthful splitting of the continental crust. Of great importance are the gravity measurements recently made in this region by E. Kohlschütter (Über den Bau der Erdkruste in Deutch-Ostafrika, *Nachr. d. Ges. d. Wiss.,* Göttingen, 1911). In the majority of the grabens the mass deficiency observed is not compensated for by material of greater specific weight at the base of the graben, but on the contrary is represented by an anomaly provided by the fill of loose unconsolidated sediments. Thus we get the impression of a fracture that is open from the top but does not completely penetrate to the base of the continental crust, so that the dense SIMA has not been able to intrude it. The filling of the rift by loose material from the margins is added to by the deposits of lakes created by the closed drainage. Today there are still numerous small earthquakes, none felt very far away, which suggests a gradual settling of the graben fill. Not all the rifts in this region give the same gravity results. Those situated closest to the border of the continent (such as the Upper Pangani and Mkomassi grabens) appear to be isostatically compensated, and evidently here the dense SIMA has risen into the rifts. For the Red Sea, long ago recognized by Suess as a continuation of the East African belt, there is also isostatic compensation, as shown by Triulzi and by Hecker, which is as might be expected in view of the great width of the fracture zone.

2. Atlantic and Andes

The essential parallelism of the two sides of the Atlantic is a not inconsiderable argument for the assumption that they represent the margins of a gigantic rift. A glance at the map is enough to show that where mountains are cut off on the east, on the opposite side they continue to the west; where there are none on one side there are none on the other. In the far north the Greenland massif corresponds to that of Scandinavia; the lesser mountain chains of Europe match those of North America. Likewise the fault zones of Middle America (Caribbean) correspond to those of the Mediterranean and the great shield of South America lies opposite that of Africa. Where the west side is jagged (Greenland-North America) the east side is jagged (Europe). Where the continental border is straight (South America), it is just as straight and almost congruent, on the other side (Africa).

And still more. In those sectors that we know best, namely Europe and North America, there is almost complete agreement. Suess (in his great work *Das Antlitz der Erde,* II, p. 164; III, p. 60 and 77) discussed these relationships in several places. The most northerly zone on both sides consists of gneiss. On the European side it occurs in Lofoton (Norway) and the Hebrides (Scotland); on the west it is found extensively in Greenland, as well as to the west of Davis Strait and Baffin Bay, continuing across Cumberland Sound to Labrador and across the Strait of Belle Isle in the southeast.

The most striking observation in this connection was made by Marcel Bertrand in 1887, who noticed the presence of Carboniferous folded rocks (called Armorican by Suess) in France, which would appear to be immediately continued on the west side by the Carboniferous formations of North America. Today this belt is mainly a maturely eroded landscape with prominent planation surfaces, extending from central Europe in a westerly direction to the southwest point of Ireland and the western limit of Brittany, where it is abruptly cut off by a deeply incised (rias) coast. According to Suess "there is no evidence that this coast (between Dingle Bay and La Rochelle) represents the natural termination of the orogenic belt. Its extension lies under the waters of the Atlantic Ocean and on the other side of that ocean."

The continuation on the American side is found in the extension of the Appalachians of Newfoundland and Nova Scotia. Here there is also an orogenic belt folded in Carboniferous times, and, just as in Europe, it is overturned to the north; likewise there is a typical rias coast, also cut off by the sea. The folds trend northeast, and eventually easterly. The fauna and flora in the formations on both sides are not only identical for the Carboniferous, but also even more clearly for the earlier systems. At this point we cannot go into the literature (ref. Dawson, Bertrand, Walcott, Ami, Salter, and others).

The disruption of these "Transatlantic Altaides," as Suess called them, in matching positions, represents one of the most striking arguments for the former continuity of the two coasts. The old assumption was that the old mountain chain formerly connecting the two sides had subsided into the Atlantic, but this is difficult to accept, as Penck has shown, because there is no trace of it along the entire distance.

Farther south (i.e., Africa, etc.) the geology is insufficiently known to make a close correlation. Nevertheless, *Gentil,* for example, believed he could trace the continuation of the Atlas Mountains towards the Canary and Cape Verde islands and into the Antilles (Caribbean). After a comparison of the floras, *Engler* came to the conclusion that there may have been a former connection between Brazil, southeast of the mouth of the Amazon and the Bay of Biafra (Cameroun); Suess found that the sediments on both sides were in "striking agreement" with Engler's conclusion. But the data are necessarily very uncertain and a detailed comparison must wait for the future.

In addition to these so-called "anatomical" details, we must now consider two questions, which to the outsider may seem important, but can only be briefly discussed. The first is whether or not, on paleontological grounds, one can speak of a direct connection between America and

Europe-Africa at a certain time interval. The second, if this is true, is when the separation took place. Both questions have been long discussed and new material is constantly appearing that may correct our model. Obviously these things are independent of any assumption of either drift or subsided continents. One point makes it difficult to interpret older paleogeographic results: the transgressions. This is because parts of the same continental land area may become separated by shallow transgressions from the point-of-view of flora and flora. From this sort of evidence it is hard to distinguish a simple transgression from a rifting and spreading cycle.

With respect to South America and Africa the geologists and biogeographers are pretty well agreed that during the Mesozoic there was a broad land connection, that is, the "Brazilo-African Continent" (see Arldt, *Die Entwicklung der Kontinente und ihre Lebewelt,* Leipzig, 1907). It was called "Archhelenis" by von Ihering. Thanks to further work by these authors and others like Ortmann, Stromer, Keilhack, and Eigenmann, it is now clear that the connection was broken in the Tertiary, perhaps Eocene/Oligocene. (Haug and Kayser put it in any case before the Miocene; von Ihering, Ortmann and Stromer put it in the Eocene. Stromer and Eigenmann believed there was still a connection in the Lower Eocene.) The precise time of the break will naturally be studied closely by the paleontologists. At this moment the great North-South fractures began to rupture, and the Atlantic began to open.

Between Europe and North America an early Tertiary rupture is also accepted, severing a broad land connection that had previously permitted a general exchange of organisms. It was slowed down in the Oligocene and definitely stopped in the Miocene. We may conjecture that the line of fractures opened slowly from south to north. The last creatures were exchanged only in the far north, via Scandinavia and Greenland. Here a land bridge existed probably until the Pleistocene (Scharff, 1909, *Proc. Roy. Irish Acad.,* v. 28, pt. 1, 1–28). We shall note a few of the arguments of general interest. The common heather (*Calluna vulgaris*) is found outside Europe only in Newfoundland and in nearby provinces; many North American species are limited to western Ireland. The heather family is entirely absent from western North America and likewise in eastern Asia and could not thus have crossed by way of the Bering Strait. The garden snail (*Helix hortensis*) lives outside of continental Europe in Ireland, Greenland, Labrador, and New England. In Wegener's opinion, this whole region was a single landmass during the Pleistocene glaciation, so that the glaciation could have covered a much smaller area that would appear on the maps today.

[*Editor's note:* Omitted here is a discussion of Wegener's erroneous interpretation of the central European loess as an interglacial steppe deposit.]

An interesting situation existed between North and South America. Proposed by *Osborn* and followed up *by Scharff,* there appears to have been a land connection until the early Tertiary. It appears to have been gradually broken off in the late Tertiary (*Kayser* says Pliocene),. Previously this former landbridge was postulated in the west, perhaps via the Galapagos, but according to our model the connection could easily be made via West Africa and would be severed with the Atlantic opening that was matched by the uplift of the Andes.

Since the Andean folding was more or less contemporary with the Atlantic opening, the idea of a causal relationship arises. The American plates drifting to the west would have encountered a resistence in the Pacific crust that appears to have been less plastic, so that the thick shelf deposits of the western margin became massively folded. In this case it seems that the sialic material was more plastic and the simatic crust was less so. Nevertheless, we must assume that the SIMA yielded appreciably because the Andean folding would in no way represent the folding of a belt 4,000 km wide, equivalent to the width of the Atlantic. Bearing in mind the way the nappe theory explains the Alps with a four to eight times crustal shortening, there seems no reason not to correlate the Andean folding with the Atlantic opening. (This is a generalization. In western North America, the time of the folding tends to migrate back from west to east. Thus only the Tertiary folding could be related to the Atlantic opening.)

3. Gondwanaland

If we turn now, following the same line of reasoning, to relate continental drift to the Tertiary folding of the Himalayas, we discover some very remarkable relationships. If we "unfold" this belt of compression that includes the highest mountains in the world, we find that the southern point of peninsular India would come down next to South Africa. During the collision with Asia, northern India became a block "ringed around with folds" (*Suess*).

On paleontological grounds an elongated linear land mass, "Lemuria" formerly occupied the Indomadagascan area which for a long time was connected to the African plate until it was separated by a broad north-south fracture system in the line of the Mozambique Channel. According to *Dacqué* and others it was already active in the Triassic and separation was certainly complete by the Lower Jurassic. *Douvillé* also believed there could be no longer any connection between Madagascar and Africa even in the Triassic. If correct, this fracturing was appreciably earlier than that identified for the Atlantic. Nevertheless the collision of this peninsula with Asia seems only to have come with the Tertiary and still continues. (In geology people often speak of "one-sided" compression in orogeny, and particularly for the Himalayas, citing a push from the north, not the south. But following the laws of physics, each stress is matched by a counter-stress. Thus an orogenic belt requires a compression from both sides, and any asymmetry must relate to other factors, such as differing heights or thicknesses of the colliding blocks, differing viscosity and so on.)

The paleontological discoveries leave no doubt that Australia had a direct land connection with India, just as between Africa and South America. This megacontinent has long been known as "Gondwanaland," and we assume Australia was part of it. The connection lasted until the Permian and the separation must have been in the Triassic or Jurassic. On the other side, however, according to *Hedley, Osborn,* and others, there were connections to South America that lasted until the Quaternary, probably by way of Antarctica, which is, unfortunately, still an unknown area. Thus the connection is uncertain. Presently it seems likely that the west coast of Australia was directly connected to the coast of India, but was separated in the Triassic period, while the south coast remained firmly attached to Antarctica. During the Tertiary the latter began to push against the Pacific Plate just like the South American block, as it separated from Africa. At this moment only the fold mountains of Victorialand and Grahamland are actually known. The final rupture of Australia from Antarctica did not occur until the Quaternary and New Zealand represents a relic of the once continuous "Antarctic Andes". These correlations should, however, be regarded mainly as tentative.

The maps showing the northerly surroundings of Australia and New Guinea suggest a collision with Southeast Asia. As *Wallace* pointed out, the faunas of New Guinea relate to Australia whereas those of the Sunda Islands belong to Southeast Asia. (The "Wallace Line" of biogeography, however, does not match the continental tectonic border). It seems likely that this northward motion led to the building of the high mountains of New Guinea.

4. Permian Ice Age

One of the most striking pieces of evidence that supports the continental drift idea is the Permian Ice Age (also partly Carboniferous according to some specialists). Traces of this glaciation are found all over the Southern Hemisphere, but nowhere in the Northern Hemisphere (except for India). Without the drift hypothesis it has posed a hopeless puzzle, for the distinctive glacial striae and grooves are found in Australia (including all states and New Zealand), in South Africa (recently also Congo and possible in Togo), South America (Brazil, Argentina, Falkland Is.; ref. Kayser, 1911, p. 266), and most remarkably in peninsular India. Koken ("Indisches Perm and die permische Eiszeit," *Festband N. Jb. f. Min.*, 1907) showed, in a special publication with a map, that with the present day arrangement of the continents such a wide distribution of glaciation was completely impossible. Even if one classed the South American material as uncertain, which cannot be accepted today, and if the South Pole is placed in the most central position (in the Indian Ocean) the ice sheets would still have to reach to (paleo-) latitudes of 30–35°. With such a glaciation hardly any part of the globe would be ice-free. With that arrangement the North Pole would fall in Mexico where there is no trace of Permian glaciation, and the South American examples would fall on the equator.

For all hypotheses that reject continental drift the Permian Ice Age represents an insoluble problem. Without any other argument or evidence, it suggests (in the words of Penck, 1906: "Südafrika und Sambesifälle," *Geogr. Zeit,* v. 12, pt. 11, p. 601–611.) that "horizontal displacements of the earth's crust must be kept in mind as a serious working hypothesis."

If we reconstruct the paleogeography of the Permian time in terms of the drift theory, we find that all of the glaciated areas become arranged concentrically around the southern point of Africa and it is only necessary to postulate a South Pole here as the center of a quite reasonable area of glaciation and everything can be explained. The North Pole at that time would lie in the Pacific near the Bering Strait.

5. Atlantic and Pacific Hemispheres [p. 305]

The broad morphological contrasts between the two hemispheres, east and west, have long been recognized. Suess (in *Das Antlitz der Erde,* II, p. 256) pointed out the faulted nature of the Atlantic and Indian Ocean coasts, in contrast to the cordilleran and island-arc coasts of the Pacific, and northeast Indian Ocean. In 1903 Becke noted that the lavas of the two are distinctive. Krümmel commented that the mean depths of the Pacific are greater than those of the Atlantic and Indian oceans. Wegener sees the Atlantic-Indian coasts as opening, the Pacific ones as converging. He follows Koken, Frech, and others in regarding the Pacific as a very ancient basin, in contrast to the youthfulness of the other two. He recognized the greater depths in the Pacific as being related to the longer cooling of the simatic crust.

6. Pole Shifts [p. 306]

Relative displacements of the pole must be accepted, after careful consideration of the evidence. In the course of the Tertiary period the North Pole migrated to its present position from the region of the Bering Straits. (Wegener does not discuss the question of the drifting plates or the polar axis that moves. Transl.) Many examples are given of displaced floras, which seem to prove a relative pole shift and not merely a general change in global climate. The shift coincides with the opening of the Atlantic. The trajectory of the North Pole shift must have taken it west of Greenland during the last ice age and then moved 10° to its present site. As the continents shift, the polar axis must adjust accordingly. Schiaparelli (1889, De la rotation de la terre, sous l'influence des actions géologiques, *Ac. Imp. Sci.,* St. Petersburg, mém. sémisec.) concluded that with a quite rigid earth, geological changes (mass displacements) could only shift the polar axis a very small amount, but that with some degree of plasticity, appreciable displacements are to be expected, particularly over extended geological intervals. After the San Francisco earthquake in 1906, Hayford and Baldwin (Coast and Geod. Surv. Rept., App. III, p. 97) calculated that a crustal block of 40,000 km^2, some 118 km thick, shifted northward about 3 m; in consequence a pole shift of 0.0007 sec (about 2 mm) would result. Drifting continents would repeat this sort of shift hundreds of times; one could consider a secular movement of a hundredth of a second per year, or 1° in 360,000 yr. A Tertiary shift of 30° could be achieved in 10 million years, at this rate.

Although the relationship between pole shifts and continental drift is at present purely theoretical, it seems reasonable, but must await precise observations and calculations in the future.

III. PRESENT-DAY CONTINENTAL MIGRATION AND OSCILLATION OF THE POLAR AXIS [p. 307]

1. Greenland

It is of the greatest importance to obtain direct measurements of continental drift. If we assume that the separation of Scandinavia and Greenland occupied 50,000 to 100,000 years (that is, if it began before the last Ice Age but following the "great" Ice Age), still continues today, and the separation distance is 1,400 km, then the drift rate is 14–28 m per year. This rate can be checked against astronomical determinations of longitude. Unfortunately, only lunar observations are available, and they are of dubious accuracy. [*Translator's note:* We omit lengthy details.] Nevertheless, over 84 yr. a net shift of 950 m is indicated; this is 11m/yr, which is accepted as reasonable.

2. North America

More precise figures are available for the suspected west-drift of America, the result indicating 4 m/yr. [Details omitted.]

3. Pole Oscillations

Thanks to the work of the International Latitude Service, it is known that although the Euler perturbation for a rigid earth should be 305 days, in the observed condition (plastic), the periodicity is lengthened to 430 days. The pole of rotation describes a looping course around the geographic axis in a series of spirals. In this succession (shown by Albrecht, 1910, *Peterm. Mitt.*, July, p. 17) there is just a suggestion of a gradual shift toward the Atlantic Ocean side. If this is confirmed it will be an important point of evidence in favor of the hypothesis of continental drift.

19

Reprinted from pages 169–170 and 180–195 of *Washington Acad. Sci. Jour.*
23:169–195 (1933)

THE THERMAL HISTORY OF THE EARTH[1]

Arthur Holmes

Speculation as to the thermal behavior and cooling history of the earth has been vigorously stimulated in recent years by the discovery that the radioactive elements constitute an important source of internal heat. Despite the vast store of data accumulated since Hopkins grappled with the problem nearly a hundred years ago, the constitution of the earth and the physical properties of its materials are still inadequately known, and in consequence every worker in this field finds an uncharted margin in which the undetermined conditions governing the earth's thermal history offer a considerable range of personal choice. It is for this reason that the field is occupied by several types of competing hypotheses.

RIVAL HYPOTHESES

The time-honoured "contraction hypothesis," made familiar by Kelvin, has been ably rejuvenated by Jeffreys (23). Joly, on the other hand, departing widely from traditional lines of treatment, has introduced a "hypothesis of thermal cycles" which has aroused world-wide interest and discussion (26). Convective circulation within the earth was envisaged by Hopkins in 1839 and later by Fisher (10), and a "hypothesis of sub-crustal convection currents" (maintained by the earth's radiothermal energy) is now being actively explored by Bull (6) and myself (16).

To the onlooker this overcrowded field of speculation may seem to be in a distressingly chaotic state. The occasion of this address provides me with a welcome opportunity to attempt to tidy it up. I

[1] Lecture delivered before a joint meeting of the Washington Academy of Sciences with the Geological Society of Washington on April 7th, 1932. Received January 29, 1933.

propose to offer a critical survey of the whole field by systematically reviewing the physical assumptions involved in each type of hypothesis and by confronting the geological consequences of each with what is known of the realities of earth history.

[*Editor's Note:* Material has been omitted at this point.]

AI—THE THERMAL CONTRACTION HYPOTHESIS

Here it is assumed that the sub-crustal material is negligibly radioactive (i.e. that the heat generated within it is insufficient to prevent continuous cooling) and that the freezing-point gradient is steeper than the convection gradient throughout (Fig. 6). Crystallization therefore begins at the bottom of the Lower Layer, presumably producing a peridotite-like rock until an outer residual magma a few tens of kilometres in thickness remains. From this residuum, which must be assumed to contain nearly all the earth's store of radioactive elements, layers of basaltic and granitic composition crystallize in turn.

The consequences of the contraction hypothesis have been repeatedly discussed, and despite the damaging criticism to which the hypothesis has been exposed, many geologists still retain their faith in its tenets. In 1930, for example, Hobbs (13) expressed his belief that "this view of a shrinking earth is more firmly established today than ever before." On the other hand, Wegener regarded the hypothesis as "completely bankrupt." Both authors omitted to give sound reasons for their contradictory opinions, but at least two sets of considerations seem to me to support Wegener's characteristically sweeping judgment.

As indicated already on p. 176 the thermal limitations of the hypothesis preclude its successful application to the outstanding phenomena of igneous activity. The widespread occurrence of plateau basalts and the appearance of plutonic intrusions during orogenesis seem to be quite inconsistent with the existence of a crystalline interior (19).

Jeffreys has claimed that the compression due to contraction is of the right order to account for mountain-building, and the validity of his claim has been as persistently denied. On both sides, however, the margin of error is so great that a really convincing comparison between theory and fact is at present out of the question (17 and 24). On the other hand, the distribution of mountain-building in time, as tested by the lead-ratios of radioactive minerals, shows that the intervals between the peaks of successive orogenic revolutions have not become systematically longer, as they should have done, according to the hypothesis. Indeed, since Pre-Cambrian time the intervals have become shorter, as the following data clearly show (18).

Orogenesis	Lead-ratios	Intervals
Alpine-Laramide	0.008	
		0.022
Hercynian-Appalachian	0.030	
		0.025
Caledonian-Taconic *ca.*	0.055	
		0.030
Upper Pre-Cambrian	0.085	
		0.045[a]
End of Middle Pre-Cambrian	0.130	

[a] It is possible that this interval represents two intervals of 0.025 and 0.020 divided by the Killarnean orogenesis with a lead-ratio of 0.110.

This apparent speeding-up of the earth's orogenic activities is demonstrated in detail by Figs. 10 and 11. In Fig. 10 a first approximation to a geological time-scale is provided by plotting lead-ratios against maximum thicknesses of sediments from the Cambrian onwards. The steepening of the curve as the present is approached accords with the findings of Barrell and Schuchert to the effect that rates of denudation and sedimentation have steadily increased since Pre-Cambrian time. In Fig. 11 the orogenic episodes recognised by Stille are plotted against the time-scale. The results show at a glance that there has been a remarkable acceleration of activity.

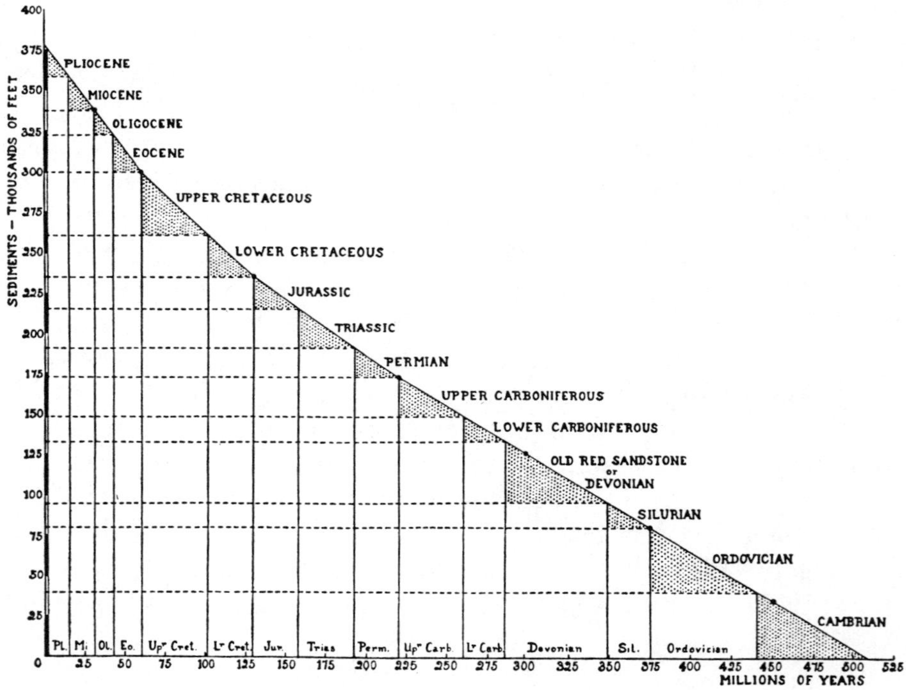

Fig. 10.—An approximate geological time-scale based on lead-ratios (plotted as abscissae) and maximum thicknesses of sediments (plotted as ordinates).

The contraction theory implies an ageing earth that should now have reached senility. The fact is that during the Tertiary period the earth was more vigorous in her behaviour than ever before during at least the last thousand million years. Many other objections to the hypothesis could be urged, but those here given seem to be ample to justify its rejection.

AII—MODIFIED FORM OF THE THERMAL CONTRACTION HYPOTHESIS

In this case (Fig. 7) the assumed distribution of radioactivity is that of AI, but the two gradients are regarded as becoming tangential at some intermediate depth. Crystallization begins at this depth and proceeds upwards as in AI. Crystallization—or possibly consolidation into the durovitreous state—also proceeds downwards into the underlying zone of convection so long as the heat brought up can be carried away by conduction through the crust. The underlying convective zone slowly thins and dies out by downward thickening of the crust as loss of heat reduces the gradient below the critical value. The bulk of the substratum is then left as a stagnant glass in which downward crystallization very slowly proceeds. The period required for complete crystallization would be very long compared with geological time. This form of the contraction hypothesis implies that the substratum was practically devoid of radioactivity from the start, since the cooling process in itself provides no mechanism for concentrating any originally dispersed radioactivity towards the surface.

Jeffreys finds that on this hypothesis crystallization may now have proceeded to a depth of a few hundred kilometres. As compared with AI a little more contraction may be available because of the volume change attending crystallization or the passage from the liquevitreous to the durovitreous state. On the other hand the gradient due to the

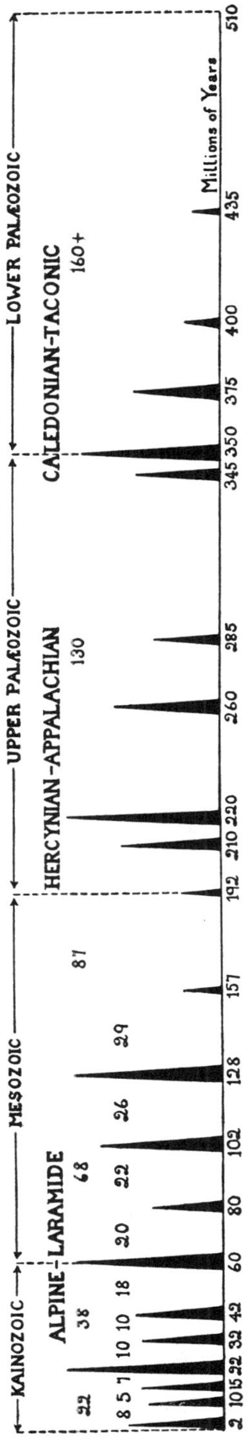

Fig. 11.—Orogenic episodes (after Stille) of Post-Cambrian time plotted on the time-scale of Fig. 10. An assessment of the relative intensities of the respective orogenies is roughly indicated by the heights of the black triangles.

increase of freezing-point with depth—which gives rise to an outflow of heat at all depths that have crystallized—is less steep than in AI, and gives a heat flow through only a shallow depth instead of all through the Earth. In this respect, therefore, there is less contraction, and on the whole AII seems to be even less capable of matching the facts than AI. Indeed, in his latest work Jeffreys clearly shows his preference for AI (25).

BI—HYPOTHESIS OF THERMAL CYCLES

The substratum is assumed to be slightly radioactive, sufficiently so for the heat generated to prevent permanent crystallization. The two assumed gradients being as in AI, crystallization begins at the bottom and proceeds towards the surface (Fig. 8). Once convection has ceased, heat is generated at all depths below a certain critical level (near the base of the basaltic layer) faster than it can be carried off by conduction. This excess of heat supplies latent heat and promotes re-fusion. Sooner or later the growing threads of fluid unite and liberate crystal individuals or aggregates which sink, unless the viscosity is sufficiently high to prevent them from doing so at any effective rate. A layer of magma thus accumulates and rises upwards, increasing in thickness as it ascends. Convection recommences within this layer, and accelerates the upward migration until the base of the basaltic layer is reached. Re-fusion of the basaltic material then proceeds. At this stage the temperature at the top of the zone of fusion is below the freezing-point of peridotite and the peridotite magma gradually solidifies as the basaltic layer is fused. Finally, under the continents, much of the granitic layer may become fused in turn before the overlying crust becomes sufficiently thin to carry off the excess of heat and bring about consolidation. During the later stages, igneous activity is inevitable in both continental and oceanic regions. Essentially the cycle consists of the ascent of a wave of fusion from great depths, thickening until it comes under the cooling influence of the surface, and then dying out as the surface is closely approached. Before the first cycle is completed another will have begun in the depths of the substratum. The earth's history is thus thought to be controlled by a succession of "thermal cycles."

It is obvious that if the maximum depth ever reached by a zone of fusion were to exceed a certain thickness (of the order of a few tens of kilometres,) it would fail to die out before breaking clean through to the surface. As this has not happened on a regional scale within known geological history there is here a clear indication that the hypothesis cannot legitimately invoke waves of fusion of more than relatively small amplitude. This limitation implies an average radioactivity in the substratum of a much lower order than that of the crust (1/100th or less).

It was assumed in the above statement that the viscosity of the substratum was not too high to prevent the effectual sinking of crystals. Actually, however, the viscosity is extremely high; Jeffreys finds a mean value exceeding 5×10^{20} c.g.s. from the 14-monthly variation of latitude. Under such conditions crystals could not sink through any appreciable distance during geological time. Thus the conditions favourable to thermal cycles would seem to be limited to within 100 km. or so from the surface. But as fusion would nevertheless steadily proceed throughout the substratum, successive cycles or groups of cycles would involve ever-increasing depths, until finally a wave of fusion would rise of such amplitude that the whole crust would be engulfed, and in part remelted. Accepting the premises, it is by no means inconceivable that such a fate may overtake the earth at some remotely future time. The danger arises from the conditions implied by the hypothesis under consideration; first, the rising convection currents maintain temperatures at the top that are well above the melting-point of the materials of the roof; and second, the heat represented by the excess of temperature over the melting point, and the latent heat of the material of the wave of fusion, have both to escape through to the surface before reconsolidation can occur.

The advantages of Joly's hypothesis are many, but unfortunately they are both dangerous and incomplete. Each period of consolidation corresponds to one of crustal contraction and mountain-building, and in this respect the hypothesis is more satisfactory than the traditional contraction hypothesis, as it provides for repeated consolidations, each time beginning afresh from conditions of fusion. Obviously, there is ample opportunity for igneous activity on the most extravagant scale. Marine transgressions and recessions are a necessary consequence of thermal cycles. Partial foundering of the lands and their invasion by epeiric seas take place during conditions of widespread fusion, followed by recovery and withdrawal of the seas during the succeeding phase of contraction and orogenesis.

Jeffreys has vigorously attacked the hypothesis on physical grounds (21 and 22). Joly requires a westerly tidal drift of the crust to drag the ocean floors over the magmatic layer and so permit the excess sub-continental heat to be discharged through the chilled and attenuated oceanic part of the crust. Jeffreys gives good reasons for thinking that this drift could not occur at the required rate. On the other hand, J. H. J. Poole (31) has developed a form of the hypothesis in which tidal drift does not appear. McCarthy (29) has pointed out that rapid loss of heat through the oceanic floors would involve abnormal heating of the oceanic waters, and this consideration undoubtedly limits the application of the hypothesis to a very thin layer of fusion. A serious objection to the hypothesis in its ideal symmetrical form is its implication that world-wide crustal tension (during fusion and expansion) has alternated in time with world-wide compression (during consolidation and contraction). The testimony of geology is that compression seems always to have been more or less active somewhere or other in the world, periodicity being a matter of climaxes in particular regions heralded by crescendos and followed by diminuendos. Moreover, tensional phenomena have often been contemporaneous with compressional phenomena elsewhere. It might be possible to modify the hypothesis so that it would become consistent with these phenomena, but to do so would necessitate the discovery of a non-uniform distribution of radioactivity in the crust on a regional scale and to an extent that the evidence so far collected fails to foreshadow. Like the thermal contraction hypothesis, this one equally fails to account for the development of geosynclines, the subsiding areas of long-continued sedimentation that formed the gathering ground of the formations out of which long belts of mountain ranges were destined to be built. It must also be pointed out that Joly's hypothesis is at variance with the rate of the apparent speeding-up of geological activities to which reference has already been made. Certainly successive cycles would be of increasing intensity if each wave arose from a greater depth than the previous one, but if so the time intervals between them would become systematically longer.

Although it seems highly improbable that the hypothesis of thermal cycles can by itself express the earth's thermal history, it would be rash to deny its applicability to certain of the problems of crustal thermodynamics. The formation, ascent, and consolidation of magmas and consequent isostatic changes of level seem to fall within the scope of its mechanism, but on a scale less grandiose than that visualized by its author.

BII—HYPOTHESIS OF SUB-CRUSTAL CONVECTION CURRENTS

Here it is assumed that the gradient in the convecting layer becomes tangential to the freezing point gradient at some intermediate depth (Fig. 9). Tangential conditions are likely to be closely approached through a zone of considerable thickness. Crystallization begins in this zone and proceeds upwards until a crust is formed over a convecting interior. Under these conditions the rising convection currents reach the overlying roof at or near the temperature of the melting point of the roof materials. Moreover, the latent heat of the substratum below the level of tangency is not liberated by crystallization and does not escape through the crust, since crystallization tends to occur only at and above the level of tangency and not at the bottom of the substratum. Only the latent heat of material fused *above* the level of tangency has to pass through the crust. From this cause there is therefore no danger of catastrophic crustal fusion on a worldwide scale after the crust has once formed. The substratum remains in a glassy state (liquevitreous of Jeffreys), characterized by rigidity, high viscosity, no permanent strength, and sufficient radioactivity to maintain a slow but ever-changing convective circulation.

It is easy to see that the zone of tangency must lie beneath the basaltic layers. It cannot be in the latter because of the marked fall of melting-point in passing upwards from peridotitic to basaltic material. On the other hand, if the zone of tangency were deep in the peridotitic layer, the crust would be too thick to carry off by conduction the heat brought up by rising convection currents. The peridotitic layer would then heat up, and its temperature gradient would steepen, until the actual zone of tangency migrated to a higher level. This implies partial fusion of the lower part of the peridotitic layer and the generation of successive waves of magma which would ascend into the crust as in the Joly cycles. The evidence of igneous activity indicates that such waves are relatively thin, whence it follows that the shell of material fused by the ascent of the zone of tangency must itself be thin. An average depth for the zone of tangency between 40 and 80 km. would seem to coordinate all the relevant data. Such a depth corresponds first with that suggested by Gutenberg and Richter (11) for the passage from the crystalline to the glassy state; and second, with the thermal necessity that conditions of fusion must normally reach a level some distance above the depth where the melting point would be attained in virtue of the radioactivity of the overlying material.

The above discussion indicates that variations in the thickness and radioactivity of the crust, together with changes in the temperature gradient of the substratum, may combine to extend very considerably the range of the shell subject to refusion. For this reason the conditions which determine whether the substratum gains on the crust by upward melting or the crust gains on the substratum by downward crystallization are likely to vary from place to place and from time to time.

To prevent convection in a layer extending far down towards the core a viscosity exceeding 10^{26} would be necessary. As we have good evidence that the actual viscosity is well below this limit there is no physical improbability about internal convection. Vertical currents ascend in certain places and on reaching the top the material promotes sub-crustal fusion and spreads out laterally, flowing horizontally to the neighboring descending currents. The horizontal flow is likely to exercise a powerful viscous drag on the lower levels of the crust, throwing the latter into tension where the currents diverge and into compression where they converge (Figs. 12 and 13). If the distance between ascending and descending currents be of continental dimensions then the drag due to horizontal flow may produce continental displacements, with distension behind and mountain-building in front. Material from the substratum ascends into the distended area, where it cools, and differentiates, while heavy compressed crustal material (eclogite from the basaltic layers) is dragged down with the descending currents.

Excess heat from the substratum is thus discharged (*a*) by cooling due to thinning of the crust and consequent development of basins,

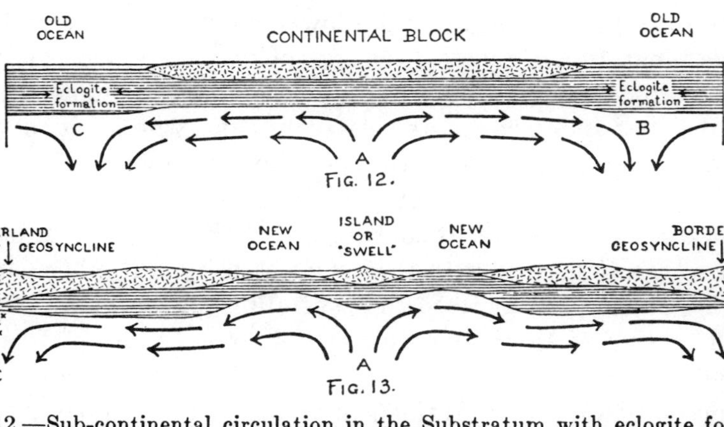

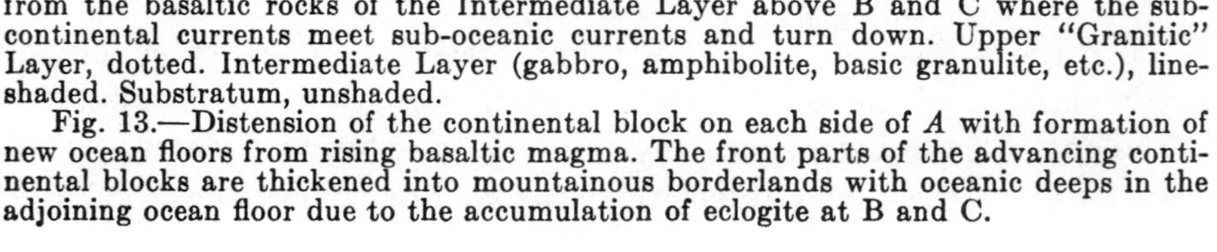

Fig. 12.—Sub-continental circulation in the Substratum with eclogite formation from the basaltic rocks of the Intermediate Layer above B and C where the sub-continental currents meet sub-oceanic currents and turn down. Upper "Granitic" Layer, dotted. Intermediate Layer (gabbro, amphibolite, basic granulite, etc.), line-shaded. Substratum, unshaded.

Fig. 13.—Distension of the continental block on each side of *A* with formation of new ocean floors from rising basaltic magma. The front parts of the advancing continental blocks are thickened into mountainous borderlands with oceanic deeps in the adjoining ocean floor due to the accumulation of eclogite at B and C.

geosynclines or ocean floors over the sites of ascending currents; (*b*) by the heating-up and remelting of crustal materials carried down in descending currents; and (*c*) by igneous activity of all kinds.

The convection hypothesis has the advantage over its rivals that it naturally and inevitably provides for simultaneous tension and compression. Good examples of this are the opening of the Uralian geosyncline at the same time as the crumpling up of the Caledonian mountains of Scandinavia, and the formation of the North Atlantic while the Rockies, Sierras, and Coast Ranges of Western North America were being formed.

According to the convection hypothesis, we should expect mountains to be formed where two series of approaching currents turn down and so compress the overlying layers of the crust (Figs. 12 and 13). Experiments by Bull (5) have shown that this kind of mechanism provides the only method yet devised for successfully imitating nappe-structures like those of the Alps. While the mountains are being built from sialic materials, the underlying basaltic material is dynamically metamorphosed into the dense eclogite facies and is sucked down by the downward-flowing currents. After the folding, some of this material will in many places remain beneath the orogenic belt, and its inevitable fusion must tend to heave up the overlying mountain range. Where it happens, however, that the fused material is squeezed out sideways, the mountain chain will break down and parts of it may even subside beneath sea-level. Other parts of a folded belt may never have risen above the sea.

Jeffreys has pointed out that many old peneplained mountain ranges are still represented by tracts of folded sediments, despite the fact that if they had been worn down to this extreme degree by denudation alone, all the sediments and much of the underlying granitic layer should have been removed (23, p. 296). He suggests that these tracts have in part been lowered by "outflow of matter from the intermediate layer, with compensating inflow in the lower layer." Such an effect would be brought about equally well by the passage of a sub-crustal current, for the latter would tend to "erode" the roots, carrying away the corroded material in the direction of movement, just as ice-sheets have ground down the protuberances of the land surface beneath them and spread out the debris in the lowlands beyond.

The distribution of the latest of the earth's mountain systems (Figs. 14 and 15) shows that they form interrupted rings surrounding the great northern land masses of N. America and Eurasia (known

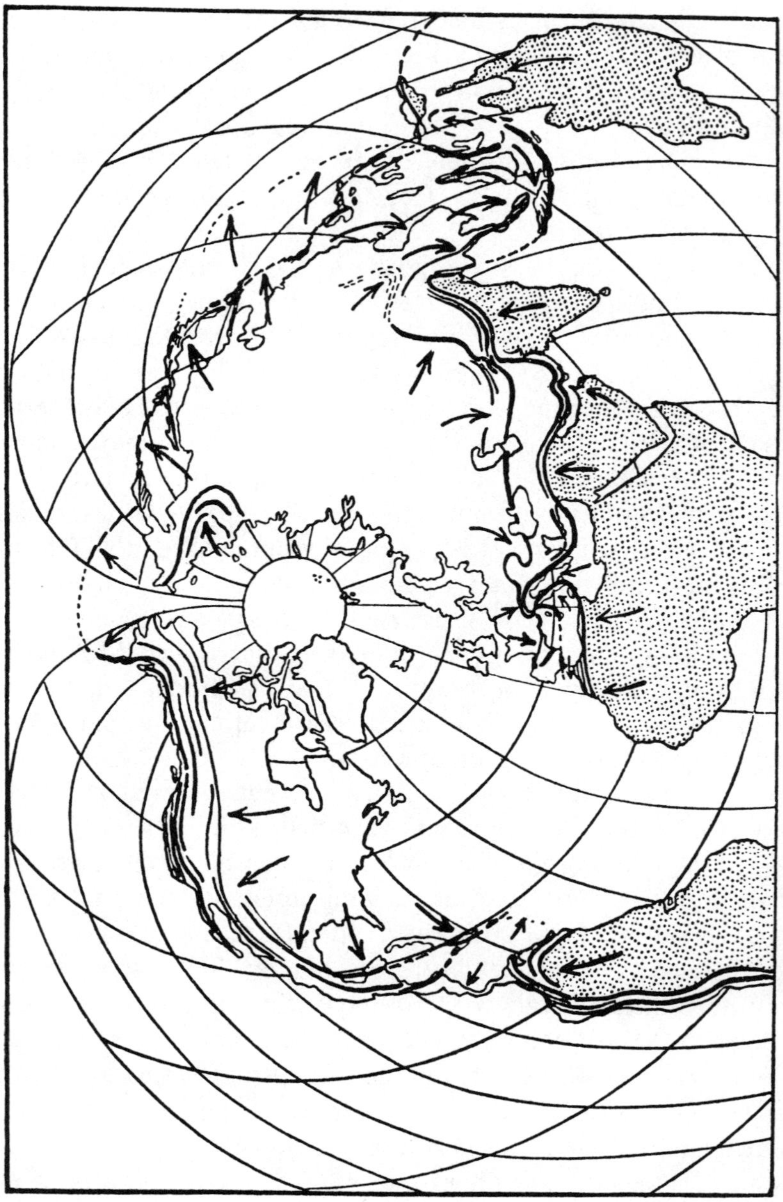

Fig. 14.—Orogenic ring peripheral to Laurasia with continental movements directed outwards towards the Pacific and the Tethys. (The adjoining blocks of Gondwanaland are dotted.) Scale: 1: 170×10^6.

as Laurasia) and the southern lands of S. America, Africa, India, Australia, and Antarctica (known as Gondwanaland). Each great land mass seems to have been the centre of a system of currents directed outwards towards the Pacific and towards the former Mediterranean—the Tethys—that lay between Gondwanaland and Laurasia. Peripheral ranges of mountains were raised up in front of each advancing continental block—and in some places, such as the East and West Indies, the process seems still to be in operation. Earthquakes and volcanoes are specially numerous in and adjacent to these peripheral rings of crustal instability.

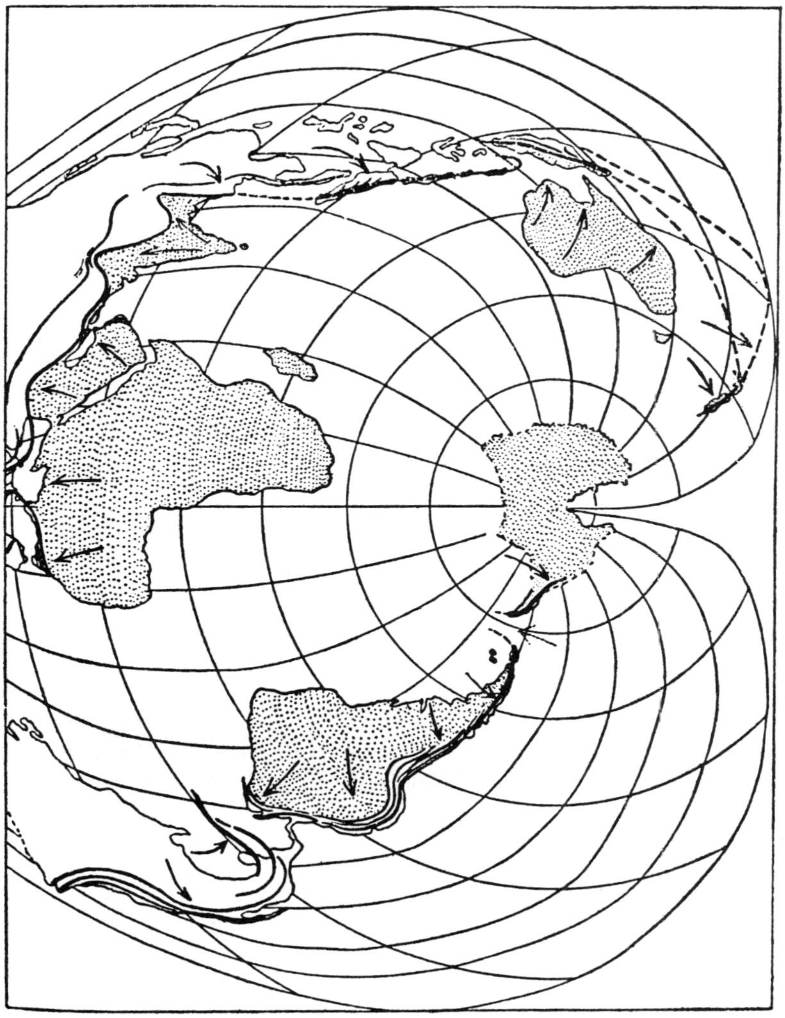

Fig. 15.—Orogenic "ring" peripheral to Gondwanaland (dotted) with continental movements directed towards the Pacific and the Tethys. Scale: 1: 210×10^6.

The formation of eclogite in front of active mountain ranges is strongly suggested by the oceanic deeps of the western margins of the Pacific. Here, too, lie the epicentres of many deep earthquakes that have originated from foci hundreds of kilometres below the surface (35). If these deep-seated disturbances are not in some way a result of currents operating far down in the zone of flowage where great masses of eclogite may be fractured as their strength is overcome, it is difficult to conceive an alternative mechanism that could be responsible for them.

The action of sub-continental currents directed towards the Pacific shores may be inferred from the echelon structure of the island festoons of Asia (34), and from the great and geologically rapid uplifts of the outer zones of many growing island-arcs (Hobbs, 12). Evidence of a different sort, suggesting the action of a sub-Pacific current, is forthcoming from a comparative study of certain coral reefs in the Fiji group made by W. M. Davis. It will suffice here to quote a single sentence from the paper referred to (8): "The changes of level in the several belts all appear to be caused by the slow westward migration of a broad and low ocean-floor anticline, preceded and followed by shallow ocean-floor synclines." Substratum currents acting on the under surface of the crust may be expected to produce migrating sub-crustal wave-forms which would be mirrored above by migrating anticlines and synclines of the kind discovered by Davis.

The remarkable belt of negative anomalies of gravity recently discovered by Meinesz in the East Indian Archipelago (30) may also be a symptom of sub-crustal currents. A powerful downward drag seems necessary to account for the astonishing departure from isostatic equilibrium that has been revealed (17). Other anomalies, however, such as those which imply that the equator of the geoid may be slightly elliptical instead of circular, have been used by Jeffreys in support of the view that the substratum has sufficient strength to withstand considerable regional departures from isostatic equilibrium (23, p. 222). If the substratum has strength, its viscosity is infinite until the strength has been overcome, and convection currents would then be impossible. The objection is not final, however, for its acceptance raises in turn a fresh difficulty. If Africa, for example, stands high to an extent implying a slight bulge in the equator, then its tendency should be to sink; yet actually its tendency has been, and geologically speaking, still is, to rise. I have therefore suggested the alternative interpretation that the "bulge" may be an expression of some dynamic process operating in the substratum at a rate slightly in advance of that of isostatic re-adjustment (16, p. 576). There is here a difference of opinion that should be amenable to mathematical treatment and decision.

Turning to evidence of a very different kind, it is worthy of notice that the maintenance and secular variations of the earth's magnetic field are regarded by Chapman (7) as possible consequences of a deep-seated internal circulation of material. Another suggestive indication of flowage in the substratum is provided by the mysterious fluctuations in the earth's rate of rotation which have been studied by de Sitter (9) and Brown (3). De Sitter points out that the sudden shortening of the day which happened in 1918 could have been produced "by a very slight adjustment of the layers of equal density in the inside of the earth, of which the effect on the dimensions of the outer surface need not exceed a fraction of a millimeter." The significance of this statement is best realised by considering that "if the whole of the Central Asian Highlands from the Himalaya to the Kwen Lun Mountains (both included) had at that time been sunk into the earth, their mean height being reduced to sea-level, the resulting shortening of the day would have been only one fourth of what is required." We have in such remarkable time-changes an indubitable indication of deep-seated widespread movements within the earth—movements which must, moreover, arise from causes originating inside the earth, quite independently of any external cause due to sun, moon, or stars.

So far, little has been said about continental drift, and space does not allow of any critical discussion of the evidence for or against it. It may be said, however, that on the thermal contraction hypothesis continental drift is manifestly impossible, whereas on the convection hypothesis a certain amount of drift is inevitable mechanically; it serves, moreover, to provide a means of escape for the heat carried up to the crust by the more powerful ascending currents. The convection process seems also to be intimately related to the acceleration of geological activities since Pre-Cambrian times, and to the cognate fact that continental drift does not appear to have operated on a considerable scale until after the Hercynian orogenesis. If eclogite is carried down by descending currents into the substratum, then some of the basaltic material must be dissolved in the substratum which, in consequence, becomes locally enriched in the radioactive elements. The tendency to circulate in convection currents should, therefore, have become more marked with the advance of geological time, and this would necessarily have provided more earth movement, more igneous activity, and in general more intense manifestations of energy in every direction. The explanation suggested implies that the radioactive layer was not concentrated from the interior, as generally assumed, but that conversely, the interior has been steadily gaining radioactivity at the expense of the lower levels of the crust. If so, the earth has an exciting geological history in store.

Although the hypothesis invoking sub-crustal convection currents cannot be regarded as established, it is encouraging to find that it is consistent with a wide range of geological and geophysical data. Moreover, it is by no means independent of the best features of the other hypotheses. It requires the local operation of thermal cycles within the crust, and it necessarily involves contraction in regions where crustal cooling takes place. It is sufficiently complex to match the astonishing complexities of geological history, and sufficiently startling to stimulate research in many new directions. And therein lies its chief merit, whether it fails or succeeds. Jeffreys (24) can find nothing inherently impossible in the hypothesis, but he thinks that "its validity would be a remarkable accident." I agree; but then I think the earth is no less a remarkable accident. It is impossible to be a geologist without realising that—in the dim light of the knowledge we have so far gained—the earth we live on is a strange and most improbable planet.

Of the figures which illustrate this paper Figs. 1–13 have been drawn for me by my friend Miss Doris L. Reynolds and in conclusion I wish to record my grateful thanks for her generous coöperation. The two maps, Figs. 14 and 15, are reproduced from an earlier paper (16) by permission of the Council of the Geological Society of Glasgow.

LITERATURE CITED

1. ADAMS, L. H. *The temperature at moderate depths within the earth.* This JOURNAL **14**: 459. 1924.

2. BRIDGMAN, P. W. *The physics of high pressure.* (London, 1931). p. 398.

3. BROWN, E. W. *The evidence for changes in the rate of rotation of the earth and their geophysical consequences.* Trans. Ast. Obs. Yale Univ. **3**: 209–235. 1926.

4. BULL, A. J. *Some aspects of the mountain building problem.* Proc. Geol. Assoc. **38**: 145. 1927.

5. BULL, A. J. *Further aspects of the mountain building problem.* Proc. Geol. Assoc. **40**: 105. 1929.

6. BULL, A. J. *The convection current hypothesis of mountain building.* Geol. Mag. **68**: 495–498. 1931.

7. CHAPMAN, S. *Cosmical magnetic phenomena.* Nature. **124**: July, 19. 1929.

8. DAVIS, W. M. *A migrating anticline in Fiji.* Amer. Journ. Sci. (5) **14**: 333–351. 1927.

9. DE SITTER, W. *On the rotation of the earth and astronomical time.* Nature. **121**: 99–106. 1928.

10. FISHER, O. *Physics of the earth's crust.* (London, 1881).
11. GUTENBERG, B., and RICHTER, C. F. *On supposed discontinuities in the mantle of the earth.* Bull. Seis. Soc. Am. **21**: 216–222. 1931.
12. HOBBS, W. H. *The unstable middle section of the island arcs.* In Gedenkboek van Dr. R. D. M. VERBECK: Geol. Mijnbouwk. Genootschap, v. Nederland en Kolonien, Geol. Ser. **8**: 219. 1925.
13. HOBBS, W. H. *Stress conditions within the lithosphere as revealed by earthquakes.* Bull. Geol. Soc. Am. **41**: 739–746. 1930.
14. HOLMES, A. *Radioactivity and the earth's thermal history. Pt. II. Radioactivity and the Earth as a cooling body.* Geol. Mag. (6), **2**: 102. 1915.
15. HOLMES, A. *Radioactivity and the earth's thermal history. Pt. IV. A criticism of parts I, II, and III.* Ibid. **62**: 504. 1925.
16. HOLMES, A. *Radioactivity and earth movements.* Trans. Geol. Soc. Glasgow. **18**: 559–606. 1928–29.
17. HOLMES, A. *Problems of the earth's crust.* Geog. Journ. **78**: pp. 445 and 541. 1931.
18. HOLMES, A. *Radioactivity and geological time.* In, *The age of the earth.* Bull. Natl. Research Council No. **80**: Washington 124–459. 1931.
19. HOLMES, A. *The origin of igneous rocks.* Geol. Mag. **69**: 543. 1932.
20. JEFFREYS, H. *The cooling of the earth.* Nature. **115**: 876. 1925.
21. JEFFREYS, H. *On Prof. Joly's "Theory of earth history."* Phil. Mag. (7) **1**: 923. 1926.
22. JEFFREYS, H. *Prof. Joly and the earth's thermal history.* Phil. Mag. (7) **5**: 208. 1928.
23. JEFFREYS, H. *The Earth.* (2nd ed.) Cambridge, 1929.
24. JEFFREYS, H. *Problems of the earth's crust.* Geog. Journ. **78**: 451. 1931.
25. JEFFREYS, H. *On the variation of melting point within the earth.* Monthly notices: Roy. Ast. Soc. Geophys. Suppl., **3**: 6–9. 1932.
26. JOLY, J. *The surface history of the earth.* Oxford, 1925. (2nd ed.) 1930.
27. JOLY, J. *Dr. Jeffreys and the earth's thermal history.* Phil. Mag. (7) **4**: 338. 1927.
28. LAWSON, R. W. *Radioactivity and the heat of the earth.* Nature. **119**: 277. 1927. and **119**: 703. 1927.
29. MCCARTHY, G. *Radioactivity and the floor of the oceans.* Geol. Mag. **63**: 301. 1926.
30. MEINESZ, F. A. VENING. *Gravity anomalies in the East Indian Archipelago.* Geog. Journ. **77**: 323–337, *Pl.* 1931.
31. POOLE, J. H. J. *The thermal instability of the earth's crust.* Sci. Proc. Roy. Dublin Soc. **19**: (N. S.) 385–408. 1930.
32. STRUTT, R. J. *On the distribution of radium in the earth's crust and on the earth's internal heat.* Proc. Roy. Soc. A. **77**: 472. 1906, and **78**: 150. 1906.
33. TAMMANN, G. *Aggregatzustände,* Leipzig. 97–98. 1923.
34. TOKUDA, S. *On the echelon structure of the Japanese Archipelago.* Trans. Jap. Journ. Geol. and Geog. **5**: 41. 1926.
35. TURNER, H. H. *Thirty-fifth Report of Committee on Seismological Investigations.* Report Brit. Assoc., 1930, Bristol, 1931, 245–247 and map.
36. WILLIS, BAILEY. *Radioactivity and theorizing.* Am. Jour. Sci. (5) **23**: 193–226. 1932.

Part V

REGIONAL GEOLOGY

Editor's Comments on Papers 20, 21, and 22

20 COLLET
The Alps and Wegener's Theory

21 NOBLE
Excerpt from *The San Andreas Rift and Some Other Active Faults in the Desert Region of Southeastern California*

22 BERNAUER
Iceland and the Question of Continental Drift

It has often been said that the traditional methods of field geology were incapable of contributing significantly to solving the problem of continental drift and that they clouded the issue by swamping the theorist with vast amounts of contradictory and inconclusive data. Standing in stark opposition to this pessimistic view, however, are the papers reproduced here, which are representative of numerous regional studies that paved the way for acceptance of continental drift theory by providing clear-cut evidence for the type and amount of movement of rock masses required by continental drift.

Among the most influential (and difficult) of the relevant areas studied were the Alps, where European geologists (beginning with Hans Conrad Escher von der Linth and his son Arnold) discovered and named the nappe structures that are so typical of parts of that range (see Pavoni, 1967), and developed the idea that an enormous thrust of Africa against the European continent was the immediate cause of the formation of the Alps. Among the leaders in this exciting revolution of tectonic ideas were many of the great European geologists of the late nineteenth and early twentieth centuries, including Eduard Suess (1875), Marcel Bertrand (1884), Hans Schardt (1893), Maurice Lugeon (1901–1902), Albert Heim (1908), Emile Argand (1924), Rudolf Staub (1924), and Leopold Kober (1923). Certainly not the least of this eminent company was Leon Collet whose short summary of Alpine structure (reprinted here as Paper 20) was presented to the Royal Geographical Society on 14 December 1925. As the eminent English geologist E. B. Bailey said after Collet's presentation:

> if we take the current Swiss view of the structure of the Alps, and the similar view, worked out by that giant of the north, Tornebohm, for the Scandinavian chain, and again the similar view attained by Scottish geologists in their Highlands, we find that some sort of continental drift becomes a matter not of interpretation but of observation (p. 309).

Another type of horizontal movement of rocks is represented by that archetype of strike-slip faults, the San Andreas fault of California. Efforts to understand the nature of the San Andreas began in 1893 when Andrew Lawson identified the structural significance of the fault and (after the 1906 San Francisco earthquake) reported that as much as 5 meters of strike-slip motion took place along portions of the fault during that earthquake. Lawson indicated that the fault could be traced for at least 1000 kilometers. In Paper 21, L. F. Noble suggested that as much as 38 kilometers of strike-slip motion may have taken place along the fault since the Eocene. In 1933, H. F. Reid suggested that continuous right-lateral slip might be occurring along the fault, and in 1948, C. A. Whitten reported that triangulation surveys showed that continued right-lateral slip of 5 centimeters per year has been taking place along the fault since 1882.

In the very next year, R. E. Wallace (1949) concluded that a total of 120 kilometers of strike-slip motion may have occurred in the last 30 million years and in 1953, M. L. Hill and T. W. Dibblee, reported their finding that 280 kilometers of movement has occurred since the early Miocene. Finally, the culmination of all this work came in 1965 when J. T. Wilson developed the concept of transform faults and identified the San Andreas as a dextral, ridge-ridge transform fault connecting the East Pacific Rise in the Gulf of California with the Juan de Fuca Ridge.

Still a third type of horizontal motion (rifting or spreading) is discussed in Paper 22 by F. Bernauer. Bernauer studied the 100-kilometer-wide belt of young volcanoes and gaping parallel fractures that transects Iceland and concluded that they represent "a rifting apart by magmatic undercurrents" and that Europe and North America may well have rifted apart in the same manner. His conclusions thus grew out of and supported the Icelandic work of his predecessors N. Nielsen (1929, 1933) and W. Iwan (1938), and they support F. B. Taylor's (1910) concept of the Mid-Atlantic Ridge as the place where the opposing Atlantic continents were once joined and subsequently separated. In this regard Bernauer's thinking can be seen as reminiscent of Arthur Holmes' (1928) convection-current theory of sea-floor thinning and as anticipating the later theories of B. Heezen (1959) and H. H. Hess (1962).

REFERENCES

Argand, E., 1924, La tectonique de l'Asie, *C. R. XIII Internat. Geol. Cong.* **1:**171-372.

Bertrand, M., 1884, Rapports de structure des Alpes de Glarus et du bassin houiller du Nord, *Soc. Geol. France Bull.,* 3rd ser., **12:**318-330.

Heezen, B. C., 1959, Paleomagnetism, Continental Displacements, and the Origin of Submarine Topography, in M. Sears, ed., *International Oceanographic Congress Preprints,* American Association for the Advancement of Science, Washington, D. C., pp. 26-28.

Heim, A., 1908, *Der bau der Schweizeralpen,* Zuercher & Furrer Zurich, 26p.

Hess, H. H., 1962, History of Ocean Basins, in A. E. J. Engel et al., eds., *Petrological Studies—A Volume in Honor of A. F. Buddington,* Geological Society of America, New York pp. 599-620.

Hill, M. L., and T. W. Dibblee, Jr., 1953, San Andreas, Garlock and Big Pine Faults, California, *Geol. Soc. Amer. Bull.* **64:**443-458.

Holmes, A., 1928, Radioactivity and Earth Movements, *Geol. Soc. Glasgow Trans.* **18:**559-606.

Iwan, W., 1938, Die Bardartalverwerfung in Nordisland, *Gesellschaft fur Erdkunde zu Berlin Zeitschraft* **9-10:**350-362.

Kober, L., 1923, *Bau und Entstehung der Alpen,* Borntraeger, Berlin, 283p.

Lawson, A. C., 1893, The Post-Pleistocene Diastrophism of the Coast of Southern California, *Calif. Univ. Dept. Geol. Sci. Bull.* **1**(4):115-160.

Lawson, A. C. (Chairman) et al., 1908, The California Earthquake of April 18, 1906, *Carnegie Inst. Washington Pub. 87* **1:**1-451.

Lugeon, M., 1901-1902, Les grandes nappes de recouvrement des Alpes du Chablais et de la Suisse, *Soc. Geol. France Bull.,* 4th ser., **1:**2-723.

Nielsen, N., 1929, Islands tektonik og Wegener-Theorien, *Skand. Naturforskermode,* **18:**4.

Nielsen, N., 1933, Contributions to the Physiography of Iceland, with particular Reference to the Highlands West of Vatnajokull, D. Kgl., *Danske Vidensk. Selsk. Skrifter, Naturvidensk. Og. Mathem.,* Afd., 9 Raekke, IV, 5, pp. 183-286.

Pavoni, N., 1967, Tectonophysics, in *The Development of Geodesy and Geophysics in Switzerland,* J. C. Thams, ed., Berichthaus, Zurich, pp. 42-67.

Reid, H. F., 1933, The Mechanics of Earthquakes, in *Physics of the Earth,* v. 6, Seismology, National Research Council, pp. 87-103.

Schardt, H., 1893, Sur l'origine des Alpes du Chablais et du Stockhorn, en Savoie et en Suisse, *Acad. Sci. C. R.* **117:**707-709.

Staub, R., 1924, Der Bau der Alpen, *Beitr. Geol. Karte Schweiz,* Neue Folge, Lief. 52, 272p.

Suess, E., 1875, *Die Entstehung der Alpen,* W. Braumuller, Vienna, 168p.

Taylor, F. B., 1910, Bearing of the Tertiary Mountain Belt on the Origin of the Earth's Plan, *Geol. Soc. America Bull.* **21:**179-226.

Wallace, R. E., 1949, Structure of a Portion of the San Andreas Rift in Southern California, *Geol. Soc. America Bull.* **60:**781-806.

Whitten, C. A., 1948, Horizontal Earth Movement, Vicinity of San Francisco, California, *Am. Geophys. Union Trans.* **29:**318-323.

Wilson, J. T., 1965, A New Class of Faults and Their Bearing on Continental Drift, *Nature* **207:**343-347.

20

Reprinted from *Geog. Jour.* **67**:301-312 (1926)

THE ALPS AND WEGENER'S THEORY

Prof. Leon W. Collet, Hon. Corr. Member R.G.S., Dean of the Faculty of Science and Professor of Geology in the University of Geneva.

Read at the Afternoon Meeting of the Society, 14 *December* 1925. *Folding Plate following p.* 304.

FROM north to south the principal subdivisions of the Western Alps are as follows :

(1) *The Pre-Alps*, a very distinctive zone stretching from the Lake of Thun to Lake Geneva and the River Arve.

(2) *The High Calcareous Alps*,* which with their greater height and accompanying glaciers are readily distinguished from (1). South of them rise the following crystalline Hercynian massifs (3) and (5).

* The principal peaks which belong to the High Calcareous Alps are, from south-west to north-east, the Range of the Fis, the Mont Buet, the Peak of Tenneverdze, the Mont Ruan, the Tour Sallière and the Dents du Midi, the Dents de Morcles, the Diablerets, the Wildhorn, the Wildstrubel, Balmhorn, Doldenhorn, Blumlisalp, Gspaltenhorn, Jungfrau, Monch, Eiger, Wetterhorn, Titlis, Urirotstock, Pilatus, Tödi, Glärnisch, Churfirsten, Säntis.

From a geological point of view, the High Calcareous Alps belong to the Helvetian Folds. These represent the sedimentary cover of the Foreland which has been folded owing to the travelling northwards of the pennine recumbent folds (see Figs. 11 and 13).

(3) *Aiguilles Rouges of Chamonix and Gastern Massif.*
(4) *The sedimentary Zone of Chamonix* and *The Zone of the Paragneiss of the Lötschental.*
(5) *Mont Blanc* and the *Massif of the Aar;* and south of these
(6) *The Pennine Alps* (crystalline), with across the watershed
(7) *The Zone of Canavese* or *The Zone of Roots;* and finally
(8) *The Calcareous Alps of the South* or *Dinarides.*

Let us look now at the evidence which has led to the conclusion that the Alps are fashioned out of a great pile of recumbent folds or nappes.

Long before Heim, Escher, during the years 1840–1850, conceived the idea of overthrusting. It was not, however, until 1870 that the famous section of the Glarus Double Fold was published with Heim as its warmest adherent. The subject is too familiar to detain us. All that need be said is that in 1883 Marcel Bertrand, following upon his researches in Provence and his visit to the North-West Highlands, substituted a single for a double fold. Suess in 1892 supported Bertrand's interpretation, and in 1903 Heim at last followed suit.

While discussion still continued regarding the double or single folding of Glarus, Schardt studied the Pre-Alps. During the five years 1890–1894 he realized that in this extensive district Trias is continually to be found overlying Tertiary sediments; and guided by this relationship, and by a just appreciation of facies considerations, he announced that the Pre-Alps are constituted of a succession of far-travelled thrust-masses of southern origin. Lugeon during 1894–1895 combated this idea, until, won over in 1896 by the accumulated force of his own observations, he began to demonstrate the fundamental *rôle* of nappes not only in every region of the Alps, but also in the Carpathians, the Apennines, and Sicily. And now to Argand,* a pupil of Lugeon, we owe a synthesis of the formation of the Western Alps, based upon a knowledge of the subject such as belongs only to genius.

Beginning in 1905 Argand, as a result of his researches in Piedmont, was led to study the Pennine Alps, where he determined the existence of six gneissic recumbent folds or nappes. The investigation was long and difficult, but it threw light on the mode of formation of the whole chain. We shall make use of Argand's sections and explanations:

Argand started from the notion of the geosyncline, so splendidly developed by Haug and destined to remain for all time one of the basal conceptions of tectonics. His equipment included a very detailed stratigraphical knowledge, and armed with this he has succeeded in straightening out the recumbent folds, and in thus reconstituting the Alpine region at various stages of its development when the general geosynclinal depression was subdivided by geanticlinal ridges. In the diagram (Fig. 1) showing an Alpine chain in embryonic stage, we may note:

* E. Argand, "Sur l'arc des Alpes Occidentales," *Eclogæ Geologicæ Helvetiæ*, 14, 1, Juillet 1916.

1. The Foreland, with granitic batholiths of Upper Palæozoic (Hercynian) Age.

1′. A swelling at the inner margin of the Foreland determined by the push of the nappes. This swelling we may name the Helvetian Geanticline.

2. The Epicontinental Sea.

3. The Foredeep, or Geosyncline of the Valais.

4. The Frontal Cordillera or Geanticline of Briançon.

5. The Intermontane Depression or Geosyncline of Piedmont.

6. The Second Cordillera or Geanticline of Dolin.

g′ shows where Hercynian granites involved in recumbent folds are in part transformed by movement and recrystallization into ortho-gneiss.

b shows the place of origin of certain grits and conglomerates found in the Carboniferous, Permian, Lower Trias, Liasic and Tertiary deposits of the Alps.

b′ shows another similar position and illustrates the intrusion phenomena of the Greenstones, guided by the dislocations upon which the Geanticline of Dolin advanced.

v′ shows how the basic magma of the Greenstones escaped at places into the Piedmont Geosyncline, where it consolidated in large measure as pillow lavas.

The diagram (Fig. 1) shows each geanticline of the embryonic Alps as a recumbent fold. Behind the last geanticline should be found what Suess calls the Hinterland of the chain. The *Foreland and Hinterland constitute the boundaries of the great Alpine Geosyncline diversified by its included geanticlines. Together they recall the two jaws of a vice.* Termier, director of "the Service geologique de France," has shown how the approach of the two jaws has led to the compression of the geosyncline, and thus to the development of the Alpine chain. All this will be realized more clearly if we examine Argand's representations of the evolution of mountain systems.

At the close of the *Middle Carboniferous* the beginnings of the two principal nappes show themselves as simple Hercynian anticlines in the great geosyncline (Fig. 2).

In the *Middle Trias* these two embryonic nappes locally rise out of the waters, but the Piedmont depression, 5, continues tolerably deep, as shown by the character of its sediments (Fig. 3).

In the *Lias* (Fig. 4) a strong horizontal push is felt. The embryos develop, while the depressions deepen. This is the period of accumulation of the well-known Breche du Telegraphe (see *b*, Fig. 1).

In the *Middle Jurassic* the Briançon embryo 4 emerges (Fig. 5).

In the *Upper Jurassic* a lessening of horizontal pressure leads to a general slight submergence (Fig. 6).

In the *Middle* and *Upper Cretaceous* the nappes of the Southern Alps, including the upper nappe of the Pre-Alps, start on their journey (their Hercynian cores are marked 8 in Fig. 7, their Mesozoic envelopes 8′).

In the *Middle Nummulitic* and *Lower Oligocene* a vigorous push makes itself felt (Fig. 8).

In the *Middle Oligocene* an organic paroxysm supervenes in which five principal phases may be distinguished :

St. Bernard Phase (Fig. 9).—The Briançon Cordillera makes its main advance, developing into the complex nappe of the Great St. Bernard, 4. The sediments of the Valais Foredeep are driven forward to yield the nappes of Simplon and Ticino, 3, and also the lower nappes of the Pre-Alps, 3′. The Foredeep itself is correspondingly displaced and begins to receive the Mollassic sedimentation characteristic of the Swiss Plateau, 3‴. The Dolin Cordillera passes into the Dent Blanche Nappe, 6. The upper nappes of the Pre-Alps, 8″, begin to separate from their roots and to be carried forward on the backs of the Pennine Nappes, an inclusive term embracing the nappes of Great St. Bernard and Dent Blanche.

Dent Blanche Phase (Fig. 10).—It is now the turn of the Dent Blanche Nappe, 6. The upper nappes of the Pre-Alps, 8″, are concomitantly discharged in front of the Pennine assemblage on the top of the lower Pre-Alpine nappes, 3′; from these latter a portion is detached and driven forward to constitute 3″, the external zone, as it is called, of this wonderful group of mountains.

Monte Rosa Phase (Fig. 11).—Born of the Piedmont Geosyncline, the Nappe of Monte Rosa, 5, develops strongly. Concurrently there is considerable back-folding in the Nappe of Great St. Bernard. The Simplon and Ticino nappes, 3, attain their maximum, while beneath them the Helvetian Nappes, 5′, come into being.

Phase of Adriatic Subsidence.—The root region assumes a more or less vertical posture (Fig. 12).

Phase Insubrienne (Fig. 13).—A fan arrangement of the roots is next developed. *The Alps and Dinarides thus face in opposite directions*. The Morcles Nappe and similar parautochthonous folds (8‴ above 1′, Fig. 13) originate beneath the Helvetian Folds (5′, Figs. 11, 12).

CONCLUSION.—*The southern jaw of our vice* (*Indo-Africa of Argand*) *has approached the northern jaw* (*Europe*) *overriding it, and has led to the contraction of the great intervening geosyncline along with the two geanticlines which modified its initial simplicity. These two geanticlines have developed into the two dominant nappes of the Alps, namely, the Nappes of Great St. Bernard and Dent Blanche.* To begin with, these two nappes, both of them Pennine Nappes according to current classification, existed alone. As time went on they found themselves moving in a throng. The Great St. Bernard Nappe set in motion the Simplon assemblage, while the Dent Blanche Nappe operated the Monte Rosa group.

THE FORMATION OF THE WESTERN ALPS:

Showing the principal stages of development.

From the paper by Professor Emile Argand, "Sur l'arc des Alpes Occidentales," Eclogæ Geologicæ Helvetiæ, xiv. 1, July 1916.

13. *"Phase insubrienne."*
12. *Phase of Adriatic subsidence.*
11. *Monte Rosa Phase.*
10. *Dent Blanche Phase.*
9. *St. Bernard Phase.*
8. *Middle Nummulitic and Lower Oligocene.*
7. *Middle and Upper Cretaceous.*
6. *Upper Jurassic.*
5. *Middle Jurassic.*
4. *Lias.*
3. *Middle Trias.*
2. *Middle Carboniferous.*
1. *Embryonic state of a chain of Alpine type.*

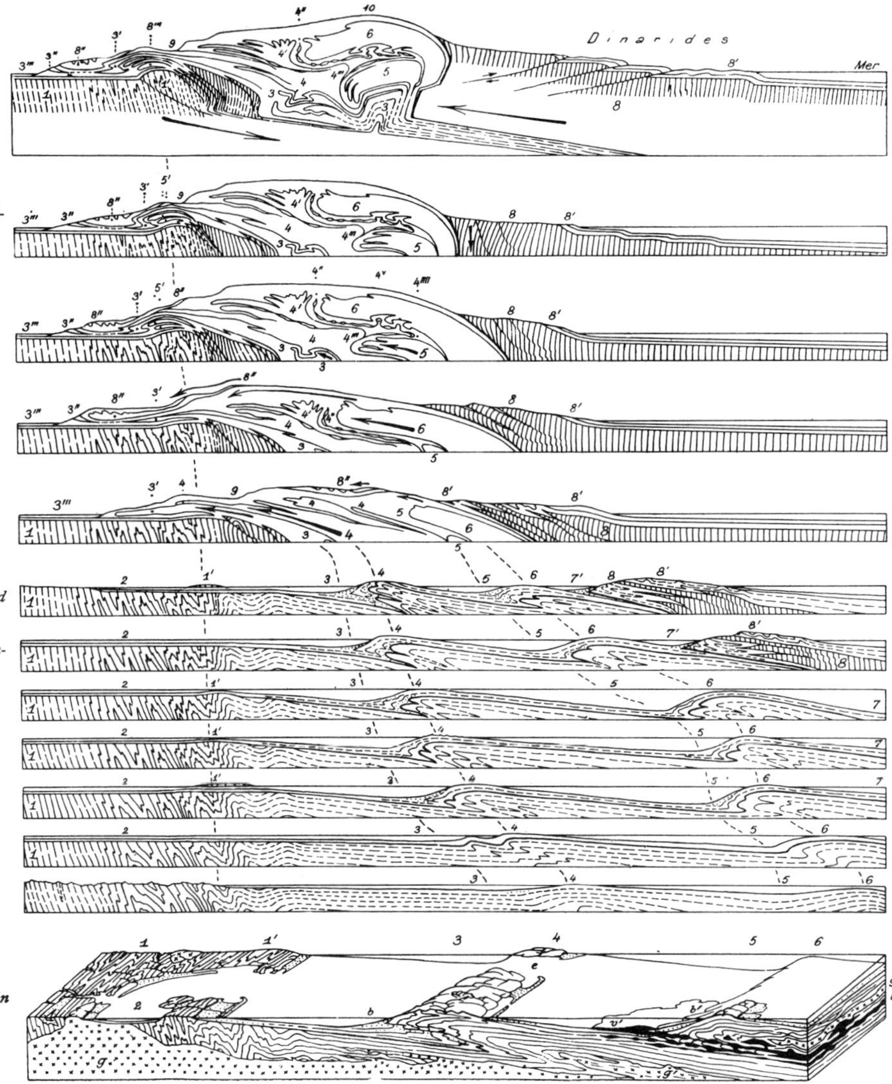

COLLET: ALPS

The *nappes of the High Calcareous Alps* (based upon 1′, Fig. 13) *owe their existence to the forward drive of the Pennine Alps.*

Suess had already shown that the "Alpides" were due to the approach of Hinterland and Foreland. But in his idea there were two directions of thrusts; one towards the north and the other towards the south. Argand shows, in 1912, that we have to deal with only one great push towards the north, with an underthrusting at the end of it. We find just the same, on a much larger scale, in the Himalayans, where the northern part is an overthrust northwards and the southern part an overthrust southwards. They are nevertheless parts of the great northwards movement. (E. Argand, "Le rythme du proplissement pennique et le retour cyclique des encapuchonnements," *Procès Verbaux Bulletin Société Vaudoise des Sciences Naturelles*, t. 48, 20 Mars 1912; and E. Argand, "La Tectonique de l'Asie," *C.R. Congrès Géologique International Belgique*, 1922.)

If we arrange now the different parts of the Alps from a tectonical point of view, we arrive at the following groups:

1. The Massifs of the Aiguilles Rouges of Chamonix and of Gastern. 2. The sedimentary Zone of Chamonix and the Zone of the Paragneiss of the Lotschental. 3. The Massifs of the Aar and of Mont Blanc. 4. The High Calcareous Alps (Helvetian Folds).	The Foreland.
The Pennine Alps.	The Geosyncline.
The Austrides of Dinarides.	The Hinterland.

The Massifs of the Mont Blanc, Aar, Aiguilles Rouges of Chamonix, Gastern are of Hercynian age. Farther to the south-west the Foreland is represented by the Massifs of the Pelvoux and Belledonne, then by the Massif of the Mercantour (Argentera of the Italians), all of Hercynian age, with a folded sedimentary cover.

Prof. Argand showed that the austro-alpine nappes belong to the Hinterland: that is, to Africa or Gondwanaland. Therefore the higher Préalpes, that can be seen from Geneva, Lausanne, and Berne, represent a small part of Africa resting on Europe or Eurasia. These important views sketched by Argand in his famous paper "Sur l'Arc des Alpes Occidentales," stated precisely at the Brussels Congress, have been accepted by Staub in his valuable memoir 'Der Bau der Alpen.'

On the base of Argand's results Dr. R. Staub found in the north-eastern part of the Swiss Alps the pennine elements covered by six higher nappes belonging more to the type of the "thrust-masses" of the north-western Highlands of Scotland than to the type of the recumbent folds of the Pennine Alps. This series of nappes has been named by Staub the Austrides, for they form the main part of the Austrian Alps.

Staub's "Austrides" represent what Termier and Argand had called, in a much more general way, the austro-alpine nappes.

For the Swiss geologists the forward drive of the Hinterland towards the Foreland cannot be denied; it is almost a matter of fact, that is splendidly summarized in Staub's geological map of the Alps.

The intrusion of basic magma in the dislocations upon which the Dent Blanche Nappe advanced (see Fig. 1) is also a matter of fact. I have seen myself the evidences on the field, guided by my friend Argand, and I have shown them to many English geologists during my summer course in the Alps in 1924.

I must state explicitly that all these results have been obtained independently of Wegener's hypothesis. That is why I think that they are a great support to Wegener's theory.

CONCLUSION.—A southern continent (Africa or Gondwanaland), made of Sial, is separated from a northern continent (Europe or Eurasia), also made of Sial, by a geosyncline or the sea of Tethys of Suess. The southern continent being drifted northwards, the floor of Tethys thinly coated with Sial has been folded (geanticlines) northwards, thus forming a mountain chain (the Alps). Sima has been injected into the laminated reversed limb of the geanticlines. The southern continent not only encountered the obstacles formed by the northern continent, but its frontal part has been thrust over it.

By the employment of Wegener's ideas we could go a good deal further and—as shown by Argand and Staub—arrive at accepting a northward drive of Europe, producing a distension to which the Mediterranean is due. But my intention this afternoon is to insist on the travel of the Hinterland towards the Foreland of the Alps, and to point out the importance of the gabbros of the reversed limb of the Dent Blanche Nappe.

As shown by Lord Rayleigh, many facts make it necessary to abandon Kelvin's classical theory of the cooling of the Earth. The work done by Prof. Joly in this direction is of great value, and it will be soon impossible to resist Wegener's attractive ideas.

Before the paper the CHAIRMAN (Mr. DOUGLAS W. FRESHFIELD) said: A week ago I had to protest to our evening meeting that I am no geologist. I do not even pretend to be able to understand many of the esoteric terms of the modern Continental school of geologists. It is therefore plain that in the position in which I find myself to-day I am one of what Mr. Thomas Hardy might call "Life's little Ironies." Still I am glad to be here, for although I cannot say much on the theories we are to discuss I can promise to introduce to you a lecturer who is a master of his subject. Professor Collet is Dean of the Faculty of Sciences in the University of Geneva; he has made a long study of the Alps and knows all the latest theories, which probably will not be the last, about their formation. I am glad also to be able to record a debt I owe to Professor Collet. Nearly forty years ago I observed that the streams that came out of the glaciers in winter were not the result of glacier-melting, but came from mountain springs whose sources were under the ice. Nobody took any notice; at any rate, the text-books went on writing about the glaciers melting in winter, until Professor Collet analyzed the water of the Grindelwald Glaciers and found that it was pure fountain-water from mountain springs rising under the bed of the ice. I will not detain you any longer, but ask Professor Collet to give his lecture.

Professor Collet then read the paper printed above, and a discussion followed.

Mr. E. B. BAILEY (Geological Survey, Edinburgh): It is a very great privilege indeed for British scientists to receive at first hand from Professor Collet an account of the wonderful renaissance of Alpine tectonics—a renaissance which dates back to 1893, when, as we have heard to-night, Professor Schardt produced his interpretation of the pre-Alps. I may say that for the next fifteen years—and this it is very important for us to realize, because it enables us to understand how thoroughly the new interpretation has been tested—practically every Alpine geologist who was worthy of the name stood up against Schardt, and produced this criticism and that criticism. Our lecturer was, fortunately for us, just old enough to take part in the initial criticism of Schardt's work. I think it is seventeen or eighteen years since I first heard Professor Collet lecturing in Edinburgh on the subject—at that time an infidel, wavering, but still one who had not been converted to the interpretations of the Alps to which we have listened to-night. Of course that is all to our advantage, because we have heard from one who was brought up in unbelief; and, as I say, all other Swiss geologists that we know, men like Heim and Lugeon, were about that time attacking Schardt. Since then every one has been convinced, and has pointed out where the error of his own criticism has lain, and what has brought him to realize the absolute truth of Schardt's interpretation of the pre-Alps. I dwell on the pre-Alps, because they still remain the most wonderful feature in the interpretation of the Alps and at the same time, in a way, the most comprehensible.

When next I heard Professor Collet on the subject of the pre-Alps, we were very much better situated, because we were actually among the pre-Alps; for in 1909 Professor Collet very kindly led a party of three Scottish geologists, of which I was privileged to be one, through the pre-Alps and the surrounding country. The pre-Alps stand up beyond Geneva, and it was explained to us, and we could follow absolutely on the ground, why they are interpreted as a travelled mass, entirely foreign to the district in which they stand. We could examine their marginal relations—could see that they were lying upon the top of the local Alps and stretching out beyond the rest upon the rocks of the great Swiss plain. Not only that, but the pre-Alps are built up of the same geological formations as they overlie. In them there are fossils easily determinable, and these fossils tell us that the pre-Alps are made up of Palæozoic, Mesozoic, and Tertiary formations, just as is the foundation upon which they lie. Not only this, but there is the extraordinary fact that we have had emphasized to-night: that although there are Mesozoic rocks in the pre-Alps and in the foundation, these Mesozoic rocks would never have been compared, and it would never have been realized that they were rocks of the same age, if it had not been for the fossils. The rocks themselves are absolutely different in character. The only possible explanation is that these overlying masses have been developed through sedimentation in another country, and have travelled, have invaded, as it were, and are now resting, a great foreign mass in the heart of Switzerland. They are just as obviously erratic as, say, a boulder of Shap granite found upon the coast of Yorkshire ; but the scale is rather different. The pre-Alps are very much more extensive than the Lake District of England, and, as you have seen in the first photograph, in altitude they are Alps—something much bigger than we have in this country. So you can imagine the bulk of this erratic mass, and that is the very essence of the story of Alpine interpretation.

Behind the pre-Alps, Professor Collet took us to see the great recumbent folds of the Rhône valley that have been interpreted, after or during his conversion, by Professor Lugeon. These folds are in a sense so very simple, that is after they have been worked out by a man of genius, that one merely has to go and see them. They are there on the hill-sides. You take the map and go over them and see that they are there.

Behind the Rhône valley, after a long gap of years, two years ago, I think it was, I was one of a party, and there are others here, who were led by Professor Collet through the much more difficult region of the Pennine range, which had been interpreted by Lugeon and Argand. And here we were on very different ground indeed. It is not, as a matter of fact, difficult to see the great folds. One sees enormous recumbent folds if one walks about, and one can follow them out with the map ; but the difficulty in this case is that the rocks have been so tremendously altered by movement and heat that they are, for the most part, in the condition of crystalline schists. There may be some British geologists who would be shocked at the very idea of calling rocks Mesozoic that have garnets in them—that have lost their fossils and have become crystalline schists —but I may say that all Swiss geologists are satisfied that the *schistes lustrés*, even where they have become thoroughly metamorphic rocks, are of Mesozoic age. And when one looks at those great folds in the Pennine Alps with their Palæozoic cores and their Mesozoic envelope—one realizes that here are folds made in the Tertiary movement that has developed the Alps.

Well, then, I must admit that I for one feel absolutely convinced as to the general truth of the Alpine interpretation as it stands to-day. Our Chairman has said that there will be other interpretations. Well, that is true. There will always be improvements, always modifications, and sometimes, possibly, periods of retrogression in the interpretation of the Alps. But the main features that have been worked out by the present generation from Schardt onwards, including many other great workers—and Professor Collet has said nothing to-night of what he himself has done on Mount Buet, and on the Jungfrau, where one has to be not only a geologist but an alpinist of the highest order—all this work has come to stay.

Then of course we have just heard from Professor Collet of Professor Argand's fascinating interpretation of the history of the Alps, starting from the first days long before the great period of Tertiary movement. Of course there is a good deal in this interpretation that is much more speculative than the mere tracing of the structure of the country ; but a lot that Argand has told us about the development of the Alps will last for ever. This attempt to build up the history of the growth of the mountain chain is not new ; it goes back for about one hundred years. We can find it in the early writings of Studer on the Alps ; it is a delightful idea, that of the mountain chain travelling forward, and the pebbles going forward from the chain as it grows, and then the chain riding over its own débris. One finds that in the early writings of Studer, and also in the story told by Argand ; and already, difficult as the subject is, I feel convinced that there is a very great measure of truth in what Argand has given us. Professor Argand is, perhaps, at the moment the most brilliant of all the brilliant school of Alpine tectonists.

Finally, if we take the current Swiss view of the structure of the Alps, and the similar view, worked out by that giant of the north, Törnebohm, for the Scandinavian chain, and again the similar view attained by Scottish geologists in their Highlands, we find that some sort of continental drift becomes a matter not of interpretation but of observation.

Mr. PHILIP LAKE (Cambridge): Like others, I have listened with the greatest interest to Professor Collet's account of the remarkable structure of the Alps as developed by the Swiss geologists. I remember well the appearance of Schardt's first paper and the discussion to which it gave rise, and have since watched the gradual conversion of every Swiss geologist, one after another, to Schardt's views. It has always seemed to me to show great open-mindedness on the part of men like Heim and Termier, that they were able to give up their own views and accept the later interpretations. No geologist can now doubt that the Alps have been pushed forward over the foreland and that the hinterland has moved towards the foreland. We have the same thing in other places. The Himalayas were probably formed by Central Asia being pushed over India. But these movements are very different from those imagined by Wegener. The great difficulty in Wegener's theory is to explain lateral movements at all. No force of any magnitude has been suggested to produce such movements. The suggestions which have been made are of very minute forces, which possibly might push the blocks of Sial through the Sima but could hardly produce folding such as we see—folding against the force of gravity. A small force may slowly overcome a viscous resistance, but it cannot overcome a force greater than itself however long it may be applied.

According to Wegener, the movement in the Alpine area should be from north to south; according to Professor Collet it was from south to north. I do not think that matters much, for it is a question of relative movement; but it is a difference. Further, the account which Professor Collet has given us hardly agrees with Wegener's ideas or with his evidence on other points. Wegener fits the two sides of the Atlantic together and from the coincidences which he observes he concludes that the probability is a million to one that his theory is correct. Now, in fitting the two together he supposes that in Carboniferous times Europe and Africa had the same relative positions that they have now. Professor Collet contends that they were farther apart. If they fit with Europe and Africa as they are now, they cannot have fitted when Europe and Africa were farther apart.

But the point which appeals to me most is that the masses of Sial are supposed to be about 100 kilometres in thickness, extending deep down into the Sima, with 4·8 kilometres rising above the latter. The whole thickness of Africa was not pushed over Europe, but only the uppermost layer. I cannot see how that could result from masses 100 kilometres thick pushing against one another. It is a difficulty on any hypothesis, even on the old hypothesis of a contracting interior. Mountain ranges are, apparently, very superficial phenomena, and it is only the outermost layer which is involved in the folds. Wegener's theory certainly offers no better explanation than the old contraction theory.

Mr. T. C. NICHOLAS (Cambridge): It is a great pleasure to me to be present to-night in order to express, even if in an inadequate manner, the great debt which I owe to Professor Collet for the memorable experience of being led by him through a part of the Alps during the summer before last. In the course of that excursion, the pleasure of which was shared with me by Mr. Bailey and other British geologists, we saw evidence for a good deal of what we have heard from the lecturer this evening, particularly with regard to the structural phenomena of the Pennine Alps. And for my own part I have no hesitation in accepting the general truth of the views now current among Alpine geologists upon the formation of the pre-Alps, High Calcareous Alps, and Pennine Alps by recumbent folding. I think it is quite incontestable that the hinterland has moved for great distances towards the foreland, and that in doing so its long-continued pressure upon the intervening geosyncline has given rise to the immense recumbent folds which Professor Collet has demonstrated to us and of which he has shown us photographs.

But though in my opinion the evidence for the movement of the hinterland towards the foreland is complete and overwhelming, I am not so sure that this evidence can be said to confirm the theory of Wegener, and I should prefer for the present to keep facts and theory separate.

Mr. J. M. WORDIE : As the hour is getting late I wish to make only one remark, and that is to ask Professor Collet if it is absolutely necessary, in the formation of the original geosyncline, to have the foreland and the hinterland pulled apart ? Could it not be imagined as formed in some other way ? When he drew the diagram he said that according to Wegener's supposition the foreland and the hinterland were pulled apart and the geosyncline formed between them. It seemed to me that there may be other possible explanations.

Dr. J. W. EVANS : I have not very much to add to the discussion, especially after the eloquent speech which we have heard from Mr. Bailey. I was very interested in Mr. Lake's criticism of the references to Wegener. Personally, I do not think that anybody except Wegener himself is prepared to accept everything that he contended for. Wegener's book is not written by a geologist ; it is written by a geodesist ; * but he dealt with some important questions from an entirely new point of view. He can be criticized in detail in many particulars, but we have only to read of the criticisms and discussions to which Wegener's hypothesis has given rise to realize the debt we owe him.

There have always been some people who have believed in great translations of continental masses. A namesake of mine, Sir John Evans, about sixty years ago suggested that the whole crust of the world might move so as to bring some countries under the North Pole and others under the Tropics ; that then a further shift might occur and these positions reversed. Only a couple of years after his first paper, published, I believe, in the *Philosophical Transactions*, there was a small book by a cleric named John Kirk, which was certainly rather heterodoxic in many respects from the geological standpoint, but in it he suggested that the tropical climate of this country in the Eocene period was due to the fact that it then lay under the Tropics and that it had been shifted in the interval from the Tropics to the temperate region. A reviewer of this book, writing in the *Geological Magazine* in 1869, thought that he only had to state this proposition for it to be rejected with ridicule. But that is the very supposition which was brought forward by Dr. Staub in that wonderful book of his on the structure of the Alps. In that book you will find substantially what we have heard to-night ; Dr. Staub goes very far in his ideas of the drift of continents. He says that Africa has been moved, roughly speaking, over 3000 miles to the north. As it moved it pushed Europe in front of it. Hence the fact that we are not enjoying a tropical climate at the present time. But in pushing Europe it badly smashed the south of that continent, and was driven over, as you have heard to-night, a great portion of Switzerland. The

* Perhaps I should have said a geophysicist, but he was, I believe, responsible for the topography of the expedition that traversed Greenland in 1912-13.

result, according to Dr. Staub, has been that there has been a contraction of the Earth's crust in Switzerland of something like 1000 miles. There are two kinds of drifting of continental masses: a drifting with the whole of the rest of the crust of the Earth, and a relative drift between the different continents. The general movement of the crust of the Earth over the interior is easier to explain than the movement of one portion of the crust relative to another, which involved the piling up of mountain ranges. Where did the force come from which produced it? Wegener rejects that due to the contraction of the Earth, but mathematicians tell us that it is the only force which is adequate to produce the result. But much smaller forces are capable of moving the whole crust of the Earth relative to the interior, and I see no reason why they should not be due to some of the minor causes which Wegener and others have suggested.

Col. Sir GERALD LENOX-CONYNGHAM: I had no intention, when I came here, of joining in this discussion, because I am no geologist, but I have just heard Dr. Evans say that Wegener is not a geologist but a geodesist. I do not think that we can accept him in the rank of geodesists. He has never written anything which gives serious geodetic support to his theory, and I thought geology was his special subject, because the foundation of his theory is mainly geological.

I should like to make one remark with reference to the most interesting lecture we have had from Professor Collet, namely, that I do not think that the isogamme chart of Professor Niethammer is secure evidence on which to base conclusions with regard to the underlying geology. I think, from the look of the chart, that the observations have not been reduced isostatically and that they are largely dependent upon the height of the different stations above sea-level. Until you have disentangled the results from the height above sea-level you cannot draw geological conclusions from them. I do not know whether Professor Collet will accept this criticism, but at present I feel that the chart does not allow one to draw geological conclusions.

The CHAIRMAN: I will ask Professor Collet to reply to the various remarks on his paper, and if there is any member present who has not had time to speak, if he will forward his remarks to the Secretary, they will be printed in the report of the proceedings in the *Journal*.

Professor LÉON W. COLLET: I quite agree with Sir Gerald Lenox-Conyngham. This map based on Niethammer's preliminary work has been published by Albert Heim in his well-known 'Geologie der Schweiz.' If I have shown it here, it is only to point out that the defects in gravity seem to correspond to the Pennine Alps and to the Austrian Alps, where we find the great recumbent folds formed into a geosyncline or on its southern border. Professor Niethammer has published, in 1921,* the results arrived at by reductions, in two maps, the one for

$$g_0'' - \gamma_0$$

and the other for

$$g - \gamma, \text{ with a depth of 120 km.}$$

These maps show the same general features shown on the map I have here. If these results do not allow to draw geological conclusions, they are interesting from a geological and geographical point of view.

* Th. Niethammer, 'Die Schwerebestimmungen der Schweizerischen Geodaetischen Kommission und ihre Ergebnisse. Verhandl. der Schweizer. Naturforsch-Gesell. Schaffausen,' 1921, pp. 50–64.

Mr. Wordie seems to have another explanation for the formation of the geosyncline, if I understood him well?

Mr. J. M. WORDIE: I asked whether you could not imagine some other way of getting the space between the foreland and the hinterland instead of by pulling apart. Could you not, for instance, have bending down of the strata?

Professor LÉON W. COLLET: If the strata had been bent down, we should have to deal with a great thickness of Sial at the bottom of the geosyncline, which is against the geological evidences seen on the field. If, on the contrary, we suppose that the Sial was very thin, it is possible to understand how the basic magma has come into it. To explain this thin cover of Sial the best is to suppose that the foreland and the hinterland were pulled apart and the geosyncline formed between them.

Professor Argand has gone much further in using Wegener's ideas, but what I wanted to show you to-night is that we have in the Alps the evidence of a movement towards Europe. This movement seems to me to be a support of Wegener's theory. It is too late to go further into details. Those who are interested in this question might read the very valuable paper by Professor Argand on the tectonics of Asia.*

It has been a great pleasure to me to come here to speak about the Alps. I am always glad to be able to pay a "wee bit" of my debt of gratitude towards this country, for I owe the best of myself to British science.

The CHAIRMAN: If a child was taken to the mountains the first question asked would be: "How are mountains made?" Well, I think we grown-up children have all been thinking about that, even earlier than Professor Collet suggested; in fact, since the days of Leonardo da Vinci, who, if he did not go very far, made interesting speculations with regard to the Alps. To-night, whether the experts thoroughly agree with all that has been said or not, we shall all agree in feeling that we have seen a step forward taken; that we have had proof that our generation is doing its share, and more, in elucidating this great problem. We shall also agree in hailing Professor Collet as one of the most distinguished of the pioneers in this branch of science. We have a feeling of very great gratitude to him for having come over at this season of the year—we only wish he had brought a little of Africa with him—to our bleak climate, which has long been moved north from the Tropics, to talk to us. We wish him a very happy return to his own country, where England is always liked and Englishmen, as I know from personal experience, are received with the utmost kindness. I ask you to give a hearty round of applause to Professor Collet for his lecture.

21

Reprinted from pages 415-422 of *Carnegie Inst. Washington Yearbook 25*, pp. 415-428 (1926)

THE SAN ANDREAS RIFT AND SOME OTHER ACTIVE FAULTS IN THE DESERT REGION OF SOUTHEASTERN CALIFORNIA[1]

L. F. Noble

SEISMOLOGY

REPORT OF THE ADVISORY COMMITTEE

(For previous reports, see Year Books Nos. 20–24.)

The year 1925–6 has been one of noteworthy progress in the study of the zones of movement in California, undertaken by the Carnegie Institution of Washington and the powerful agencies cooperating with it. The purpose of these efforts and the definite plan which has been followed hitherto have been adequately set forth in the earlier reports of this Committee. It is appropriate that this report should indicate some of the more conspicuous results which have thus far been attained, for it is through these that the promise for the future is best appraised.

All of the Annual Reports of the Committee have contained brief mention of the progress of the detailed study of the southern portion of the San Andreas rift, the master fault of California, undertaken by Dr. L. F. Noble, of the U. S. Geological Survey, and his associates. This deep fault, active, as Dr. Noble has shown, through long periods of geologic time, may prove to be the ultimate source and center of the widespread group of movements which this Committee has set itself to study. The report is therefore of fundamental importance to any study of earth movements in this region and of immediate interest to all American seismologists.

[1] By Dr. L. F. Noble. Printed by permission of the Director of the U. S. Geological Survey.

A report is presented herewith of the progress of the author's work during the past six years in mapping and studying some of the faults in the desert region of southeastern California that exhibit evidence of recent movements. This work has included two projects. The first is a study of the San Andreas rift in southern California, and begins with that part of the rift which lies along the southern border of Mohave Desert near the base of the San Gabriel Range and follows the course of the rift southeastward across a part of that range into Cajon Pass. From Cajon Pass it follows the course of the rift along the southern base of the San Bernardino Range and through San Gorgonio Pass into the Salton Sink. The second project is a reconnaissance of the Mohave Desert, Death Valley, and Panamint Valley regions eastward and northeastward from the San Andreas rift to the Nevada boundary for the purpose of locating and studying the faults which have been active in Quaternary time in this little known region.

The first project was undertaken in 1920, and work was continued on it intermittently for three years. In that time an areal geologic map of about 50 miles of the rift northwest of Cajon Pass was completed. The map covers a strip averaging 6 miles in width. The rift southeast of Cajon Pass has not yet been mapped and studied. It is planned to do this work in the spring of 1927, and later to prepare a report upon the San Andreas rift for publication by the Geological Survey.

In 1923 attention was shifted to the study of the faults in the desert region east of the San Andreas rift and this study has been continued intermittently since that time, several trips being made into the region each year. On three of these trips the author was accompanied by Professor H. E.

Gregory of Yale and on two by Professor W. M. Davis of Harvard. This work has not only yielded much information concerning the active faults of the region, but has resulted in the discovery of many other interesting geologic features, an account of which will be written later. This investigation is still in progress.

The San Andreas Rift

The area of the San Andreas rift already studied in this investigation lies along the southern border of Mohave Desert and extends across a part of the San Gabriel Mountains into Cajon Pass. It reaches from the Southern Pacific track in the vicinity of Palmdale to the Santa Fé track in Cajon Pass. It is shown on parts of the Elizabeth Lake, Tujunga, Rock Creek, San Antonio and Hesperia topographic sheets of the Geological Survey.

Southeasterly across the area runs a continuously traceable chain of scarps, trough-like depressions, and ridges which afford clear and unmistakable evidence of recent earth-movements. This line of topographic features, which is so straight that one may sight along it for distances of 25 miles or more, marks the position of a profound fault in the underlying rocks. This straight and continuous fault is the San Andreas fault—unquestionably the master fault. Bordering the master fault is a belt of roughly parallel branching and interlacing fractures which in places attains a width of 6 miles. This belt of faults roughly parallel with the main fault will be referred to as the San Andreas fault-zone. The San Andreas fault and the fault-zone together will be referred to in a general way as the San Andreas rift.

Recent Features

The scarps, ridges and depressions, which form a continuously straight line on the master fault and constitute the so-called "fault-trace," differ in no respect from the similar recent topographic features that mark the San Andreas fault where it has been described in the region north and south of San Francisco. Many of them involve the very youngest alluvial fans and are so fresh that they are entirely unmodified by erosion—in all probability some date from the Fort Tejon earthquake of 1857. It is characteristic of these features that they change abruptly from place to place along the strike of the fault. Within half a mile a scarp may give place to a ridge and the ridge to a trough; or a scarp facing in one direction may die out and change to a scarp facing in the opposite direction. Other ridges, sinks and scarps may appear at any place parallel with the main fault-trace. At places these parallel features may occupy a belt as much as half a mile in width. The features differ greatly in size. Some depressions are trenches a few feet wide and a foot or two deep; others are as much as 100 feet deep and many hundreds of feet wide; many of the depressions are undrained and contain ponds. Some of the features are in rock, others in alluvium. Some ridges are composed entirely of shattered rock—essentially a fault gouge; others are composed of alluvium. At places a ridge may run straight across the mouth of a canyon and block the drainage. The features exhibit all stages of modification by erosion, from those that are still fresh to those that are almost obliterated. In some parts of the rift many subjacent features are entirely buried under alluvial deposits; in other parts they are being exhumed by erosion. Some features involve both the recent alluvium and the older rocks, others involve only older dissected Quaternary beds and older rocks, and still others involve only early Quaternary and older rocks. It is therefore evident that the features are the product of movements that have taken place repeatedly throughout Quaternary time, and that the movements are still in progress. In general the older features are on a much larger scale than the recent ones, indicating that the older movements were of greater magnitude. Some of the oldest features occur several miles away from the main fault, although in general parallel with it.

Some recent features afford clear and convincing evidence of horizontal movement along the San Andreas fault. At a place about 3 miles southeast of Cajon Pass four deep ravines, that descend the steep San Bernardino Mountain slope east of the fault, are offset abruptly at the fault-trace, each ravine appearing again west of the fault at a point at least 150 feet northwest of the place where it is offset. This feature was discovered by Mr. Peters of Devore, California, who pointed it out to Professor Davis and the author.

The Fault-Zone

The belt of branching, interlacing, and roughly parallel faults that borders the main San Andreas fault and constitutes the San Andreas fault-zone exhibits a structure so complex that a simple statement concerning it is difficult to make. In general, the zone is a fault mosaic of elongated sliver-like blocks whose longer axes trend parallel with the strike of the main fault, but at many places the rock masses are so intricately shattered and different formations so mixed together that it is impossible to map them or to determine their relations and age. The evidence of the nature of the fault-movements is somewhat conflicting; the dominant structure is a sort of slicing which appears to be mainly the result of horizontal shear along the San Andreas fault. No exposures of the fault-plane are obtainable, but the straight course of the fault across an uneven topography suggests that it is nearly vertical. On the other hand, the planes of a large number of faults in the fault-zone are obviously thrust planes, nearly all of which dip southwest at angles ranging from 55° to nearly vertical. These structures are the result of thrusting brought about by compressive forces acting from the southwest. Normal faults occur at a number of places in the fault-zone. Some are of considerable magnitude and have high scarps. Graben structures are common in some parts of the zone. The normal faulting, however, is a relatively unimportant feature, probably the result of vertical adjustments within this zone of profound weakness.

One noteworthy feature of the fault-zone is the thorough crushing and shattering of the harder rocks everywhere along the main fault, usually through a belt several hundred yards wide. In this belt the rocks are at most places so ground up that they are practically gouge. At many places the rocks are thoroughly decomposed. Granite, for example, at most places along the fault is so rotten that when wet it becomes plastic like clay.

Another noteworthy feature of the zone is the occurrence of long narrow blocks of granite which form inverted prisms inset in soft Tertiary or early Quaternary beds by faulting in such a way that the fault-planes, which separate the granite from the sedimentary beds, dip under the granite prism from both sides, the apex of each granite prism pointing downward. Some of these granite blocks are a mile long and a quarter of a mile wide. This structure suggests that the granite blocks have been squeezed up by compression. Exposure of granite lying upon Tertiary or early Quaternary beds and separated from them by thrust planes that are almost flat are common in some parts of the fault-zone.

An interesting feature of some of the branching faults of the San Andreas rift is the occurrence along them of narrow strips of steeply dipping Tertiary sediments bordered on both sides by vastly older crystalline rocks. One of these ribbons of Tertiary sediments pinched between the crystalline rocks runs 12 miles through the highest part of the San Gabriel Range and is at most places less than 100 yards wide. The beds at places are nearly vertical.

They lie in a great thrust fault that branches from the San Andreas fault zone at a low angle and runs nearly parallel with it for many miles. Several cross-faults within the San Andreas fault-zone are marked by similar ribbons of nearly vertical sediments. It is obvious that faults characterized by this structure are profound, indeed.

Although the San Andreas fault runs for some distance near the northern base of the San Gabriel Range, it does not actually bound the range, for it crosses the range at an elevation of 6,800 feet between Rock Creek and Cajon Pass. The initial uplift of the range did not take place upon the main San Andreas fault, but, rather, upon faults that branch from the San Andreas fault-zone. The planes of these branching faults, wherever exposed to observation, are thrust planes which dip steeply southwestward into the mountain mass in such a way that the older crystalline rocks of the range override younger sedimentary rocks at the base of the range.

Areal Geology

Eleven different rock-formations have been recognized in the area studied; 2 of them are probably pre-Cambrian, 1 is Paleozoic, 1 Mesozoic, 4 Tertiary and 3 Quaternary.

The two formations which are believed to be pre-Cambrian are a series of schists and a series of gneisses. Both are largely recrystallized and have undergone profound metamorphism. The schist series is chiefly of sedimentary origin and the gneiss series is of igneous origin. Which is the older is not known.

The schist series is composed chiefly of quartz-sericite-albite schist, which at most places exhibits a distinctive bluish-gray color due to the presence of innumerable tiny flakes of graphite, but it includes many beds of chlorite schist and actinolite schist, and, in places, beds of metamorphosed limestone (marble) and sandstone (quartzite). This schist series is the so-called "Pelona schist" of Hershey. In all probability it is the correlative of the Rand schist of the Randsburg district, with much of which it is lithologically identical.

The gneiss series is an intricate assemblage or "complex" of banded and contorted gneisses which range in composition from granite and quartz-monzonite gneiss to diorite gneiss. A gray quartz-diorite gneiss is the prevailing rock, but the series includes much amphibolite and in places large bodies of crushed and altered pegmatite and aplite. Both the gneiss series and the schist series are injected by Mesozoic granite and, at places, by quartz porphyry of unknown age.

The only Mesozoic rock represented in the area, so far as known, is granite. This granite extends eastward from the San Andreas rift for great distances over the Mohave Desert region and is part of a great batholith which is probably of the same age as the Sierra Nevada batholith and may be connected with it. At many places along the rift this granite contains masses of limestone and schist, some of them several miles in extent. All these masses are inclusions in the granite; many apparently represent roof pendants of the batholith.

The Paleozoic rocks are the masses of limestone and schist just referred to. They are always surrounded by and injected with the granite or granite gneiss and are thoroughly altered by contact metamorphism. A few obscure fossils have been found in them by the author—chiefly crinoid stems and corals. The fossils are too badly altered to be specifically determinable, but they are believed to represent Carboniferous forms.

The oldest Tertiary formation is a series of ten sandstones, conglomerates and dark shales which represent the Martinez formation of Eocene age.

Nearly 5,000 feet of these beds are exposed on Rock Creek, where they contain typical Martinez fossils. This formation is the youngest marine formation exposed in the area studied, and in fact anywhere in the Mohave Desert, so far as known. All succeeding formations are of terrestrial origin.

What appears to be the next younger Tertiary formation is a series of andesitic and rhyolitic lavas, breccias and hard red sandstones that may represent the Sespe formation of Oligocene (?) age. No fossils have been found in it and all contacts with other Tertiary formations in the rift are faults. Consequently, its stratigraphic position is unknown.

Overlying the Martinez formation with marked angular unconformity is a series of fanglomerates, arkose sandrocks and shales, which, near Rock Creek, have an exposed thickness of about 5,000 feet. These beds are particularly well exposed in the Devil's Punchbowl near Rock Creek and in Cajon Pass, where they form great tilted ledges of sandrock, that are familiar to all travelers through the Pass. Vertebrate fossils have been found in them which suggest a stage of the upper Miocene not far from that represented by the Mint Canyon formation of Kew. Accordingly, the beds are here assigned tentatively to the Mint Canyon formation.

The three Tertiary formations just described are in general well consolidated and are profoundly disturbed. Their areal distribution has no relation to the present topography. Obviously, they were deposited before the initial uplift of the San Gabriel Range and before the movements along the San Andreas rift that resulted in the network of older faults in the fault zone.

Above the Mint Canyon formation, and separated from it by an angular unconformity, are beds of fine arkose sand, clay and gravel which may be of late Pliocene age, or of early Quaternary age, or both. Probably they are roughly equivalent to the Saugus formation of Kew. They are well exposed at Cajon Pass, where they overlie the Mint Canyon formation, and at many places northwest of Cajon Pass along the desert border and in the rift. In contrast with the older Tertiary formations they are all poorly consolidated and in general are much less disturbed. They are believed to have been deposited after the initial uplift of the San Gabriel Range and after the older faulting along the San Andreas rift referred to in the preceding paragraph, but before uplifts took place that raised the mountains to their present height. Their deposition appears to have followed a period of erosion which had reduced the land to a fairly mature topography, for the basal beds are composed chiefly of fine material. Along the rift they commonly fill valleys formed by the earlier faulting. A few obscure fragments of vertebrate bones found in the basal beds at Cajon Pass suggest Pliocene forms, but the evidence is not certain. At many places along the rift it is difficult, and, at some places, impossible to separate the beds just described from older Tertiary and from younger Quaternary deposits, owing to the lack of fossils, to the complex structure of the fault-zone, and to the fact that the beds are lithologically identical with beds in the underlying Mint Canyon and overlying Quaternary. Degree of consolidation is not a reliable guide in correlating beds in the rift, for although in a general way the Mint Canyon formation is more firmly consolidated than the later formations, many Mint Canyon beds are as loosely consolidated as Quaternary beds.

The Quaternary deposits comprise at least three formations, all of which are unconsolidated. The oldest formation consists chiefly of fanglomerate, but includes some arkose sand and sandy clay. It overlies the Pliocene or early Quaternary formation just described. In places it is tilted and moderately faulted, but on the whole it is not greatly disturbed. It was deposited after the San Gabriel Range had attained something like its present configuration and height. The next younger formation consists of dissected alluvial fans and terrace remnants and is undisturbed except along the San Andreas fault zone. Its dissection records the latest uplift in the San Gabriel Range. The youngest formation is the recent alluvium—fan deposits, boulder wash and soil. It is undisturbed except where involved in the recent movements along the San Andreas fault.

Distribution of Rocks along the San Andreas Rift

Very broadly speaking, the San Andreas fault zone in the area studied is a belt made up of representatives of all the rock-formations just described, bordered on both sides by pre-Tertiary crystalline rocks. The prevailing rocks north of the fault-zone are Mesozoic granites with inclusions of limestone; those south of it are pre-Cambrian schists and gneisses. The Tertiary formations lie chiefly within the fault-zone, where, together with masses of the older and younger formations, they form the complex mosaic of long sliver-like blocks already described. However, for a distance of 15 miles northwest of Cajon Pass, only crystalline rocks occur in the fault-zone; and at and northwest of Cajon Pass and near Palmdale, Tertiary and Quaternary rocks extend beyond the zone into Mohave Desert. All Quaternary formations except a few patches of recent alluvium occur either within or north of the zone.

Scarcely anywhere in the fault-zone are the rocks on opposite sides of the master fault similar. This relation is particularly marked in the pre-Tertiary rocks. For example, in a stretch 50 miles long extending from a point 6 miles southeast of Rock Creek to the Crafton Hills east of Redlands, the fault is bordered continuously on the south by the series of pre-Cambrian schists already described, whereas the rocks north of the fault are Mesozoic granites and pre-Cambrian gneisses. These relations not only show that the movements along the fault have been of very great magnitude, but they strongly suggest that the movements began in pre-Tertiary time.

The distribution of certain Tertiary rock masses along the master fault affords a suggestion that a horizontal shift of many miles has taken place along the rift. On the north side of the fault, near Cajon Pass, a small block of strata lithologically similar to beds in the Martinez formation at Rock Creek is associated with Mint Canyon beds. The only other exposure of Martinez associated with Mint Canyon beds anywhere in the region lies on the opposite side of the fault at Rock Creek, 24 miles northwest of the locality in Cajon Pass. At both localities the beds of both formations are so intricately faulted that a displacement of any magnitude is conceivable, and it thus appears possible that horizontal movements along the fault have dragged the rock-masses north of the fault to the southeast in relation to those south of the fault, or have dragged the masses south of the fault to the northwest in relation to those north of the fault. The evidence just cited, however, is not convincing, and it certainly is not definite enough to amount to proof. It is to be hoped that the study of the rift in the region south of Cajon Pass which is planned for this winter will throw more light on the problem.

Broader Topographic Features

The San Andreas fault-zone is in striking topographic contrast with the areas bordering it on the north and south. North of the fault-zone a gently sloping alluvial plain stretches far out into the Mohave Desert. This plain is made up of alluvial fans built out from the mountainous region south of the rift. South of the fault-zone lies the high San Gabriel Mountain Range, a part of which is crossed by the rift. Everywhere south of the rift these mountains are characterized by exceedingly rugged, youthful topography. The drainage is insequent, or dendritic, ridges are serrated and irregular, and spurs are knife-edged and sprawling.

The mountain peaks are of different heights and no flat-topped ridges exist. Within the fault-zone the topography is very different. Long, narrow ridges and long narrow valleys trending parallel with the San Andreas fault are the dominant topographic forms. The ridges have even, gently sloping tops. These even tops of the ridges, many of which are composed of heterogeneous rocks of complex structure, are believed to represent remnants of a peneplain. Near Rock Creek, where the ridges are of granite, they present fine examples of the pan-fan or near pan-fan stage of desert erosion. Some of this pan-fan topography is in process of being exhumed from a cover of Quaternary alluvium.

Even where the rift crosses the San Gabriel Range it is characterized by the parallel longitudinal valley and ridge topography just described. The high divide where the master fault crosses the range is a gap, or col, that lies midway between Cajon Pass and Rock Creek. The elevation of this gap is a little under 6,800 feet. The recreation camp of the recently-opened Los Angeles County Park is situated in this gap. It is reached by a magnificent road that follows the master fault all the way from Rock Creek to Cajon Pass and affords a splendid opportunity to study many recent features along the fault. The divide, or gap just described, is on the axis of a broad upwarp which has bowed up this part of the San Gabriel Range in late Quaternary time. Northwest of this axis of warping the fault-zone slopes northwestward toward Rock Creek, the long parallel valleys and the crests of the long even ridges descending steadily in that direction until, northwest of Rock Creek, many of the older rift features plunge beneath a cover of Quaternary alluvium. Southeast of this axis the fault-zone slopes southeastward toward Cajon Pass, the rift ridges and valleys descending steadily all the way to Cajon Creek. One ridge finally passes under the alluvial filling of San Bernardino Valley.

This downwarp, or rather failure of uplift, in the area about Cajon Pass is one of the factors to which the Pass owes its existence. The other factors are faulting, soft rocks and erosion. Faulting along the San Andreas fault-zone has inset a great mass of soft sedimentary rocks in the hard crystalline rocks about the pass in such a way that the soft sedimentary rocks extend from Mohave Desert through to the main San Andreas fault almost at the edge of San Bernardino Valley. Only a narrow belt of crystalline rocks south of the fault separates these soft rocks from San Bernardino Valley. The downwarp, or absence of strong uplift, by keeping depressed this narrow belt of crystalline rocks already breached obliquely by the San Andreas fault and rendered weak by crushing along the fault, has enabled Cajon Creek easily to maintain its course across the crystalline belt, and the softness of the sedimentary rocks has enabled the creek quickly to excavate in them by headwater erosion the huge amphitheater under the rim of Mohave Desert at the head of the pass. The amphitheater includes all the upper valley of Cajon Creek and its rim is a line of steep southward-facing bluffs, 8 miles in length. From the summit of the bluffs an alluvial plain slopes northeastward many miles into Mohave Desert. This alluvial plain is the surface of a gigantic alluvial fan built out northeastward from the high part of the San Gabriel Range, afterward tilted and somewhat disturbed in its upper portion, and then dissected by wide and deep-stream channels running down its slope. No greater surprise awaits the geologist who travels up this fan toward Cajon Pass from the desert than to find that, instead of resting against a mountain slope, the upper part of the fan drops suddenly away at the amphitheater just described—the amphitheater has completely beheaded the fan. The stream valleys on the fan are all truncated headward by the bluff along the head of the amphitheater, so that they appear in cross-section along the summit of the bluff as deep notches. The great width and depth of some of the valleys at the point of truncation is eloquent testimony of the former great extent of the fan beyond the point where it is beheaded. Inasmuch as the fan drains into Mohave River its beheading must have had far-reaching consequences in the Quaternary history of that stream. Altogether, Cajon Pass affords one of the most interesting and spectacular geologic studies in the west.

In brief, the topographic expression of the San Andreas rift in the area studied is not a great scarp bounding a mountain range, but, rather, a belt of long narrow valleys and ridges which at one place lies along the base of a mountain range but at another place lies across a part of the range; this part of the range has recently been broadly uplifted across the path of the rift.

Geologic History

Until a study of the rift southward from Cajon Pass shall have been completed it would be premature to attempt a systematic outline of the geologic history. However, an incomplete summary of the probable sequence of events may be given from the evidence at hand.

The distribution of the pre-Tertiary rocks along the rift indicates that movements took place upon it as far back as late Mesozoic or early Tertiary time. The first movements whose date can be established approximately, however, took place at some period between late Miocene and early Quaternary time, because they involve the upper Miocene Mint Canyon formation but not a formation that is believed to be either late Pliocene or early Quaternary or both. These movements were of great magnitude. Apparently they were largely horizontal, although definite proof of their nature is not yet established. Many of them certainly were thrusts. The initial uplift of the San Gabriel Range is believed to have taken place at some time during this period, not upon the main San Andreas fault, but upon faults that branch from the San Andreas; it was brought about by thrusting from the south, southwest, or west. These movements were followed by a period of erosion which appears to have resulted in peneplanation. Remnants of the peneplain are preserved on the flat-topped ridges of the San Andreas fault zone. Probably the peneplain is to be correlated with the Ricardo peneplain of Baker. Beds which may be either late Pliocene or early Quaternary, or both, were deposited in depressions in this more or less mature surface. As deposition went on, new depressions were constantly being formed along the rift by faulting. The beds thus deposited continued to be disturbed by fault-movements along the rift, and, at some time in the early Quaternary, the uplift of the San Gabriel Range was renewed. At some period after this uplift, alluvial deposits washed from the slopes of the range may have covered parts of the San Andreas fault-zone and these deposits may have helped preserve the flat tops of the ridges by burying them. Still later in Quaternary time, but before the Glacial Period, the northern part of the San Gabriel Range was bowed up locally in the present high region about Baldy and North Baldy Peaks. This movement was independent of the San Andreas rift, for the rift is gashed diagonally across the axis of uplift and has been raised with it. As a result of the erosion following this last uplift, many of the ridges in the rift in this high country have been considerably dissected.

The different stages of erosion exhibited by the recent, sub-recent, and older topographic features along the rift are proof that faulting has been taking place at intervals all through Quaternary time. The movements are still in progress.

[*Editor's Note:* The remainder of the article has been omitted.]

22

ICELAND AND THE QUESTION OF CONTINENTAL DRIFT

Ferdinand Bernauer

This article was translated expressly for this Benchmark volume by Rhodes W. Fairbridge from "Island und die Frage der Kontinentalverschiebungen," *Geol. Rundschau,* **30:**357–358 (1939)

1. In Iceland, in the 100-200 km wide belt of youthful volcanoes, there is a series of parallel fractures that are wide open and up to 50 m across, so wide in fact that they have not been filled in or covered by volcanic material.
2. Judging from the width of the openings in many places, one can calculate that there has been a postglacial extenison of the surface of 1-3 percent. Thus, assuming an average age for the surface rocks of 5,000 years, this would make an average annual extension of 2–6 mm/km. Older fractures are almost certainly present but are buried.
3. This activity continues up till the present day, as is shown by historical data and by direct observations of earthquakes and volcanic eruptions, as well as by 1–2 m wide gaping fractures of the same type within the historical lava streams.
4. Their only reasonable interpretation is to be found in a relative stretching of the underlying material, which by a friction couple leads to extension fractures in the overlying brittle crust. A natural model for this process can be observed where active modern lava flows develop a thick crust of slag that becomes broken up into strips and separated into islands by the fast-moving undercurrent (e.g., Laki, Sveinagja, Eldgja). It is noteworthy that kinks and bends in the fractures are reflected by the adjacent blocks and only die out over considerable distances.
5. I have not seen any indications in the Bau Islands that could support the idea of the Icelandic tension zone being a central graben over a crustal "tumor" in the sense of Haarmann and Cloos. Rather, I could visualize a system of magmatic undercurrents that spread out in a flat sheet, carrying a crust passively above it. Here, that crust is exceptionally thin.
6. As a consequence of this flat distribution, subsidence of the crust occurs in tensional areas. In this way the ocean can invade the subsided areas and thus reach the fresh magma exposed in the bottom of the fractures and in their future extenisons. This leads to the formation of rocks of palagonite facies, within which there are numerous basalt plugs, dikes, and minor intrusions marked by glassy edges and a microfracturing in columnar form, suggestive of quenching in water or in wet muddy sediment. The water-saturated tuffs suggest submarine explosions of highly liquid magma.

7. From a regional point of view, there is evidence in the overlying crust of a gradual loading by this magma and a progressive rise in gravity. Its high density counterbalances the gravity differential, which is to be expected due to the exceptionally light palagonite formation in the younger volcanic zone. (Actually, Schleusner, in a gravity profile normal to the Bardar Valley, showed that there were only minor variations at the boundaries of plateau basalts and the palagonite formation.) Indeed, on account of the inertia (or sluggish isostatic readjustment), initial uplifts still show a tilt of the uplifted land forms into blocks and ridges; this can be seen both on the large scale (e.g., the whole Reykjanes Peninsula) and in detail (e.g., Herdubreid, Sveifluhals, etc.).
8. The continuation of the Icelandic volcanic belt is to be found along the equally volcanic Reykjanes Ridge and farther south in the Mid-Atlantic Ridge. It seems likely that they have a similar structure. Furthermore, the similarity between this trend and those of the two opposite continental margins, in my opinion, suggests a common origin, perhaps a rifting apart by magmatic undercurrents of the sort described above. The continents would thus have drifted apart but not in the manner that Wegener had assumed; rather one might expect to find many larger or smaller strips and blocks of the former crust left behind in the floor of the Atlantic.The movements in Iceland provide perhaps an order of magnitude for these motions; further observations then appear to be important.

Part VI

A DOCUMENTED CHRONOLOGY OF PLATE TECTONICS

CONTENTS

PREFACE

During the period 1965-1970, the science of geology experienced a revolution more profound than anything that had happened to the field since it first became a science in the late 18th Century, about two hundred years ago. Not even the remarkable synthesis of organic evolution conceived by Charles Darwin and Alfred Russell Wallace in the middle of the 19th Century, the development of quantitative age-determination techniques in the early 20th Century or the late 20th Century extension of geology to other planets required such drastic changes in our thinking about the earth. In fact, there is hardly a branch of geology that remains unaffected by the implications of plate tectonics.

Despite the fact that this revolution was a very long time in the making, (some parts of it were developed in 1908 by F.B. Taylor, in 1912 by Alfred Wegener, in 1928 by Arthur Holmes, and in 1960 by Harry Hess) most geologists remained unconvinced until the middle and late 1960s when in a few short years the implications of certain crucial observations were finally realized and additional key elements in the general theory were conceived. The result was that, in a very short time, most geologists were convinced of the validity of the new theory. Some writers have even claimed that the conversion can be traced to a series of three meetings that took place in the six-month period between November, 1966 and April, 1967. Whatever the time span, the switch in opinion was almost total. Only a handful of geologists opposed the new theory and even their small number declined during the early 1970s as a crushing weight of new evidence came in from deep-sea drilling, paleontology, paleomagnetism, seismology, vulcanology and bathymetric and heat-flow studies.

The purpose of the Chronology, as the title suggests, is to trace the development of plate tectonics and the lines of evidence that led to its development and acceptance from their early beginnings up to about 1975, to do so in a distinctly chronological format as contrasted to the topical approach taken by other histories of the theory, and to document each step of that development with readily accessible, comprehensive citations to the literature. An additional feature of the Chronology that will be useful to readers and researchers is its extensive name and subject index.

In preparing the Chronology my most important source was, of course, the original papers themselves and the lists of references they provide. My basic procedure was to start with important recent works and to move backward through the citation tree, selecting what seemed important and eliminating what turned out to be not so important. I also made extensive use of the general histories of plate tectonics by Hallam, Marvin and Sullivan and of many less-comprehensive historical accounts. (See the list of references at the end of the Preface.) The judgement as to what was significant was entirely my own. This is not to say that I had no help. Many individuals gave generously of their time, ideas and knowledge. Particular thanks are due the late E.C. Bullard who read an early draft, took the time to make many detailed suggestions for changes or additions and helped me track down numerous leads. Others who read drafts and offered comments were R.W. Fairbridge, J.T. Wilson and Susan Schlee. H.J. Harrington graciously provided me with copies of his Pacific Basin annotated bibliographies and also helped with specific problem entries. The following individuals helped out with comments on one or more specific points: F.W. Aumento, S.W. Carey, W.W. Hay, B.C. Heezen, J.R. Heirtzler, U.B. Marvin, W.C. Pitman III, E.I. Smith, B.W. Troxell, and R.E. Wallace. I emphasize, however, that the views reflected by general content, by inclusion or exclusion of certain works, and by the wording of specific entries are entirely my own. I also express my appreciation to the University of Wisconsin - Parkside Library Staff, and to the Wisconsin Inter-Library Loan Service for the unfailing help they provided. Without their assistance, the Chronology would not have been written.

James H. Shea

University of Wisconsin - Parkside

REFERENCES

Bell, P.M., ed., 1975, United States National Report to the International Union of Geodesy and Geophysics: Review of Geophysics and Space Physics, v. 13, no. 3, 1107 p.

Bullard, E.C., 1964, Continental drift: Geological Society of London Quarterly Journal, v. 120, p. 1-33.

Bullard, E.C., 1968, Reversals of the earth's magnetic field: Royal Society of London Philosophical Transactions, v. A263, p. 481-524.

Bullard, E.C., 1975, The emergence of plate tectonics: a personal view: in Donath, F.A., and others, eds., Annual review of earth and planetary sciences, v. 3, Palo Alto, Calif., Annual Reviews, Inc., p. 1-30.

Bullard, E.C., 1975A, William Maurice Ewing 1906-1974: Royal Society of London Biographical Memoirs Fellows, v. 21, p. 269-311.

Carozzi, A.V., 1970, New historical data on the origin of the theory of continental drift: Geological Society of America Bulletin, v. 81, p. 283-86.

Cox, Allan, 1973, Plate tectonics and geomagnetic reversals: San Francisco, W.H. Freeman and Co., 702 p.

Davison, Charles, 1927, The founders of seismology: Cambridge, Cambridge University Press, 240 p.

Du Toit, A.L., 1937, Our wandering continents, an hypothesis of continental drift: New York, Hafner Pub. Co., 366 p.

Hallam, A., 1973, A revolution in the earth sciences: Oxford, Clarendon Press, 127 p.

Harrington, H.J., 1976, Tectonics of the Pacific Basin and the circum-Pacific rim; an annotated bibliography for the interval 1967-1975: Armidale, New South Wales, Australia, University of New England, 212 p.

Harrington, H.J., 1972, An annotated bibliography of the circum-Pacific rim and the Pacific Ocean basin (with special reference to continental drift, and to publications of the interval 1960 to mid-1967): Armidale, New South Wales, Australia, Department of Geology, University of New England, 119 p.

Hume, W.F., 1948, Terrestrial theories - a digest of various views as to the origin and development of the earth and their bearing on the geology of Egypt: Cairo, Geological Survey, Government Press, 731 p. (Note: This book was completed in 1939 but its publication was delayed by World War II.)

Irving, E., 1964, Paleomagnetism and its application to geological and geophysical problems: New York, John Wiley & Sons, 399 p.

Kasbeer, Tina, 1972, Bibliography of continental drift and plate tectonics: Geological Society of America Special Paper 142, 96 p.

Kasbeer, Tina, 1975, Bibliography of continental drift and plate tectonics, v. 2: Geological Society of America Special Paper 164, 151 p.

Marvin, U.B., 1973, Continental drift - the evolution of a concept: Washington, D.C., Smithsonian Institution Press, 239 p.

Nagata, Takesi, 1961, Rock magnetism: Tokyo, Maruzen Co. Ltd., 350 p.

Rupke, N.A., 1970, Continental drift before 1900: Nature, v. 227, p. 349-350.

Schlee, Susan, 1973, The edge of an unfamiliar world - a history of oceanography: New York, E.P. Dutton & Co., 398 p.

Sullivan, Walter, 1974, Continents in motion - the new earth debate: New York, McGraw-Hill Book Co., 399 p.

Tarling, Don and Maureen, 1971, Continental drift - a study of the earth's moving surface: London, G. Bell, 140 p.

Wegener, A.lfred, 1924, The origin of continents and oceans (translated by J.G.A. Skerl from the 3rd German edition): New York, E.P. Dutton & Co., 212 p.

Wertenbaker, William, 1974, The floor of the sea - Maurice Ewing and the search to understand the earth: Boston, Little Brown & Co., 275 p.

1749

1 P. Bouguer concluded that the deflection of a plumb line by the Andes Mountains is much less than expected from a consideration of their apparent mass.

Bouguer, Pierre, 1749, La figure de la Terre: Paris, C. Jombert, 394 p.

1775

2 L. Euler showed that any displacement of a portion of a rigid spherical surface on that surface can be considered as a rotation about a fixed axis. Such axes came to be known as "Euler poles," and their geometry proved very useful in understanding plate motions.

Euler, Leonard, 1775, Formulae generales pro translatione quacunque corporum rigidorum: Novi Comentarii Academiae Scientiarum Imperialis Petropolitanae (now Akademiia Nauk SSSR, Leningrad), v. 20, p. 189-207.

1795 - 1846

3 Hans Conrad Escher von der Linth and his son Arnold discovered the repeated age sequence of rocks at various places in the Alps, conceived the idea of giant overthrust sheets, named these structures "nappes", and showed that nappes are abundant in the Alps.

Escher von der Linth, H.C., 1795, Geognostische Nachrichten uber die Alpen: in Briefe aus Helvetien, Erster Brief, Neues Bergmann. J., I, Freyberg, p. 116-160.

___ 1807, Geognostische Nachrichten uber die Alpen: in Briefe aus Helvetien, Erster und Zweiter Brief, Alpina II, Freyberg, p. 1-58.

Escher von der Linth, Arnold, 1841, Geologische Carte des Kantons Glarus und seiner Umgebungen, nebst Profilen: Verh. Schweiz Naturforsch. Ges., Zurich.

___ 1846, Gebirgskunde, mit Karte und Profilen: in Heer, O., und Blumer, J.J., Der Kanton Glarus, VII Band der Gemalde der Schweiz.

1828

4 A. Brongniart named the genus Glossopteris and described several species (including G. browniana) from the Glossopteris flora from Australia and India.

Brongniart, Adolphe, 1828, Histoire des vegetaux fossiles, ou recherches botaniques et geologiques sur les vegetaux renfermes dans les diverses couches du globe (2 vols.): Paris, Dufour and D'Ocange (reprinted 1965, A. Asher, Amsterdam). (See p. 222-226 and plates 61-63.)

1831 - 1963

5 Various authors suggested that Earth contraction offers the best explanation of Earth tectonics.

de Beaumont, L.E., 1831, Researches on some of the revolutions which have taken place on the surface of the globe . . . : Philos. Mag., v. 10, p. 241-264.

Le Conte, Joseph, 1872, A theory of the formation of the great features of the earth: Amer. Jour. Sci., v. 104, p. 345-355, 460-472.

Dana, J.D., 1873, On some results of the earth's contraction from cooling, including a discussion of the origin of mountains, and the nature of the earth's interior: Amer. Jour. Sci., v. 105, p. 423-443, 474-475; v. 106, p. 6-14, 104-115, 161-172, 304, 381-382.

Suess, Eduard, 1885-1909, Das Antlitz der Erde: Prague and Vienna, F. Tempky, 5 volumes totalling 2936 p.
Hobbs, W.H., 1923, The Asiatic arcs: Geol. Soc. America Bull., v. 34, p. 243-252.
Jeffreys, Harold, 1924-1976 , The earth (six editions): Cambridge, Cambridge Univ. Press.
___ 1931, On the mechanics of mountains: Geol. Magazine, v. 68, p. 435-442.
Bucher, W.H. 1933, The deformation of the earth's crust: Princeton, N. Jersey, Princeton Univ. Press, 518 p.
Nolke, Friedrich, 1933, Gibt es eine geologische Universaltheorie?: Scientia, Bologna, v. 53, no. 249, p. 20-30.
De Lury, J.S., 1945, Outline of an hypothesis to explain crustal movements: Royal Soc. Canada Trans., Ser. 3, v. 39, sec. 4, p. 1-11.
Landes, K.K., 1952, Our shrinking globe: Geol. Soc. America Bull., v. -240.
Lees, G.M., 1953, The evolution of a shrinking earth: Geol. Soc. London Quart. Jour., v. 109, p. 217-257.
Wilson, J.T., 1954, The development and structure of the crust: in Kuiper, G.P., ed., The earth as a planet, Chicago, Univ. of Chicago Press, p. 138-214.
___ 1959, Geophysics and continental growth: Amer. Scientist, v. 47, p. 1-24.
Scheidigger, A.E., 1963, Principles of geodynamics (2nd ed.): Berlin, Springer.

1839

6 W. Hopkins conceived the concept of convection currents in the earth's interior.

Hopkins, W. 1839, Researches in physical geology: Royal Soc. London Philos. Trans., v. 129, p. 381-385.

1840

7 Sir James Clark Ross made a sounding of 4,438 m in the south Atlantic at 27 degrees 26 minutes S latitude, 17 degrees 29 minutes W longitude, thereby ushering in the era of lead and line sounding. In 1968 this area was resounded by the United States Coast and Geodetic Survey ship Discoverer which measured a depth of 3,843 m. However, since the area is one of rough topography and there is uncertainty as to the precise position of the Ross sounding, the determination was probably relatively accurate.

Ross, J.C., 1847, A voyage of discovery and research in the southern and Antarctic regions during the years 1839-43: London, John Murray, 2 vols., 366 and 477 p. (Reprinted in 1969, New York, A. M. Kelley Pubs.)
Dietz, R.S., and Knebel, H.J., 1968, Survey of Ross's deep sea sounding site: Nature, v. 220, p. 751-753.

1842

8 C. Darwin theorized that fringing reefs (earliest stage), barrier reefs (intermediate stage), and atolls (latest stage) are all developmental stages in the history of reef growth on gradually subsiding volcanic islands.

Darwin, Charles, 1842, The structure and distribution of coral reefs: London, Smith Elder and Co.

1843

9 W.B. and H.D. Rogers reported that the structure of the Appalachians of Virginia and Pennsylvania consists in part of gigantic overthrust blocks that were pushed toward the northwest.

Rogers, W.B., and Rogers, H.D., 1843, On the physical structure of the A palachian chain, as exemplifying the laws which have regulated the elevation of great mountain chains generally: Assoc. Amer. Geologists report, p. 474-531.

1846

10 James Dwight Dana advanced the idea that the continents and ocean basins are permanent features of earth's topography.

Dana, J.D., 1846, On the volcanoes of the moon: Amer. Jour. Sci., v. 52, p. 335-355.

___ 1847, On the origin of continents: Amer. Jour. Sci., v. 53, p. 94-100.

___ 1847, Origin of the grand outline features of the earth: Amer. Jour. Sci., v. 53, p. 381-398.

___ 1881, The continents always continents: Nature, v. 23, p. 410.

(See also)

Willis, Bailey, 1910, Principles of paleogeography: Science, v. 31, p. 241-260.

Stille, Hans, 1945, Die tektonische Entwicklung der pazifischen Randgebiete: Geotektonische Forschungen, Heft 7/8.

___ 1948, Ur und Neuozeane: Abh. Deutsch Akad. Wiss. Berlin, Jahrgang 1945-46, no. 6.

1847

11 J.D. Hooker suggested that similarities in the fossil plants of the southern continents demonstrate that these continents were once connected with each other.

Hooker, J.D., 1847, Flora Antarctica: Part I of the botany of the Antarctic voyage of H.M. discovery ships Erebus and Terror in the years 1839-1843, under the command of Captain Sir James Clark Ross: London, Reeve Brothers.

___ 1853, Introductory essay to the flora of New Zealand: London, 39 p. (A summary of this essay appeared in Amer. Jour. Sci., v. 17, p. 241-252 and 334-350.)

1848

12 A United States Navy expedition led by Lieutenant W.E. Lynch determined the maximum depth of the Dead Sea to be almost 400 m.

Jenkins, J.S., 1855, Expedition to the Dead Sea and the River Jordan: Part II of Voyage of the United States Exploring Expedition commanded by Captain Charles Wilkes of the U.S. Navy in 1838-1842, New York, J.C. Derby, p. 460-517.

1848 - 1886

13 H.F. Blanford, W.T. Blanford, W.B. Clarke, R.D. Oldham and T. Oldham conclusively demonstrated the Permo-Carboniferous age of the Glossopteris flora and extensively documented the remarkable similarity of the fossils of this flora from India and Australia.

Clarke, W.B., 1848, On the genera and distribution of plants in the Carboniferous rocks of New South Wales: Geol. Soc. London Quart. Jour., v. 4, p. 60-66.

Oldham, Thomas, 1860, On the geological relations and probable geological age of the several systems of rocks in central India and Bengal: Geol. Survey India Mem., v. 2, p. 299-341.

Clarke, W.B., 1861, On the relative positions of certain plants in the coal-bearing beds of Australia: Geol. Soc. London Quart. Jour., v. 17, p. 354-362.

Oldham, Thomas, 1861, Additional remarks on the geological relations and probable geological age of the several systems of rocks in central India and Bengal: Geol. Survey India Mem., v. 3, p. 197-213.

Blanford, H.F., 1875, On the age and correlations of the plant-bearing series of India, and the former existence of an Indo-Oceanic continent: Geol. Soc. London Quart. Jour., v. 31, p. 519-542.

Blanford, W.T., 1876, Note on the geological age of certain groups comprised in the Gondwana series of India, etc.: Geol. Survey India Record, v. 9, pt. 3, p. 79-86.

Clarke, W.B., 1878, Remarks on the sedimentary formations of New South Wales (4th ed.): Syndey, T. Richards, Govt. Printer, 165 p.

Blanford, W.T., 1884, (Presidential address to Section C (Geology)): British Assoc. Adv. Sci. Rept., 54th Ann. Mtg., p. 691-711. (Also in Geol. Survey India Record, v. 18, pt. 1, 1885)

___ 1886, On additional evidence of the occurrence of glacial conditions in the Palaeozoic Era, and on the geological age of the beds containing plants of Mesozoic type in India and Australia: Geol. Soc. London Quart. Jour., v. 42, p. 249-260.

Oldham, R.D., 1886, Memorandum on the correlation of the Indian and Australian coal-bearing beds: Geol. Survey India Record, v. 19, pt. 1, p. 39-47.

1849

14 J.D. Dana concluded that there has been a regular progression in the extinction age of the Hawaiian chain volcanoes, with the northwesternmost dying out first, followed by those further southeast.

Dana, J.D., 1849, Geology: v. 10 of U.S. Exploring Expedition 1838-1842, New York, Putnam, 756 p.

1849 - 1861

15 M.F. Maury conducted a systematic program of measuring the depths of the oceans. An accurate, simple and inexpensive method of depth determination was developed, many U.S. Navy ships were equipped with the necessary equipment and the results were progressively incorporated into a long series of maps and charts issued by the Navy's Depot of Charts and Instruments. Maury discovered the Mid-Atlantic Ridge (variously referred to as the Telegraph Plateau, Middle Ground, and the Dolphin Rise) and showed it on his bathymetric maps of the Atlantic Ocean. He also tested a sonic method of depth determination where an explosion was set off at shallow depth and observers listened for the return of the sound wave reflected off the bottom. However, the return could not be heard with the unaided human ear so no results were obtained. One of Maury's assistants, Midshipman J.M. Brooke, designed a modification of the basic sounding weights (cannon balls) which returned many samples of sediment from great depths.

Maury, M.F., 1855, The physical geography of the sea: New York, Harper and Brothers, 287 p.

Williams, F.L., 1963, Matthew Fontaine Maury, Scientist of the sea: Brunswick, New Jersey, Rutgers Univ. Press, 720 p.

1853

16 M. Melloni showed by experiment that cooling rocks of certain kinds acquire a magnetic alignment parallel to that of the earth at the time they cooled.

Melloni, Macedonio, 1853, Sur l'aimentation des roches volcaniques: Acad. Sci. Comptes Rendus, v. 37, p. 229-231.

1855

17 G.B. Airy hypothesized that the mass deficiency of mountains is due to a downward extension of part of the earth's low-density crust into the more dense mantle.

Airy, G.B., 1855, On the computation of the effect of attraction of mountain masses, as disturbing the apparent astronomical latitude of stations in geodetic surveys: Royal Soc. London Philos. Trans., v. B145, p. 101-104.

18 J.H. Pratt calculated that the Himalaya Mountains must have a substantial mass deficiency at depth.

Pratt, J.H., 1855, On the attraction of the Himalaya Mountains and of the elevated regions beyond them upon the plumb-line in India: Royal Soc. London Philos. Trans., v. B145, p. 53-100.

1856

19 J.M Safford reported that the Appalachian structure of eastern Tennessee is comprised of eight large overthrust fault blocks which are piled upon one another like "thick slates or tiles on a roof."

Safford, J.M., 1856, A geological reconnaissance of the state of Tennessee: Nashville, Mercer, 164 p.

1857

20 In his presidential address to the American Association for the Advancement of Science, J. Hall reported that his studies of Paleozoic rocks in New York and adjacent states led him to conclude that 12 km of generally shallow-water sediments had been deposited in an elongate, shallow basin in that area and that the weight of the accumulating sediments had caused the gradual subsidence. The sediments were said to have been folded during subsidence and the deeper ones metamorphosed and intruded by igneous rocks. Eventually, the rocks were uplifted to form the modern Appalachians while the source area to the east sank beneath the sea.

Hall, James, 1883, Contributions to the geological history of the American continent: Am. Assoc. Adv. Sci. Proc., 31st mtg., p. 29-69.

1858

21 Antonio Snider suggested that at the time of the Biblical deluge the continents separated from each other and took their present positions as a result of an outpouring of material from within the earth. As evidence he suggested that the continents (particularly those bordering the Atlantic Ocean) fit together like the pieces of a jigsaw puzzle.

Snider, Antonio, 1858, La creation et ses mysteres devoiles: Paris, A. Franck and E. Dentu, 487 p.

1859

22 W.T. Blanford, H.F. Blanford and W. Theobald described evidence from the lower part of the Gondwana Group for glaciation of India during the late Paleozoic.

Blanford, W.T., Blanford, H.F., and Theobald, W., 1859, The geological structure and relations of the Talchir coal-field in the district of Cuttach: Geol. Survey India Mem., v. 1, p. 33-88.

23 R.N.Rubidge and R. Tate reported the discovery of Glossopteris and associated plant fossils in the Karoo beds of South Africa and directed attention to the close similarity of the South African flora to fossil floras of Australia and India.

Rubidge, R.N., 1859, On some points in the geology of South Africa: Geol. Soc. London Quart. Jour., v. 15, p. 195-198.

Tate, Ralph, 1867, On some secondary fossils from South Africa: Geol. Soc. London Quart. Jour., v. 23, p. 139-175.

1859 - 1888

24 R. Daintree, T.W.E. David, E.J. Dunn, R.L. Jack, R.D. Oldham, T. Oldham, A.R.C. Selwyn and C.S. Wilkinson discovered, described and conclusively established the existence of late Paleozoic glacial beds associated with a Glossopteris - Gangamopteris flora in the provinces of New South Wales, Victoria and Queensland, Australia. The great similarity of the rock types and fossils with those of India and South Africa was apparent to these workers since several of them had worked in South Africa or India as well as Australia.

Selwyn, A.R.C., 1859, Geological notes of a journey in South Australia from Cape Jervis to Mount Serle: Adelaide Parliamentary Papers, South Australia 20, 4 p.

Oldham, Thomas, 1861, Additional remarks on the geological relations and probable geological age of the several systems of rocks in central India and Bengal: Geol. Survey India Mem. 3, p. 197-213.

Daintree, R., 1866, Report on the geology of the District of Ballan: Melbourne, folio, J. Ferres, Govt. Printer.

Selwyn, A.R.C., 1866-67, Notes on the physical geography, geology and mineralogy of Victoria: Melbourne.

Jack, R.L., 1879, Report on the Bowen River coal-field: Brisbane, 44 p. + map.

Wilkinson, C.S., 1879, Notes on the occurrence of remarkable boulders in the Hawksbury rocks: Royal Soc. New South Wales Jour., v. 13, p. 105-107.

Oldham, R.D., 1886, Memorandum on the correlation of the Indian and Australian coal-bearing beds: Geol. Survey India Record, v. 19, p. 39-47.

David, T.W.E., 1887, Evidence of glacial action in the Carboniferous and Hawksbury series, New South Wales: Geol. Soc. London Quart. Jour., v. 43, p. 190-195.

Dunn, E.J., 1888, Notes on the occurrence of glaciated pebbles and boulders in the so-called Mesozoic conglomerate of Victoria: Royal Soc. Victoria Trans. and Proc., v. 24, p. 44-46.

1860

25 J. Marcou prepared the first world paleogeographic map. The map showed the earth during Jurassic time and depicted an immense "Americo-Africo-Australie" continent which included eastern North America , South America, Africa, parts of Europe, Arabia, India, the East Indies and Australia. The distribution of fossils, particularly ammonites, was the basis for the paleogeography.

Marcou, Jules, 1857-1860, Lettres sur les roches du Jura et leur distribution geographique dan les deux hemispheres: Paris, F. Klincksieck, 364 p.

1861

26 C.J.F. Bunbury described a collection of fossil plants from central India that included several species of Glossopteris and called attention to its strikingly close similarity to that of the "coal-formation of Australia."

Bunbury, C.J.F., 1861, Notes on a collection of fossil plants from Nagpur, central India: Geol. Soc. London Quart. Jour., v. 17, p. 325-346.

27 W.E. Logan reported that the Appalachian front in Canada consists of a giant overthrust belt where older rocks have been thrust northwestward over younger rocks. His line showing the western boundary of the fault is known as "Logan's Line."

Logan, W.E., 1861, Remarks on the fauna of the Quebec group of rocks and the primordial zone of Canada (in a letter addressed to Mr. Joachim Barrande of Paris): Amer. Jour. Sci., v. 81, p. 216-220.

___ 1862, Considerations relating to the Quebec group and the upper copper-bearing rocks of Lake Superior: Amer. Jour. Sci., v. 83, p. 320-327.

28 J. Nicol reported finding overthrust faults in the Northwest Highlands of Scotland and compared them to similar structures in the Alps. The article started a long controversy among British geologists, a controversy that was effectively settled by the work of C. Callaway, C.T. Clough, W. Gunn, L.W. Hinxman, J. Horne, C. Lapworth and B.M. Peach which fully confirmed Nicol's finding.

Nicol, James, 1861, On the structure of the Northwestern Highlands and the relations of the gneiss, red sandstone and quartzite of Sutherland and Ross-Shire: Geol. Soc. London Quart. Jour., v. 17, p. 85-113.

Callaway, Charles, 1883, The age of the newer gneissic rocks of the northern Highlands: Geol. Soc. London Quart. Jour., v. 39, p. 355-414. (Discussion: p. 414-422)

___ 1883, The Highland problem: Geol. Mag., new ser., dec. 2, v. 10, p. 139-140.

Lapworth, Charles, 1883, The secret of the Highlands: Geol. Mag., new ser., dec. 2, v. 10, p. 120-128, 193-199, 337-344.

Peach, B.N., Horne, John, Gunn, W., Clough, C.T., and Hinxman, L.W., 1907, The geological structure of the Northwest Highlands of Scotland; with petrological chapters and notes by J.J.H. Teal, edited by Sir Archibald Geikie, Geol. Survey Great Britain Mem., 668 p.

29 J.H. Pratt hypothesized that the mass deficiency of mountains is due to density differences within the crust.

Pratt, J.H., 1861, On the Indian arc of Meridian: Royal Soc. London Philos. Trans., v. B151, p. 579-594.

1864

30 P.L. Sclater concluded from his studies of mammalian distribution that Africa, South America, India and Madagascar had formerly been united in one large continent for which he coined the name "Lemuria."

Sclater, P.L., 1864, The mammals of Madagascar: Quart. Jour. Sci., v. 1, p. 213-219.

1865

31 P. Gervais described a small aquatic-reptile fossil that had been found more than 30 years previously in Griqualand, South Africa, and named it Mesosaurus Tenuidens.

Gervais, Paul, 1867-1876, Zoologie et paleontologie generales, Ser. I: Paris, A. Bertrand. (See p. 223-228 and plate 42.)

1868

32 P.C. Sutherland described evidence from the Dwyka Series (the lowermost part of the Karoo Group) for a late Paleozoic glacial epoch in southern Africa.

Sutherland, P.C., 1868, On the geology of Natal: Pietermaritzburg.
___ 1870, Notes on an ancient boulder clay of Natal: Geol. Soc. London Quart. Jour., v. 26, p. 514-517.

1869

33 L.M.H. Lartet suggested on the basis of his field studies that the Jordan Valley - Dead Sea fault has undergone 160 km of strike slip.

Lartet, L.M.H., 1869, Essai sur la geologie de la Palestine . . . : Paris, V. Masson et fils, 292 p.

1869 - 1895

34 W. Carruthers described three new species of fossil plants from Rio Grande do Sul Province in Brazil. The plants were said to belong to genera which occur in the British coal measures, thus establishing their Paleozoic age. In 1895 R. Zeiller correctly identified some of the specimens as genera (including Gangamopteris) which are part of the Glossopteris flora of India.

Carruthers, W. 1869, On the plant remains from the Brazilian coal beds, with remarks on the genus Flemingites: Geol. Magazine, v. 6, p. 151-156.
Zeiller, R. 1895, Sur quelques empreintes vegetales des gisements houillers du Bresil meridional: Competes Rendus Acad. Sci. Paris, v. 121, p. 961-964.
___ 1895, Note sur la flore fossile des gisements houillers de Rio Grande do Sul: Soc. Geol. France Bull., Ser. 3, v. 23, p. 601-629.

1870 - 1888

35 W.T. Blanford, F. Fedden, R.B. Foote, C.I. Griesbach, R.D. Oldham, T. Oldham, W. Waagen, H. Warth and A. B. Wynne conclusively established the glacial origin of the Talchir boulder beds of India by finding numerous faceted and striated boulders in the unit and by showing that the boulder bed rests in places on a striated pavement of Vindhyan limestone.

Foote, R.B., 1870, Notes on the geology of the neighbourhood of Madras: Geol. Survey India Record 3, p. 11-17.
Blanford, W.T., 1872, Description of the geology of Nagpur and its neighbourhood: Geol. Survey India Mem. 9, 36 p.
Oldham, Thomas, 1872, (note): in Blanford, W.T., Description of the geology of Nagpur and its heighbourhood: Geol. Survey India Mem. 9, p. 10, 28-31.
Fedden, F., 1875, On the evidence of "ground ice" in tropical India during the Talchir period: Geol. Survey India Record 7, p. 16-18.
Griesbach, C.L., 1880, Geology of the Ramkola and Tatapani coal-fields: Geol. Survey India Mem. 15, Art. 2, 64 p.
___ 1880, Geological notes: Geological Survey of India Record 8, p. 83-94.
Wynne, A.B., 1881, Travelled blocks of the Punjab: Geol. Survey India Record 14, p. 153-154.
Oldham, R.D., 1884, Some rough notes for the construction of a chapter of the history of the earth: Asiatic Soc. Bengal Jour., v. 53, pt. 2, no. 3, p. 187-198.
Blanford, W.T., 1886, Notes on a smoothed and striated boulder from a pre-Tertiary deposit in the Punjab Salt Range: Geol. Mag., v. 3, p. 494-495.
___ 1886, On additional evidence of the occurrence of glacial conditions in the Paleozoic era and on the geological age of the beds containing plants of Mesozoic type in India and Australia: Geol. Soc. London Quart. Jour., v. 42, p. 249-260.

Wynne, A.B., 1886, On a striated and faceted fragment from Chel Hill Olive conglomerate, Salt Range, Punjab: Geol. Mag., v. 3, p. 492-494.
Warth, H. 1887, On the identity of the Olive Series in the east with the Speckled Sandstone in the west of the Salt Range in the Punjab: Geol. Survey India Record 20, pt. 2, p. 117-119.
Waagen, W., 1888, The Carboniferous ice age: Geol. Survey India Record 21, pt. 3, p. 89-130.
Warth, H., 1888, A facetted pebble from the Boulder Bed (Speckled Sandstone) of Mount Chel in the Salt Range in the Punjab: Geol. Survey India Record 21, pt. 1, p. 34-35.

1871

36 G.W. Stow directed attention to many similarities among fossil plants and animals from southern Africa, India and Australia and concluded that those lands must have been part of an enormous continent, portions of which subsequently sank beneath the sea.

Stow, G.W., 1871, On some points in South-African geology: Geol. Soc. London Quart. Jour., v. 27, p. 497-548. (See: "On the probable existence of an ancient southern continent", p. 546-548.)

1872

37 E.J. Dunn firmly established the glacial origin of the Dwyka (Ecca) boulder beds of South Africa by discovering striated, grooved and faceted boulders in the beds near the Orange River.

Dunn, E.J., 1888, Notes on the occurrence of glaciated pebbles and boulders in the so-called Mesozoic conglomerate of Victoria: Royal Soc. Victoria Trans. & Proc., v. 24, p. 44-46.

38 H.B. Medlicott coined the term "Gondwana (meaning land of the Gonds) system" for a set of rock units that are characteristic of central India. The name was first published by O. Feistmantel in 1876. The stratigraphy and paleontology of the Gondwana system were worked out by Medlicott and Feistmantel and by W.T. Blanford and T. Oldham who determined that it is largely non-marine and that it ranges in age from late Carboniferous to at least Jurassic. It was also realized that extremely similar sequences of rocks occur in Australia and South Africa.

Feistmantel, Ottokar, 1876, Notes on the age of some fossil floras in India: Geol. Survey India Record, v. 9, pt. 6, p. 115-144.
Feistmantel, Ottokar, Oldham, T., and Morris, J., The fossil flora of the Gondwana system: Paleontologica Indica, Series II, XI, and XII, v. 1-4.
Medlicott, H.B., Blanford, W.T., and Oldham, R.D., 1893, A manual of the geology of India, 2nd ed.: Calcutta, Govt. of India, 543 p.
Fox, C.S., 1931, The Gondwana system and related formations: Geol. Survey India Mem., v. 58, 241 p.

38A E. Reclus postulated that the continents "are slowly traveling around the circle of the globe."

Reclus, Elisee, 1872, The earth: New York, Harper and Brothers, 567 p. (English translation by B. Woodward).
Berkland, J.O., 1979, Elisee Reclus - neglected geologic pioneer and first(?) continental drift advocate: Geology, v. 7, p. 189-192.

1873

39 J.D. Dana coined the term "geosynclinal" (later almost always shortened to "geosyncline") for the elongate basins that were filled with great thicknesses of shallow-water sediments and that were later uplifted to form fold-mountain belts.

Dana, J.D., 1873, On some results of the earth's contraction from cooling, including a discussion of the origin of mountains, and the nature of the earth's interior: Am. Jour. Sci., 3rd ser., v. 5, p. 423-443, 474-475; v. 6, p. 6-14, 104-115, 161-172, 304, 381-382.

1873 - 1876

40 The British research vessel Challenger began the era of modern oceanographic exploration by making a four-year, around the world voyage for the purpose of investigating the world ocean. The vessel sampled and mapped the distribution of deep-sea sediments, sounded several trenches, obtained the deepest sounding made up to that time (in the Marianas Trench), and used soundings and deep-water temperatures to demonstrate that the Mid-Atlantic Ridge bisects the Atlantic Ocean, is 2,000 m high and is continuous from about 50 degrees N to about 40 degrees S latitude.

Thomson, C.W., and Murray, John, 1885, Report of the scientific results of the voyage of H.M.S. Challenger during the years 1873-1876: v. 1, pt. 1, Narrative, printed and sold for Her Majesty's Stationery Office by publishers in London, Edinburgh, and Dublin, 509 p. (Reprinted in 1975 by Johnson Reprint Corp.)

Thomson, C.W., 1895, Report of the scientific results of the voyage of H.M.S. Challenger 1873-1876: A summary of the scientific results, pt. 1, printed and sold for Her Majesty's Stationery Office by various publishers in London, Edinburgh, Glasgow and Dublin, 796 p.. (Reprinted in 1975 by Johnson Reprint Corp.)

___ 1878, The voyage of the Challenger, The Atlantic, a preliminary account of the general results: 2 vols., New York, Harper & Bros., 391 & 340 p.

1874

41 H.F. Blanford concluded on the basis of a variety of evidence that (1) the Gondwana Series of India ranges in age from early Permian to latest Jurassic, (2) the Gondwana glaciation occurred in early Permian time, (3) India, South Africa and Australia were part of an immense "Indo-Oceanic continent" in the Permian and the first two remained connected until the end of the Miocene, and (4) the Laccadives, Maldives and Chagos Islands are remnants of this continent.

Blanford, H.F., 1875, On the age and correlations of the plant-bearing series of India and the former existence of an Indo-Oceanic continent: Geol. Soc. London Quart. Jour., v. 31, p. 519-542.

42 The Japan trench was discovered by a sounding taken from the USS Tuscarora.

Anonymous, 1874, Deep-sea soundings in the Pacific Ocean: Nature, v. 10, p. 356-357.

1875 - 1935

43 European geologists, led by E. Argand, S. von Bubnoff, M. Bertrand, L.W. Collet, E. Haug, F. Heritsch, A. Heim, L. Kober, M. Lugeon, H. Schardt, R. Staub, G. Steinmann, E. Seuss, P. Termier and W. von Seidlitz, worked out the geology of the Alps and associated ranges. Chief among the structural features they described were the great overthrusts

and perhaps hundreds of kilometers of crustal shortening brought about, they said, by a "push" of Africa against Europe.

Suess, Eduard, 1875, Die Entstehung der Alpen: Vienna, W. Braumuller, 168 p.
Bertrand, Marcel, 1884, Rapports de structure des Alpes de Glarus et du bassin houiller du Nord: Soc. Geol. France Bull., 3rd ser., v. 12, p. 318-330.
Schardt, Hans, 1893, Sur l'origine des Alpes du Chablais et du Stockhorn, en Savoie et en Suisse: Acad. Sci. Comptes Rendus, v. 117, p. 707-709.
___ 1893, Sur l'origine des Prealpes Romandes: Archives Sci., 3rd ser., v. 30, p. 570-583.
Bertrand, Marcel, 1897, Structure des Alpes Francaises et recurrence de certains facies sedimentairies: Comptes Rendus, Cong. Geol. Internat., pt. 3, p. 163-177.
Schardt, Hans, 1898, Les region exotiques du versant Nord des Alpes Suisses (Prealpes du Chablais et du Stockhorn et les klippes): Soc. Vaudoise Sci. Nat. Bull., v. 34, no 128, p. 113-219.
Lugeon, Maurice, 1901-1902, Les grandes nappes de recouvrement des Alpes du Chablais et de la Suisse: Soc. Geol. France Bull., 4th Ser., v. 1, 723 p.
Termier, Pierre, 1903, Les nappes des Alpes orientales et la synthese des Alpes: Soc. Geol. France Bull., 4th ser., v. 3, p. 711-766.
Steinmann, Gustav, 1906, Geologische Probleme des Alpengebirges: Zeitschrift Deutsch. - Ost. Alpenver, 3f bd.
___ 1907, Alpen und Apennin: Monatsber. Deut. Ges., v. 59, p. 117-185.
Heim, Albert, 1908, Der bau der Schweizeralpen: Zurich, Zuercher & Furrer, 26 p.
Argand, Emile, 1916, Sur l'arc des Alpes Occidentales: Eclogae Geol. Helvetiae, v. 14, p. 145-191.
von Bubnoff, Serge, 1921, Die Grundlagen der Deckentheorie in den Alpen: Stuttgart, E. Schweizerbart'sche Verlagsbuchandlung, 149 p.

Heim, Albert, 1921, Geologie der Schweiz, Band 2, Die Schweizer Alpen: Leipzig, Tauchnitz, 1018 p.
Kober, Leopold, 1923, Bau und Entstehung der Alpen: Berlin, Borntraeger, 283 p.
Argand, Emile, 1924, La tectonique de l'Asie: Comptes Rendus, Cong. Geol. Internat. XIII, v. 1, p. 171-372.
Staub, Rudolf, 1924, Der Bau der Alpen: Beitr. Geol. Karte Schweiz, Neue Folge, Leif. 52, 272 p.
Haug, Emile, 1925, Contribution a une synthese stratigraphique des Alpes Occidentales: Soc. Geol. France Bull., 4th ser., v. 25, p. 97-244.
Collet, L.W., 1926, The Alps and Wegener's theory: Geog. Jour., April, p. 301-312.
von Seidlitz, W., 1926, Entstehen und Vergehen der Alpen: Stuttgart, F. Enke, 267 p.
Heritsch, Franz, 1927, Die Deckentheorie in den Alpen: Berlin, Gebruder Borntraeger, 210 p.
Kober, Leopold, 1931, Das Alpine Europa und sein Rahmen - ein geologisches Gestaltungsbild: Berlin, Borntraeger, 310 p.
Collet, L.W., 1935, The structure of the Alps (2nd ed.): London, E. Arnold, 304 p. (Reprinted in 1974 by R.E. Krieger Pub. Co., New York)

1879

44 G.H. Darwin hypothesized that the moon originated by fissioning from the earth as a result of the interaction of rotation and the tidal effects of the sun.

Darwin, G.H., 1879, On the precession of a viscous spheroid, and on the remote history of the earth: Royal Soc. London Philos. Trans., v. 170, pt. 2, p. 447-538.

1881

45 C. Darwin suggested that a test of his reef theory could be made by drilling several hundred feet down into a coral atoll and extracting cores.

Darwin, Charles, 1881, Letter to Alexander Agassiz dated 5 May: in Darwin, Francis, ed., The life and letters of Charles Darwin including an autobiographical chapter, New York, Basic Books, 1959, v. 2.

1882

46 O. Fisher suggested that separation of the continents was caused by flow towards a 65 km-deep scar (the Pacific basin) left by the fissioning of the moon from the earth.

Fisher, Osmond, 1882, On the physical cause of the ocean basins: Nature, v. 25, p. 243-244.

___ 1889, Physics of the earth's crust (2nd ed.): London, Macmillan.

1883

47 J.D. Everett summarized the work of a British Committe on terrestrial heat flow which had determined an average value of .054 watts per square metre, a value which is in very close agreement with modern estimates of continental heat flow. The work was flawed, however, by the fact that estimated rock conductivities were used rather than measured values from the rocks whose temperatures were measured.

Everett, J.D., 1883, British Association for the Advancement of Science Report for 1882: p. 72-90.

48 A.E. Tornebohm concluded that parts of the Caledonian Mountains of Scandanavia are characterized by great overthrusts. By 1896 he had documented a record of at least 130 km of overthrusting.

Tornebohm, A.E., 1883, Om Dalformationens geologiska alder: Geol. Foren. Stockholm Forhandl., v. 6, p. 622-661.

___ 1888, Om Fjallproblemet: Geol. Foren. Stockholm Forhandl., v. 10, p. 328-336.

___ 1896, Grunddragen af det centrala Skandinaviens bergbyggnad: Kgl. Svenska Vetensk. Akad. Handl., v. 28 (5), 212 p.

1884

49 F. Galton noted that there is a great "depression or trough" that extends continuously from the Dead Sea at one end to Lake Tanganyika at the other.

Galton, F., 1884, (untitled comment on a report to the Royal Geog. Soc. by J. Thomson): Royal Geog. Soc. Proc. (now Geog. Jour.), v. 6, p. 711.

50 F.R. Helmert concluded that the mass of continents above sea-level is compensated at depth by a low-density crust that extends below sea level.

Helmert, F.R., 1884, Die Tiefe der Ausgleichsflache bei der Prattschen Hypothese fur das Gleichgewicht der Erdkruste und der Verlauf der Schwerestorung vom Innern der Kontinente und Ozeane nach den Kusten: Sitzungsberichte der Koniglich Preussischen Akademie der Wissenschaften, v. 48, p. 1192-1198.

1884 - 1896

51 W.T. Blanford delivered presidential addresses to the Geology Section of the British Association for the Advancement of Science (Montreal, 1884) and the Geological Society

of London (1890) and wrote several important papers in which he (1) directed attention to striking similarities in the geology of Australia, India, southern Africa and South America and speculated that these continents had once been connected, (2) attacked the theory of "permanence" of continents and ocean basins on the basis of a great variety of evidence, and (3) described detailed evidence for the Paleozoic age of the Glossopteris - Gangamopteris flora and the associated glaciation.

Blanford, W.T., 1884, (Presidential address to Section C (Geology)): British Assoc. Adv. Sci. Rept., 54th Ann. Mtg., p. 691-711.

___ 1886, On additional evidence of the occurrence of glacial conditions in the Paleozoic Era, and on the geological age of the beds containing plants of Mesozoic type in India and Australia: Geol. Soc. London Quart. Jour., v. 42, p. 249-260.

___ 1890, The permanence of ocean basins (Anniversary address of the President): Geol. Soc. London Quart. Jour. Proc., v. 46, p. 59-109.

___ 1896, On the papers by Dr. Kossmat and Dr. Kurtz, and on the ancient geography of "Gondwana-land": India Geol. Survey Record 29, p. 52-59.

1885 - 1909

52 E. Suess coined the name "Gondwanaland" for an immense southern land mass which he conceived on the basis of fossil evidence. Large portions of this super-continent were said to have subsided to form parts of the modern ocean basins, leaving isolated remnants behind as modern Australia, South America, Africa, Madagascar and India. Suess conceived the idea that the continents consist of less dense, granitic SAL (later modified by Wegener to SIAL) and that oceanic crust is composed of denser, basaltic SIMA. He also suggested that the African rift valley system formed in response to tensional stresses in the crust.

Suess, Eduard, 1885-1909, Das Antlitz der Erde: Prague and Vienna, F. Tempky, 5 volumes totalling 2936 p.

___ 1904-1924, The face of the earth (translated by H.B. Sollas): Oxford, Clarendon Press, 5 volumes totalling 2403 p.

___ 1891, Die Bruche des ostlichen Afrika: Denk. K. Akad. Wiss., Wien (Math. Nat.), v. 58, p. 555-584.

1886

53 B. Lotti described an association of rocks from the Island of Elba which consists of serpentenized ultramafic rocks overlain by argillite, pillow lavas, gabbro and chert.

Lotti, B., 1886, Descrizione geologica dell'Isola d'Elba: Mem. Descrit. Carta Geol. Ital., v. 2.

1886 - 1922

54 E. Blackwelder, M.R. Campbell, C.L. Dake, R.A. Daly, C.R. Longwell, G.R. Mansfield, R.G. McConnell, R.W. Richards, A.R. Schultz, A.C. Veatch, and B. Willis discovered and mapped the great overthrust sheets of the western United States and Canada, including the McConnell, Lewis, Absaroka, Crawford, Muddy Mountain, Willard, Bannock, Darby and Hart Mountain faults.

McConnell, R.G., 1887, Report on the geological structure of a portion of the Rocky Mountains, accompanied by a section measured near the 51st parallel: Geol. Survey Canada Annual Rept., 1886, v. 2, Rept. D, p. 5-41.

Willis, Bailey, 1902, Stratigraphy and structure, Lewis and Livingston ranges, Montana: Geol. Soc. America Bull., v 13, p. 305-352.

Veatch, A.C., 1907, Geography and geology of a portion of southwestern Wyoming: U.S. Geol. Survey Prof. Paper 56, 178 p.
Blackwelder, Eliot, 1910, New light on the geology of the Wasatch Mountains, Utah: Geol. Soc. America Bull., v. 21, p. 517-542.
Daly, R.A., 1912, Geology of the North American Cordillera at the 49th parallel: Geol. Survey Canada Mem. 38, pt. 1, 546 p.
Richards, R.W., and Mansfield, G.R., 1912, The Bannock overthrust, a major fault in southeastern Idaho and northeastern Utah: Jour. Geology, v. 20, p. 681-709.
Campbell, M.R., 1914, The Glacier National Park: a popular guide to its geology and scenery: U.S. Geol. Survey Bull. 600, 54 p.
Schultz, A.R., 1914, Geology and geography of a portion of Lincoln County, Wyoming: U.S. Geol. Survey Bull. 543, 141 p.
Dake, C.L., 1918, The Hart Mountain overthrust and associated structures in Park County, Wyoming: Jour. Geology, v. 26, p. 45-55.
Longwell, C.R., 1922, The Muddy Mountain overthrust in southeastern Nevada: Jour. Geology, v. 30, p. 63-72.

1887

55 M. Bertrand called attention to similarities between the geosynclinal - folded belts of eastern North America and a belt of similar age in northwestern Europe and noted their truncation by the Atlantic Ocean.

Bertrand, Marcel, 1886-1887, La chaine des Alpes et la formation du continent European: Soc. Geol. France Bull., 3e Ser., v. 15, p. 442.

56 E. Rudolph reported that the Mid-Atlantic Ridge is seismically active.

Rudolph, E. 1887, Ueber submarine Erdbeben und Eruptionen: Gerlands Beitr. Geophysik, v. 1, p. 133-373.

57 C.B. Warring hypothesized that, early in the history of the earth, all of the land masses had been clumped together into one great triangular continent. This great mass "moved slowly westward, we may think of it as a great cake of ice floating on a river of very moderate velocity . . ." Subsequently, it broke into parts which then became the modern continents. As evidence for his theory he cited the "remarkable parallelism between the shores of the Atlantic" along its full length.

Warring, C.B., 1887, The evolution of continents: Trans. Vassar Brothers Inst., v. 4, p. 256-274.

1888

58 O.A. Derby concluded from evidence found near Sao Paulo, Brazil that South America experienced a late Paleozoic ice age.

Derby, O.A., 1888, Uber Spuren einer carbonen Eizeit in Sudamerica: Neues Jahrb. Mineral. Geol. Palaontol., v. 2, p. 172-176.

1888 - 1973

59 O. Ampferer, V.V. Beloussov, W.H. Bucher, L.U. De Sitter, P.J.C. De Wijkerslooth de Weerdesteijn, E. Haarmann, H.J. McCunn, E. Reyer, and R.W. van Bemmelen developed general tectonic hypotheses based on the idea that vertical movements of the earth's crust are the most important kind of movement and that such limited horizontal movements as do take place are caused by primary vertical movements.

Reyer, Eduard, 1888, Theoretische Geologie: Stuttgart, Schweizerbart (E. Koch), 867 p.
___ 1892, Geologische und geographische Experimente, Heft I: Deformation und Gebirgsbildung: Leipzig, Engelmann, 52 p.
___ 1894, Geologische und geographische Experimente, Heft III: Rupturen, Heft IV: Methoden und Apparate: Leipzig, Engelmann, 32 p.
Ampferer, O., 1906, Uber das Bewegungsbildung von Faltengebirgen: J.d. K.K. Geol. Reichanstalt (Wien), bd. 56.
Haarmann, Erich, 1930, Die Oszillationstheorie: Stuttgart, F. Enke, 260 p.
van Bemmelen, R.W., 1932, Die undatie Theorie: Natuurk Tijdschr. Nederl. Indie, v. 92, p 85-242, 373-402.
___ 1933, The undation theory of development of the earth's crust: Internat. Geol. Congr., Washington, v. 2.
De Wijkerslooth de Weerdesteijn, P.J.C., 1934, Bau und Entwicklung des Apennins, besonders der Gebirge Toscanas: Amsterdam, Geologisch Inst., 426 p.
Bucher, W.H., 1956, Role of gravity in orogenesis: Geol. Soc. America Bull., v. 67, p. 1295-1318.
de Sitter, L.U., 1956, Structural geology: New York, McGraw-Hill, 552 p.
Beloussov, V.V., 1962, Basic problems in geotectonics: New York, McGraw-Hill, 816 p.
van Bemmelen, R.W., 1966, On mega-undations: a new model for the earth's evolution: Tectonophysics, v. 3, p. 83-127.
McCunn, H.J., 1973, Vertical uplift explanation for plate tectonics: Amer. Assoc. Petroleum Geologists Bull., v. 57, p. 1644-1657.

1889

60 C.E. Dutton rejected the contraction hypothesis as a cause of folded mountains because contraction would lead to compressive strain that "would act in every direction equally" and "could not form long, narrow belts of parallel folds." The term "isostasy" was proposed "for this condition of equilibrium of figure, to which gravitation tends to reduce a planetary body."

Dutton, C.E., 1889, On some of the greater problems of physical geology: Washington Philos. Soc. Bull., v. 11, p. 51-64.

61 R. Mantovani called attention to various geologic similarities along the coastlines of continents bordering the Indian Ocean and suggested that those continents had been torn apart and scattered to their present locations.

Mantovani, R., 1889-1890, Les fractures de l'ecorse terrestre et la theorie de Laplace: Soc. Arts et Lettres de la Reunion (St. Denis) Bull., p. 41-53.

62 C.D. White described the discovery of evidence for a late Paleozoic ice age in southern Africa, India, Australia and South America, discussed the problematic fossil evidence for the age of the glaciation, and concluded that those areas probably were united in an enormous continent, though the evidence from South America was yet fragmentary and inconclusive.

White, C.D., 1889, Carboniferous glaciation in the southern and eastern hemispheres - with some notes on the Glossopteris flora: American Geologist, v. 3, p. 299-330.

1891 - 1896

63 G. Bodenbender, A. Hettner, F. Kurtz and R. Zeiller reported the discovery in Brazil and Argentina of fossil plants of the Glossopteris flora.

Hettner, A., 1891, Das sudlichste Brasilien (Rio Grande do Sul): Zeitschrift Gesell. Erdkunde Berlin, v. 26, p. 85-144.
Kurtz, F., 1894, Sobre la existencia del Gondwana inferior en la Republica Argentina: Revista del Museo de la Plata, v. 6, p. 125-139 + plates.
Bodenbender, G., 1895, Sobre la edad de algunas formaciones Carboniferas de le Replica Argentina: Revista del Museo de la Plata, v. 7, p. 129-148.
Zeiller, R., 1895, Sur quelques empreintes vegetales des gisements houillers du Bresil meridional: Acad. Sci. Paris Comptes Rendus, v. 121, p. 961-964.
___ 1896, Note sur la flore fossile des gisements houillers de Rio Grande do sul (Bresil meridional): Soc. Geol. France Bull., v. 23, p. 601-629 + plates.
___ 1896, Remarques sur la flore fossile de l'altai a proppos des dernieres decouvertes paleobotaniques de mm. les Drs. Bodenbender et Kurtz dans la Republique Argentine: Soc. Geol. France Bull., v. 24, p. 466-487.

1892

64 E. von Rebeur-Paschwitz made the first verified recording of a distant earthquake by detecting, at Potsdam, Germany, waves later determined to have originated at an epicenter in Japan.

Rebeur-Paschwitz, E. von, 1894, Horizontalpendel-Beobachtungen auf der Kaiserlichen Universitats-Sternwarte zu Strassburg 1892-1894: Gerlands Beitr. Geophysik, v. 2, p. 211-535.
___ 1895, Europaische Beobachtungen des grossen japanischen Erdbebens von 22 Marz 1894, und des venezuelanischen Erdbebens von 28 April 1894, nebst Untersuehungen uber die Fortpflanzungsgeschwindigkeit dieser Erdbeben: Petermann's Metteilungen, v. 41, p. 13-21, 39-42.

1893

65 A.C. Lawson recognized the San Andreas fault as an important structural element in western California.

Lawson, A.C., 1893, The post-Pleistocene diastrophism of the coast of southern California: Calif. Univ. Dept. Geol. Sci. Bull., v. 1, no. 4, p. 115-160.
(See also)
Hill, M.L., 1981, San Andreas fault: history of concepts: Geol. Soc. America Bull., v. 92, part I, p. 112-131.

66 R.D. Oldham described overthrust faults in the Himalaya Mountains of India.

Oldham, R.D., 1893, A manual of the geology of India, Stratigraphic and structural geology (2nd ed.): Calcutta, Govt. of India, 543 p.

1895

67 M. Neumayr prepared a paleogeographic map of the world based on fossil affinities which showed South America and Africa connected along their full lengths (but not displaced) and also showed Africa, Madagascar and India connected by a land bridge.

Neumayr, Melchior, 1895, Beschreibende Geologie (2nd ed.): v. 2 of Erdgeschichte, Leipzig and Vienna, Bibliographisches Inst.

1895 - 1904

68 F. de Montessus de Ballore assembled a catalog of almost 157,000 earthquakes and showed that (1) earthquakes occur mostly (95%) in two narrow, world-girdling belts (the

Alpine-New Zealand and the circum-Pacific), (2) the belts form an angle of 67 degrees with each other, and (3) the belts coincide with the great Mesozoic geosynclines that underwent orogeny during the Cenozoic to form the world's young fold-mountain belts.

Montessus de Ballore, F. de, 1895, Relation entre la relief et la seismicite: Acad. Sci. Comptes Rendus, v. 120, p. 1183-1187.

___ 1900, Introduction a un essai de description seismique du globe et mesure de la seismicite: Gerlands Beitr. Geophysik, v. 4, p. 331-382.

___ 1903, Loi generale de la repartition des regions seismique instables a la surface du globe: Berichte d. II Internat. Konferenz zu Strassburg, Gerlands Beitr. Geophysik, Erganzungsband 2, p. 325-334.

___ 1903, Sur l'existence de deux grands cercles d'instabilite sismique maxima: Acad. Sci. Comptes Rendus, v. 136, p. 1707-1709.

___ 1904, Sur la coincidence entre les geosynclinaux et les grands cercles de sismicite maxima: Acad. Sci. Comptes Rendus, v. 139, p. 686-687.

1896

69 J.W. Gregory suggested the term "rift" valley for the system extending from the Jordan Valley to Mozambique and described evidence supporting its fault origin and continuity.

Gregory, J.W. 1896, The great rift valley: London, John Murray, 422 p.

1896 - 1898

70 The Royal Society of London sponsored the drilling of a test hole on the seaward margin of Funafuti Island (Ellice Islands) in the Pacific Ocean. The hole reached a depth of 340 m and penetrated shallow-water carbonates all the way to the bottom.

Coral Reef Committee of the Royal Society, 1904, The atoll of Funafuti, borings into a coral reef and the results: London, Royal Society, 428 p. + maps.

1897 - 1903

71 J. Milne plotted graphs of elapsed time vs. distance for "large" waves, "first phase" (P) waves and "second phase" (S) waves, developed the time-difference method for determining epicenter locations and pointed out that the linear plot of the "large waves" shows that they are surface waves.

Milne, John, 1897, The highest apparent velocities at which earthquake waves are propagated: Brit. Assoc. Adv. Sci. Rept., p. 172-176.

___ 1898, On the velocity of propagation of large waves: Brit. Assoc. Adv. Sci. Rept., p. 220-226.

___ 1900, Interval between the first tremor and the maximum motion: Brit. Assoc. Adv. Sci. Rept., p. 65-67.

___ 1902, Time curves for the earthquakes recorded during four years ending in 1900: Brit. Assoc. Adv. Sci. Rept., p. 65-67.

___ 1903, The velocity of propagation of earthquake vibrations: Brit. Assoc. Adv. Sci. Rept., p. 84-85.

1898

72 A.E. von Triulzi described a positive gravity anomaly along the Red Sea.

Triulzi, A.E. von, 1898, Expedition S.M. Schiff Pola in das Rothe Meer, Nordliche Halfte, II Relative Schwerebestimmungen: Akad. Wiss. Vienna, Math. Nat., v. 65, p. 131-206 + plates.

___ 1901, Expedition S.M. Schiff Pola in das Rothe Meer, Sudliche Halfte, XII Relative Schwerebestimmungen: Akad. Wiss., Vienna, Math. Nat., v. 69, p. 143-217.

1899

73 G. Folgerhaiter studied the magnetic inclination of human artifacts such as bricks and pottery and showed that the magnetization they acquired when fired is stable for periods at least as long as 2000 years.

Folgerhaiter, G., 1899, Sur le variations seculaires de l'inclinaison magnetique dans antiquite: Jour. Phys., 3rd ser., v. 8, p. 5-16.

1900 - 1927

74 E. Haug supported the concept of sunken continents in his classical paper on geosynclines and in his widely read "Traite." His world maps showed a "Continent Africano-Brasilien" and a "Continent Australo-Indo-Malagache," both of which were said to have existed into the Mesozoic.

Haug, Emile, 1900, Les geosynclinaux et les aires continentales: Soc. Geol. France Bull., v. 28, p. 617-711.

___ 1907-1927, Traite de Geologie: Paris, Colin, 3 vols., 1396 p. (see v. 2).

1901 - 1928

75 H.T. Ferrar collected fossil plant material from the Beacon System at a locality near Taylor Glacier, west of the Ross Sea while on the British National Antarctic Expedition. Early studies of the specimens turned up nothing identifiable but W.N. Edwards later split the rock specimens and found several fragments of Glossopteris, thus establishing their Permo-Carboniferous age and demonstrating that Antarctica was a Gondwana continent.

Ferrar, H.T., 1907, Report on the field geology of the region explored during the Discovery Antarctic Expedition 1901-1904: British National Antarctic Expedition Report 1901-1904, Nat. Hist., v. 1, p. 1-100.

Seward, A.C., 1914, Antarctic fossil plants: British Antarctic Terra Nova Expedition, Geology, v. 1, p. 1-49.

Edwards, W.N., 1928, The occurrence of Glossopteris in the Beacon sandstone of Ferrar Glacier, South Victoria Land: Geol. Magazine, v. 65, p. 323-327.

1901 - 1903

76 O. Hecker made numerous gravity measurements by simultaneously measuring air pressure in two different ways (by finding the boiling point of water and by using a mercury barometer) and determined that the strength of gravity at sea is not much different from the strength on land, thereby severely limiting the kinds of crustal models that are tenable. He also found a negative gravity anomaly over the Tonga Trench and a positive anomaly over the Tonga Plateau.

Hecker, O. 1903, Bestimmung der Schwerkraft auf dem Atlantischen Ocean: Veroff k. preuss. geodat. Inst. Berlin, new ser., no. 11, 137 p.

___ 1903, Bestimmung der Schwerkraft auf dem Indischen und Grossen Ocean: Veroff. k. preuss. geodat. Inst. Berlin, new ser., no. 12, 233 p.

1902

77 F.D. Frech, in his well-known "Lethaea," supported the Gondwana concept by showing an "Indo-Afrikanischer" continent that included Australia, India, Africa, South America and most of the intervening Indian and South Atlantic Oceans.

Frech, F.D., 1902, Lethaea geognostica, Theil I, Lethaea palaeozoica, Bd. 2, Lief 3-4: Stuttgart, Die Dyas, 354 p.

1904

78 E. von Drygalski concluded from sounding information that the equatorial Mid-Atlantic Ridge is cut by an east-west gap at what is now known as the Romanche fracture zone.

Drygalski, E. von, 1904, Zum Kontinent des eisgen sudens, Deutsche sud-polar Expedition fahrten und forschungen des Gauss 1901-1903: Berlin, George Reimer.

1904 - 1906

79 B. Brunhes and P. David discovered that the fossil magnetic field direction of certain lava flows and that of their underlying baked clays is roughly the reverse of the present field and proposed that the earth's field was reversed at the time those flows were extruded.

David, P., 1904, Sur la stabilite de la direction d'aimantation dans quelques roches volcaniques: Acad. Sci. Comptes Rendus, v. 138, p. 41-42.

Brunhes, B. 1906, Recherches sur la direction d'aimantation des roches volcaniques: Jour. Physique, 4e Ser., v. 5, p. 705-724.

1905

80 E.A.N. Arber prepared a comprehensive catalog of the Glossopteris flora of the world and summarized what was known of its age and geographic distribution. The concept (said to be "generally accepted") of a gigantic Gondwana continent, which was to some degree separated from a northern continent, was endorsed.

Arber, E.A.N., 1905, Catalogue of the fossil plants of the Glossopteris flora in the Department of Geology, British Museum (Natural History): London, Brit. Mus. (Nat. Hist.), 255 p.

1906

81 Dr. Alexander Agassiz used sounding data obtained from the United States Fish Commission steamer Albatross to discover the East Pacific Rise. The Rise was named the "Albatross Plateau" and was known by that name for many years.

Agassiz, Alexander, 1906, Reports on the scientific results of the expedition to the eastern tropical Pacific, v. 5 - General report of the expedition: Mem. Harvard Museum Comp. Zoology, v. 33, 75 p. + 91 plates.

82 G. Steinmann called attention to the association in many mountain belts of the world of pillow lavas, radiolarian cherts and serpentinized ultramafic rocks, a group which he referred to as "ophiolithischen Massengesteine" and which is now known as the "ophiolite suite" or the "Steinmann trinity."

Steinmann, Gustav, 1906, Geologische Beobachtungen in den Alpen, II Die Schardtsche Uberfaltungstheorie und die geologische Bedeutung der Tiefseeabsatze

und der ophiolithischen Massengesteine: Naturf. Gessel. Freiburg, bd. 16, p. 18-66.
___ 1926, Die ophiolithischen Zonen in den Mediterranen Kettengebirgen: Comptes Rendus, Cong. Geol. Internat. XIV (Madrid), v. 2, p. 637-667.

1907

83 W.H. Pickering suggested that when the moon separated from the earth, most of the earth's crust was carried away and "the remainder was torn in two to form the eastern and western continents. These then floated on the liquid surface like two large ice floes."

Pickering, W.H., 1907, The place of origin of the moon - the volcanic problem: Jour. Geology, v. 15, p. 23-38.
___ 1924, The separation of the continents by fission: Geol. Magazine, v. 61, p. 31-35.

84 H. von Ihering concluded that Africa and South America were once connected by a land bridge which he named "Arch-Helenis." In 1927 he postulated the existence during the Eocene of three land bridges crossing the Atlantic, a large bridge crossing the north Pacific, one connecting South America with Antarctica, and one between Madagascar and India.

Ihering, Hermann von, 1907, Archhelenis und Archinotis, Gessamelte Beitrage zur Geschichte der Neotropischen Region: Leipzig, W. Engelmann, 350 p.
___ 1927, Die Geschichte des Atlantischen Ozeans: Jena, G. Fisher, 237 p.
___ 1931, Land bridges across the Atlantic and Pacific Oceans during the Kainozoic Era: Geol. Soc. London Quart. Jour., v. 87, p. 376-391.

85 D. White concluded that "the occurrence of this (Gangamopteris) flora in great uniformity, including an estraordinarily high degree of specific identity, and in relative purity, contemporaneously in India, Australia, South Africa and southern South America, leaves no recourse but to conclude that the land surfaces over which it extended were in such continuity, or intimate geographical relation, as to freely permit the migration of the flora. . ."

White, David, 1907, Permo-Carboniferous climatic changes in South America: Jour. Geology, v. 15, p. 615-633.

1908

86 A.C. Lawson and others reported that upwards of 5 m of strike slip took place along hundreds of km of the San Andreas fault during the 1906 San Francisco earthquake. The San Andreas was identified as being almost 1,000 km long and was recognized as being highly significant in the seismicity of California. The report also described two sets of markers that had been set up at Olema (Marin County) and at Crystal Springs Lake (San Mateo County) to simplify measurement of future movement along the fault.

Lawson, A.C. (Chairman), and others, 1908 (reprinted 1969), The California earthquake of April 18, 1906: Carnegie Inst. Washington Pub. 87, v. 1, 451 p.

87 J.H. MacGregor described fossils of a small aquatic reptile from the Permian of the state of Parana (Brazil) which he called Mesosaurus brasiliensis. He also said that his finds were "indistinguishable" from specimens of Mesosaurus previously found only in the Karoo beds of South Africa.

MacGregor, J.H., 1908, On Mesosaurus Brasiliensis nov. sp. from the Permian of Brasil: Brasil Commissao Estudios Minas Carvao Pedra Brasil Annual Report 1908, Part 2, p. 302-336.

88 F.B. Taylor proposed a theory of continental drift in which capture of the moon by the earth led to a gradual migration of Earth's large land masses away from the poles. The Mid-Atlantic Ridge was said to be a residual ridge along a line of parting away from which the continents moved.

Taylor, F.B., 1909, Bearing of the Tertiary mountain belt on the origin of Earth's plan: Geol. Soc. America Bull., v. 20, p. 625. (Note: Taylor's presentation is listed by title only in the record of the Geological Society of America's 1908 meeting.)

___ 1910, Bearing of the Tertiary mountain belt on the origin of the earth's plan: Geol. Soc. America Bull., v. 21, p. 179-226.

___ 1923, The lateral migration of land masses: Washington Acad. Sci. Proc., v. 13, p. 445-447.

___ 1925, Movement of continental masses under action of tidal forces: Pan Amer. Geologist, v. 43, p. 15-50.

___ 1926, Greater Asia and isostasy: Amer. Jour. Science, v. 12, p. 47-67.

___ 1928, Sliding continents and tidal and rotational forces: in W.A.J.M. van Waterschoot van der Gracht, and others, Theory of continental drift, Amer. Assoc. Petroleum Geologists, Tulsa, OK, p. 158-177.

See also:

Totten, S.M., 1981, Frank B. Taylor, plate tectonics and continental drift: Jour. Geol. Education, v. 29, p. 212-220.

Black, G.W. Jr., 1979, Frank Bursley Taylor - forgotten pioneer of continental drift: Jour. Geol. Education, v. 27, p. 67-70.

1910

89 P.L. Mercanton found that certain rocks from Spitzbergen crystallized at a time when the earth's magnetic field was reversed.

Mercanton, P.L., 1910, Etat magnetique des diabases de l'Isfjord au Spitzbergen: Acad. Sci. Comptes Rendus, v. 151, p. 1092-1094.

90 A. Mohorovicic reported discovery of the major seismic discontinuity which was later named after him. This discontinuity came to be regarded as the boundary between the crust and the mantle.

Mohorovicic, Andrija, 1910, Das beben vom 8 x 1909: in Jahrbuch des meteorologischen observatoriums in Zagreb (Agram) fur das jahr 1909, v. 9, part 4, p. 1-63.

1911

91 O. Ampferer and W. Hammer postulated that the Alps formed by an underthrusting of Europe beneath Africa and that the underthrust material was absorbed in the mantle by a process of "Verschluckung", i.e., "swallowing."

Ampferer, O., and Hammer, W., 1911, Geologischer Querschnitt durch die Ostalpen vom Allgau zum Gardsee: Jahrb. Geol. Reichsanstalt, v. 61, no. 3-4, p. 531-709.

92 H.B. Baker theorized that, during a close encounter with the earth, Venus pulled the moon from the earth and that, as a result, the continents, which had been united, separated and took up their present relative positions.

Anonymous, 1911, Origin of the moon; interesting theory advanced by Dr. Howard B. Baker: Detroit Free Press, 23 April, Feature Mag. section, p. 7.
Baker, H.B., 1912, The origin of continental forms, II: Mich. Acad. Sci. Ann. Rept., 1911-1912, p. 116-141.
___ 1912, The origin of continental forms, III: Mich. Acad. Sci. Ann. Rept., 1911-1912, p. 107-113.
___ 1913, The origin of continental forms, IV: Mich. Acad. Sci. Ann. Rept., 1913-1914, p. 26-32.
___ 1914, The origin of continental forms, V: Mich. Acad. Sci. Ann. Rept, 1913-1914, p. 99-103.

93 P. Marshall pointed out that andesitic lavas in the western Pacific are restricted to the western side of a line that runs from New Zealand to the Samoan Islands and then passes north of New Guinea and that the line coincides with a series of deep ocean trenches. This line was later called the "andesite line" by A. Born and was sometimes referred to as the "Marshall line."

Marshall, Patrick, 1911, Oceania: in Steinmann, G., and Wilckens, O., eds., Handbuch der regionalen geologie, v. 7, no. 2, Heidelberg, C. Winter Universitatsbuchhandlung, 36 p.
Born, A., 1933, Der geologische aufbau der Erde: in Gutenberg, Beno, ed., Handbuch der Geophysik, Berlin, Borntraeger, p. 565-567. (See Fig. 306)

1912

94 A. Wegener made the first public presentation of his theory of continental drift to the Geological Association in Frankfurt-am-Main on January 6th in an address entitled "The geophysical basis of the evolution of the large-scale features of the earth's crust (continents and oceans). On January 10th he made a presentation to the Society for the Advancement of Natural Science in Marburg on "Horizontal displacements of the continents." The first published versions appeared as follows.

Wegener, Alfred, 1912, Die Entstehung der Kontinente: Petermanns Geog. Mitt., v. 58, p. 185-195, 253-256, 305-309.
___ 1912, Die Entstehung der Kontinente: Geol. Rundschau, v. 3, p. 276-292.
Jacoby, W.R., 1981, Modern concepts of earth dynamics anticipated by Alfred Wegener in 1912: Geology, v. 9, p. 25-27.

95 On their ill-fated return journey from the South Pole, E.A. Wilson and H.R. Bowers of the R.F. Scott Terra Nova Expedition collected specimens of *Glossopteris* on Mount Buckley at the head of Beardmore Glacier at about 85 degrees S latitude. Within a few days the last member of the polar party died but the samples were found the following year and later described by A.C. Seward.

Debenham, F., 1913, Summary of geological journeys: in Scott, R.F., Scott's last expedition, London, Smith Elder & Co., p. 438-440.
Scott, R.F., 1913, Scott's last expedition: London, Smith, Elder & Co., v. 1, 564 p.
Seward, A.C., 1914, Antarctic fossil plants: British Antarctic Terra Nova Expedition, Geology, v. 1, p. 1-49.
Seaver, G., 1933, Edward Wilson of the Antarctic: London, John Murray, 294 p.

1914

96 J. Barrell coined the terms lithosphere and asthenosphere (sphere of weakness) and suggested that the latter yields readily to long-enduring strains of limited magnitude and plays a fundamental role in lithospheric deformation.

Barrell, Joseph, 1914, The strength of the earth's crust, Part VI, Relations of isostatic movements to a sphere of weakness - the asthenosphere: Jour. Geology, v. 22, p. 655-683.

1915

97 The first edition of A. Wegener's full theory of continental drift was published in German. In support of the theory, evidence from geodesy, geophysics, geology, paleontology, biology, paleoclimatology and bathymetry was described at length. The second edition appeared in 1920, the third in 1922 and the fourth, and final, edition in 1929. The third edition was translated into English, Spanish, French, and Russian. The current English edition is a translation by John Biram of the fourth German edition and was published by Dover in 1966.

Wegener, Alfred, 1915, Die Entstehung der Kontinente und Ozeane: Braunschweig, Germany, F. Vieweg und Sohns, 94 p.

___ 1966, The origin of continents and oceans (transl. of 4th German ed. by J. Biram): New York, Dover.

1916

98 J. Kiedel described a series of striking similarities between the Cape Range of South Africa and the Sierra de la Ventana of South America. The comparisons included lithology of both sedimentary and igneous rocks, age of folding, fossil floras and faunas, unconformities and structural trends.

Kiedel, J., 1916, La geologia de las Sierras de la provincia de Buenos Aires y sus relaciones con las montanas de sud Africa y los Andes: Republica Argentine Anales del Ministerio de Agricultura, Seccion Geologica, Tomo XI, No. 3, 78 p.

1917

99 T. Arldt surveyed the opinions of twenty authorities on the postulated existence of various land bridges at various times during geologic history and found strong support for the existence of some of these features during the Paleozoic and Mesozoic. Those included in the survey were Arldt, Burckhardt, Diener, Frech, Fritz, Handlirsch, Haug, Ihering, Karpinsky, Koken, Kossmat, Katzer, Lapparent, Matthew, Neumayr, Ortmann, Osborn, Schuchert, Uhlig and Willis.

Arldt, Theodor, 1917-1922, Handbuch der Palaeogeographie (2 v.): Leipzig, Borntraeger, 1647 p.

1921

100 H.A. Brouwer described many remarkable similarities between the igneous rocks of South America and Africa and concluded that drift must have separated the continents.

Brouwer, H.A., 1921, De alkaligesteenten van de Serra do Gericino ten Noordwesten van Rio de Janerio en de overeenkomst der eruptiefgesteenten van Brasilie en Zuid-Africa: Koninkl. Nederlandse Akad. Wetensch. Proc., Part 29, p. 1005-1020.

101 A. Heim estimated that the folds and overthrusts of the Alps add up to 200-300 km of crustal shortening.

Heim, Albert, 1921, Geologie der Schweiz (2 vols.): Leipzig, Tauchnitz.

102 E. Tams showed that seismic surface waves with periods of about 20 seconds travel faster along oceanic paths than continental and concluded that continental and oceanic crustal structures are quite different.

Tams, E., 1921, Die Seismizitat der Erde: in Wien, W., and Harms, F., Handbuch der experimental Physik, v. 25, no. 2, p. 361-437.

1921 - 1944

103 A.L. Du Toit, concentrating on southern Africa and South America, assembled a large body of data which supports the theory of continental drift. Among the lines of evidence he put together from the literature and from his own field work were: (1) similarities in lithology and sequence between stratigraphic units of the same age on different continents, (2) "fit" of opposing continental margins, (3) continuation of numerous structural trends in the positions called for by drift, (4) position of the Mid-Atlantic Ridge, (5) similarities in lithology and age of igneous bodies, (6) similarities of fossil faunas and floras, (7) synchroneity of glaciation and simplification of the pattern with drift, and (8) similarities in placement and age of unusual ore bodies. Du Toit's classic 1937 book summarized the evidence for drift and also provided a lengthy summary and evaluation of the extensive literature on drift. Among other things, Du Toit suggested that Spain has rotated counterclockwise as a result of relative movement between Africa and Europe.

Du Toit, A.L., 1921, Land connections between the other continents and South Africa in the past: South African Jour. Sci., v. 18, p. 120-140.

___ 1921, The Carboniferous glaciation of South Africa: Geol. Soc. South Africa Trans., v. 24, p. 188-227.

___ 1922, The evolution of the South African coastline: South African Geog. Jour., p. 5-12.

___ 1927, A geological comparison of South America with South Africa: Carnegie Inst. Washington Pub. 381, 158 p.

___ 1928, Some reflections upon a geological comparison of South Africa with South America: Geol. Soc. South Africa Trans., v. 31, p. 19-38.

___ 1937, Our wandering continents, an hypothesis of continental drifting: New York, Hafner Pub. Co., 366 p.

___ 1944, Tertiary mammals and continental drift - a rejoinder to George G. Simpson: Amer. Jour. Sci., v. 242, p. 145-163.

1922

104 At the opening session of the XIII International Geological Congress in Brussels, E. Argand presented a concept of global tectonics of which continental drift was the foundation. Argand's views on global tectonics are widely regarded as the definitive statement of early twentieth-century "mobilism", as opposed to "fixism."

Argand, Emile, 1924, La tectonique de l'Asie: Comptes Rendus Cong. Geol. Internat., XIII, v. 1, p. 171-372.

___ 1977, Tectonics of Asia (Transl. and ed. by A.V. Carozzi): New York, Hafner Press, 218 p.

105 M. Marti developed a continuous graphic recorder for oceanographic sounding data.

Marti, M., 1922, Oscillographe enregistreur continu a grande sensibilite et son application au sondage continu par le son: Brevet d'Invention XII 3, no. 568-075. (dated 7 Sept. 1922).

106 The first continuous bottom profile across an ocean (Newport, R.I. to Gibraltar) was made by H.C. Hayes' continuous sonic depth finder on the USS Stewart.

Whitcroft, T.H., 1943, Sonic sounding as developed by the U.S. Navy: U.S. Naval Inst. Proc., v. 69, p. 216-223.

107 E. Tams reported that the Mid-Atlantic Ridge seismic belt continues across the Arctic Ocean.

Tams, E., 1922, Die Seismischen Verhaltnisse des europaischen Nordmeeres: Centralblatt f. Mineralogie, Geologie u. Paleontologie, Jahrg., no. 13, p. 385-397.

108 H.H. Turner reported the discovery of deep-focus earthquakes.

Turner, H.H., 1922, On the arrival of earthquake waves at the antipodes, and the measurement of the focal depth of an earthquake: Royal Astron. Soc. Monthly Notices, Geophys. Supplement, v. 1, p. 1-13.

1922 - 1924

109 A. Wegener published the first English description of his theory of continental drift. The third German edition (1922) of his book was also translated into English and published.

Wegener, Alfred, 1922, The origin of continents and oceans: Discovery, v. 3, p. 114-118.

___ 1924, The origin of continents and oceans (translated from the third German edition by J.G.A. Skerl): New York, E.P. Dutton & Co., 212 p.

Schwarzbach, M., 1980, Alfred Wegener und die Drift der Kontinente: Stuttgart, Wissenschaften Verlagsgesellschaft, 160 p.

1922 - 1925

110 The British scientific journal Nature carried an important series of reviews, summaries and reports of discussions on the theory of continental drift. The English and German editions of Wegener's book were reviewed and the theory was brought to the attention of many English-speaking scientists. Another important review appeared in Geological Magazine. Most of the comment was decidedly unfavorable.

Cole, G.A., 1922, Wegener's drifting continents: Nature, v. 110, p. 798-801.

Lake, Phillip, 1922, Wegener's displacement theory: Geol. Magazine, v. 7, p. 338-346.

Wright, H.B., 1923, The Wegener hypothesis: Nature, v. 111, p. 30-31.

Anonymous, 1923, The distribution of life in the southern hemisphere: Nature, v. 111, p. 131.

Lake, Phillip, 1923, Wegener's hypothesis of continental drift: Nature, v. 111, p. 226-228.

Crook, T., 1923, The Wegener hypothesis and the origin of oceans: Nature, v. 111, p. 225-256.

Evans, J.W., 1923, The Wegener hypothesis of continental drift: Nature, v. 111, p. 393-394.

Gregory, J.W., 1925, (Review of) The origin of continents and oceans by Prof. Alfred Wegener: Nature, v. 115, p. 255-257.

Coleman, A.P., 1925, Permo-Carboniferous glaciation and the Wegener hypothesis: Nature, v. 115, p. 602.

Meyrick, Edward, 1925, Wegener's hypothesis and the distribution of microlepidoptera: Nature, v. 115, p. 834-835.

Anonymous, 1925, The origin of continents: Nature, v. 116, p. 481.

1923

111 A. Keith estimated that crustal shortening in the Appalachians totals over 300 km.

Keith, Arthur, 1923, Outlines of Appalachian structure: Geol. Soc. America Bull., v. 34, p. 309-380.

112 H. Nakano devised the technique of using azimuthal compression-rarefaction patterns to determine the "first motions" of earthquakes.

Nakano, H., 1923, Notes on the nature of the forces which give rise to the earthquake motions: Central Meteorol. Obs. Japan, Seismol. Bull., v. 1, p. 92-120.

113 F.A. Vening-Meinesz developed a two-pendulum method for determining gravity at sea in a submerged submarine to an accuracy of five mgals. The method was tested on a round-trip voyage aboard the Dutch submarine KII between Holland and the East Indies. Among the first results was a confirmation of the general isostasy of the ocean basins.

Vening Meinesz, F.A., 1925, The determination of gravity at sea in a submarine: Geog. Jour., v. 65, p. 501-521.

1923 - 1926

114 R.A. Daly developed a "landslide" theory of continental drift which held that "the continents appear to have slid down-hill, to have been pulled down, over the earth's body, by mere gravity; mountain structures appear to be the product of enormous, slow landslides." The thin, glassy, plastic-solid crust on the lava lake at Halemaumau, Kilauea, Hawaii was proposed as an analog of the larger-scale process of crustal sliding.

Daly, R.A., 1923, The earth's crust and its stability: Amer. Jour. Sci., v. 205, p. 349-371.

___ 1926, Our mobile earth: New York, Scribner's, 342 p.

1924 - 1976

115 H. Jeffreys published six editions (1924, 1929, 1952, 1959, 1970 and 1976) of his authoritative book, The Earth. Because of his eminence as a geophysicist, Jeffreys' arguments against continental drift were widely accepted as compelling. Among the points made by Jeffreys were that: (1) Wegener's reasoning as to land bridges was invalid because it requires perfect isostasy; (2) The Eotvos force invoked by Wegener as a cause for drift is much too weak; (3) The sima of the ocean floor must be about as strong as the sial and, therefore, the sial could not drift through the sima; and, (4) The fit of South America with Africa is a misfit by about 15 degrees. In the 4th edition, Jeffreys found the paleopole location studies made through 1956 to be unconvincing.

Jeffreys, Harold, 1976, The Earth: Cambridge, Cambridge Univ. Press.

1925

116 R. Chevallier measured the directions of fossil magnetism in 81 samples of Mt. Etna lava flows which could be historically dated as ranging in age from 394 BC to AD 1911 and showed that the field directions of the three youngest agree quite well with the directions recorded at the times they originated.

Chevallier, R., 1925, L'aimantation des laves de l'Etna et l'orientation du champ terrestre en Sicile du XIIe au XVIIe siecle: Ann. Phys., v. 4, p. 5-162.

117 E. Krenkel supported the concept of an extensive Gondwana continent that was said to have existed until the end of the Permian.

Krenkel, Erich, 1925, Geologie Afrikas: Berlin, Gebruder Borntraeger, v. 1, 461 p.

1925 - 1927

118 The German research ship Meteor, making the first extensive use of echo-sounding apparatus, demonstrated that the Mid-Atlantic Ridge is a range of rugged mountains and obtained bathymetric profiles that showed a deep rift valley near the center of the

Ridge. The data obtained were combined with other information and used to prepare a detailed bathymetric map of the Atlantic Ocean that showed a continuous Mid-Atlantic Ridge lying midway between opposite coasts. Among other features shown were the Romanche Trench and several subsidiary ridges such as the Rio Grande and Walvis Ridges. Basalt was dredged from the Mid-Atlantic Ridge.

Mauer, Hans, and Stocks, Theodor, 1933, Die Echolotungen des Meteor: Wissenschaftliche Ergebnisse der Deutschen Atlantischen Expedition auf dem forschungs- und vermessungsschiff Meteor 1925-1927, Band II, Part 1 (text), Part 2 (Echolotprofile), Walter de Gruyter & Co., Berlin and Leipzig.

Stocks, Theodor, and Wust, George, Die Tiefenverhaltnissed des Offenen Atlantischen Ozeans: Deutsche Atlantische Expedition 1925-1927, Band III, I Teil, Leif 1, Walter de Gruyter & Co., Berlin and Leipzig, 32 p. + map.

1926

119 The American Association of Petroleum Geologists convened a symposium in New York on "the origin and movement of landmasses, both intercontinental and intracontinental, as proposed by Alfred Wegener." Participants who supported drift or were sympathetic were W.A.J.M. van Waterschoot Van der Gracht (who wrote the introductory and longest contribution and also rebutted the opposing views point by point in the final paper), John Joly, G.A.F. Molengraff, Alfred Wegener, F.B. Taylor, and J.T. Singewald, Jr.. The opponents of drift who took part were Bailey Willis, R.T. Chamberlin, J.W. Gregory, Charles Schuchert, C.R. Longwell, William Bowie, David White, and E.W. Berry.

Van der Gracht, W.A.J.M. van Waterschoot, and others, 1928, Theory of continental drift: Tulsa, Okla., Amer. Assoc. Petroleum Geologists, 240 p. (later reissued by the A.A.P.G. in microfilm)

Holmes, Arthur, 1928, Continental drift: Nature, v. 122, p. 431-433.

120 B. Gutenberg concluded that the earth has a low velocity zone at depths of about 100-200 km and inferred that the zone has a lower rigidity than those adjacent to it.

Gutenberg, Beno, 1926, Untersuchungen zur Frage, bis zu welcher Tiefe die erde kristallin ist: Zeitschr. Geophysik, v. 2, p. 24-29.

___ 1960, Low-velocity layers in the earth, ocean and atmosphere: Science, v. 131, p. 959-965.

121 P.L. Mercanton obtained samples from widely separated localities (e.g., Spitzbergen and Australia) and found that some have normally oriented fossil magnetism and some reversed. He suggested that paleomagnetism might be used to test polar wandering and continental drift.

Mercanton, P.L., 1926, Inversion de l'inclinaison magnetique terrestre aux ages geologiques: Terr. Magn. Atmos. Elec., v. 31, p. 187-190.

___ 1926, Aimantation de basaltes Groenlandais: Acad. Sci. Comptes Rendus, v. 180, p. 859-860.

___ 1931, Inversion inclinaison magnetique aux ages geologique: Acad. Sci. Comptes Rendus, v. 192, p. 978.

122 L.F. Noble reported that two masses of lithologically similar rock on opposite sides of the San Andreas fault appear to have been separated by almost 40 km of strike slip.

Noble, L.F., 1926, The San Andreas rift and some other active faults in the desert region of southeastern California: Carnegie Inst. Washington Yearbook 25, p. 415-428.

___ 1927, The San Andreas rift and some other active faults in the desert region of southeastern California: Seismol. Soc. America Bull., v. 17, p. 25-39.

___ 1932, The San Andreas rift in the desert region of southern California: Carnegie Inst. Washington Yearbook 31, p. 355-363.

123 B. Sahni concluded from study of the geographical distribution of late Paleozoic botanical provinces and their likely climatic ranges that the Glossopteris and Gigantopteris realms must have originally been widely separated and that parts of the two realms subsequently moved toward each other by continental drift.

Sahni, Birbal, 1926, The southern fossil floras, a study in the plant-geography of the past: 13th Indian Sci. Congr. (Bombay), Geol. Sect., Pres. Address, p. 229-254.

___ 1936, Wegener's theory of continental drift in the light of paleobotanical evidence: Indian Bot. Soc. Jour., v. 15, p. 319-332.

1927

124 E.B. Bailey called attention to similarities in age, structure and sedimentary facies among certain Appalachian trends and the Caledonian and Hercynian trends of Europe and suggested that this evidence supports Wegener's concept of continental drift.

Bailey, E.B., 1927, Across Canada with Princeton: Nature, v. 120, p. 673-675.

___ 1928, The Paleozoic mountain systems of Europe and America: Nature, v. 122, p. 811-814.

___ 1929, The Paleozoic mountain systems of Europe and America: Brit. Assoc. Adv. Sci. Report (Glasgow, 1928), Section C, p. 57-76.

125 B. Gutenberg proposed his theory of "continental flow" (Fliesstheorie). A significant part of the theory was based on the idea that sialic crust 20-25 km thick underlies the Atlantic and Indian Oceans, whereas the Pacific basin was thought to have a totally simatic crust.

Gutenberg, Beno, 1927, Die Veranderung der Erdkrusts durch Fliessbewegungen der Kontinentalscholle: Gerlands Beitr. Geophysik, v. 16, p. 239-247, and v. 18, p. 281-291.

___ 1936, Structure of the earth's crust and the spreading of the continents: Geol. Soc. America Bull., v. 47, p. 1587-1610.

126 E. Tams showed that earthquakes occur in a wide band along the Mid-Atlantic Ridge.

Tams, E., 1927, Die seismischen Verhaltnisse des offenen Atlantischen Ozeans: Zeitschr. Geophysik, v. 3, p. 361-363.

1927 - 1976

127 Various authors developed expanding-earth hypotheses, attempted to find theoretical bases for them, and showed how they can explain many tectonic phenomena.

Lindemann, B., 1927, Kettengebirge: kontinentale Zerspaltung und Erdexpansion: Jena, G. Fischer.

Moschelles, J., 1929, The theory of dilation, a new theory of the origin and activity of endogenous forces: an essay review: Geol. Magazine, v. 66, p. 260-268.

Hilgenberg, O.C., 1933, Vom wachsenden Erdball: Berlin, Charlotenberg, O. Hilgenberg, 55 p.

Halm, J.K.E., 1935, An astronomical aspect of the evolution of the earth: Jour. Astron. Soc. S. Africa, v. 4, p. 1-28.

Dirac, P.A.M., 1938, A new basis for cosmology: Royal Soc. London Proc., v. A165, p. 199-208.

Dicke, R.H., 1957, Principle of equivalence and the weak interactions: Rev. Modern Physics, v. 29, p. 355-362.

Egyed, L., 1957, A new dynamic conception of the internal constitution of the earth: Geol. Rundschau, v. 46, p. 101-121.
Carey, S.W., 1958, The tectonic approach to continental drift: in Carey, S. W., ed., Continental Drift: a symposium, Hobart, Univ. of Tasmania, p. 177-358.
Dicke, R.H., 1959, Gravitation - an enigma: Amer. Scientist, v. 47, p. 25-40.
Egyed, L., 1960, On the mechanism of mountain building and folding: Geol. Rundschau, v. 50, p. 225-234.
Heezen, B.C., 1960, The rift in the ocean floor: Sci. American, v. 203, no. 4, p. 98-110.
Jordan, Pascual, 1971, The expanding earth: Oxford, Pergamon Press, 202 p.
Carey, S.W., 1976, The expanding earth: Amsterdam, Elsevier, 488 p.

1928

128 L. Kober predicted that the Mid-Atlantic Ridge would be found to be part of a continuous, world-encircling belt of oceanic ridge that extends into all the world's oceans.

Kober, L., 1928, Der Bau der Erde (2nd ed.): Berlin, Borntraeger, 499 p.

129 K. Wadati demonstrated conclusively that earthquakes occur at depths of several hundred kilometres.

Wadati, K., 1928, On shallow and deep earthquakes: Geophysical Mag., v. 1, p. 162-202.

1928 - 1933

130 A. Holmes proposed a theory of sea-floor spreading that included (1) rising convection currents beneath oceanic ridges, (2) creation of new ocean floor and crust as continents are spread, (3) rafting of continents away from spreading centers, (4) descent of convection currents below oceanic trenches and remelting of crust, (5) development of tectonic mountains where continents encounter downgoing convection, and (6) basalt-eclogite phase change and density differences as a driving mechanism. It was pointed out that convection current theory best accounts for many observed facts, including (1) simultaneous exertion of compressive and tensional forces in the crust, (2) global distribution of young compressive mountain systems, (3) existence and association of oceanic trenches, deep-focus earthquakes and strong negative gravity anomalies, (4) continued existence of and variations in the earth's magnetic field, and (5) variations in the earth's rate of rotation.

Holmes, Arthur, 1915, Radioactivity and the earth's thermal history, Part I, The concentration of the radioactive elements in the earth's crust: Geol. Magazine, v. 2, p. 60-71.
___ 1915, Radioactivity and the earth's thermal history, Part II, Radioactivity and the earth as a cooling body: Geol. Magazine, v. 2, p. 102-112.
___ 1916, Radioactivity and the earth's thermal history, Part III, Radioactivity and isostasy: Geol. Magazine, v. 4, p. 264-274.
___ 1925, Radioactivity and the earth's thermal history, Part IV, A criticism of Parts I, II, and III: Geol. Magazine, v. 62, p. 504-515.
___ 1925, Radioactivity and the earth's thermal history, Part V, The control of geological history by radioactivity: Geol. Magazine, v. 62, p. 534-544.
___ 1927, Some problems of physical geology and the earth's thermal history: Geol. Magazine, v. 64, p. 263-278.
___ 1928, Radioactivity and continental drift: Geol. Magazine, v. 65, p. 236-238.
___ 1928, Radioactivity and continental drift: Nature, v. 122, p. 431-433.
___ 1928, Radioactivity and earth movements: Geol. Soc. Glasgow Trans., v 18, p. 559-606.

___ 1933, The thermal history of the earth: Washington Acad. Sci. Jour., v. 23, p. 169-195.

___ 1944, The machinery of continental drift - the search for a mechanism: in Principles of physical geology, London, Thomas Nelson & Sons, Ltd., p. 505-509. (Reprinted in Cox, 1973)

___ 1965, Principles of physical geology: New York, Ronald Press, 1288 p.

Meyerhoff, A.A., 1968, Arthur Holmes - originator of spreading ocean floor hypothesis: Jour. Geophys. Res., v. 73, p. 6563-6565. (Replies to this article by R.S. Dietz (p. 6567) and H.H. Hess (p. 6569).

Frankel, Henry, 1978, Arthur Holmes and continental drift: British Jour. Hist. of Sci., v. 11, p. 130-150.

1929

131 Soundings taken from the non-magnetic research ship Carnegie demonstrated that the "Albatross Plateau" (now known as the East Pacific Rise) is a broad, rugged mountain range. The results of this cruise were not published until the mid-1940s and even then the significance of the sounding data was not appreciated.

Carnegie Institution of Washington, 1942-1944, Scientific results of cruise VII of the "Carnegie" during 1928-1929 (4 vols.): Washington, D.C., Carnegie Inst. Wash. Pub. 544.

132 Acting upon a motion by S.H. Haughton, Director of the Geological Survey of South Africa, the Council of the International Geological Congress established an "International Commission for the Correlation of the Karoo (Gondwana) System in its World Distribution."

International Geological Congress, 1930, Compte Rendu of the XV Session, South Africa, 1929: v. 1, p. 185.

Haughton, S.H., 1936, Report of the Commission on the Distribution of the Karoo (Gondwana) System: Internat. Geol. Congr., Rept. XVI Session, U.S.A., 1933, v. 2, p. 1115-1146.

___ 1939, Report of the Secretary of the Commission on the Distribution of the Karoo (Gondwana) System: Internat. Geol. Congr., Rept. XVII Session, U.S.S.R., 1937, v. 1, p. 121-134.

___ 1950, Proceedings of the Commission on the Distribution of the Gondwana (Karoo) System: Internat. Geol. Congr., Rept. XVII Session, Great Britain, Part 1, p. 192-206.

Teichert, Curt, 1952, A brief history of the Gondwanaland comcept and of the International Gondwana Commission: Symposium sur les Series de Gondwana, XIX Internat. Geol. Congr., p. 7-12.

133 M. Matuyama found that Japanese and Manchurian Cenozoic lavas fall into two groups, those with normal and those with reversed fossil magnetic fields. Since those having reversed fields were all early Quaternary or older, he concluded that the earth's field assumed its present orientation sometime in the middle Quaternary.

Matuyama, Motonori, 1929, On the direction of magnetization of basalt in Japan, Tyosen and Manchuria: Proc. Imp. Acad. Japan, v. 5, p. 203-205. (Reprinted in Cox, 1973)

___ 1929, On the direction of magnetization of basalt in Japan, Tyosen and Manchuria: Proc. 4th Pacific Sci. Congr., p. 1-3.

134 N. Nielsen suggested that the median rift zone of Iceland is actively undergoing spreading as a result of the action of tensional forces.

Nielsen, N., 1929, Islands tektonik og Wegener-Theorien: Det. 18, Skand. Naturforskermode, 1929, 4 p.

135 R.H. Rastall pointed out that objections to continental drift based on the absence of a driving mechanism neglect to consider that many geological processes are assumed to occur despite our inability to develop acceptable causative theories for them. As examples he cited the occurrence of ice ages and the formation of great overthrust sheets accompanied by hundreds of kilometres of crustal shortening in the world's fold mountain belts. He then pointed out that the use of land bridges to account for the distribution of organisms through time is considerably less plausible than drift and suggested that one of the fundamental problems of geology that had not yet been satisfactorily explained is why the sial is concentrated into continents that cover less than one-third of Earth's surface rather than being more uniformly spread out.

Rastall, R.H., 1929, On continental drift and cognate subjects: Geol. Magazine, v. 66, p. 447-456.

1929 - 1930

136 J.W. Gregory, in consecutive presidential addresses to the Geological Society of London, concluded that "The combined evidence of stratigraphical geology, palaeontology, and the distribution of the existing animals and plants proves that there was no ocean entitled to rank as the Atlantic during the Paleozoic and Mesozoic Eras, that its formation began at the end of the Cretaceous . . .", and that the Pacific basin was occupied in part at various times by enormous continents that reached from China to North America, for example, and from Australia to South America.

Gregory, J.W., 1929, The geological history of the Atlantic Ocean: Geol. Soc. London Quart. Jour., Proc., v. 85, p. 68-122.

___ 1930, The geological history of the Pacific Ocean: Geol. Soc. London Quart. Jour., Proc., v. 86, p. 72-136.

1930

137 K. Sagisaka showed that seismograms of deep-focus earthquakes exhibit the same pattern of first motions as shallow shocks and must, therefore, be produced by essentially the same mechanisms.

Sagisaka, K., 1930, A relation between the motion at a hypocenter and the types of seismogram: Geophys. Mag., v. 3, p. 165-176.

138 F.A. Vening Meinesz reported the discovery of very large (more than 200 mgal) negative gravity anomalies in an 8,000 km long belt in the East Indies. The belt is flanked on each side by a positive anomaly belt, with the difference between them reaching as much as 430 mgal within a horizontal distance of less than 200 km. The low-gravity belt was shown to be located beside the major oceanic trenches of the area and to parallel the belt of earthquake epicenters and volcanic activity. It was suggested that the belt may continue northwest to northern India and northeast to the Aleutians. A similar anomaly belt was reported in the West Indies. Another important conclusion reached was that oceanic ridges are isostatic. The gravity determinations were made while submerged in the Dutch submarine K XIII.

Vening Meinesz, F.A., 1931, Gravity anomalies in the East Indian Archipelago: Geog. Jour., v. 77, p. 323-337.

Vening Meinesz, F.A., Umbgrove, J.H.F., and Kuenen, P.H., 1934, Gravity expeditions at sea: Delft, Netherlands Geodetic Comm.

1930 - 1940

139 R.M. Field played a seminal role in developing interest in the geophysical exploration of the ocean basins, in finding money and ships to support the research, and in attracting capable people (e.g., E.C. Bullard, W.M. Ewing, J.B. Hersey, H.H. Hess, and J.T. Wilson) into that line of investigation. He was instrumental in getting the American Geophysical Union Committee on Geophysical and Geological Study of Oceanic Basins started and helped set up important symposia on the subject in 1932 (Washington, D.C.) and in 1937 (Philadelphia).

Field, R.M., 1932, Symposium on the application of geophysics to ocean basins and margins, Introduction: Am. Geophys. Union Trans., v. 13, p. 11-12. (Symposium papers to p. 49)

___ (Chm.), 1933, Report of Committee on Geophysical and Geological Study of Ocean Basins: Am. Geophys. Union Trans., v. 14, p. 9-16 + Appendices.

___ 1936, Recent developments in the geophysical study of ocean basins: Am. Geophys. Union Trans., v. 17, p. 20-23.

___ 1937, Structure of continents and ocean basins: Jour. Washington Acad. Sci., v. 27, no. 5, p. 181-195.

___ 1938, Geophysical exploration of ocean basins: Geol. Soc. London Quart. Jour., v 94, p. iv-vii.

Field, R.M., and others, Symposium on the geophysical exploration of the ocean bottom, arranged by the American Geophysical Union: Amer. Philos. Soc. Proc., v. 79, p. 1-144.

Field, R.M., 1938, The importance of geophysics to submarine geology: Am. Philos. Soc. Proc., v. 79, p. 1-7.

___ 1938, Report on the work of the International Commission on Continental and Oceanic Structure: Am. Geophys. Union Trans., v. 19, p. 252-254.

___ (Chm.), 1938, Report of Special Committee on Geophysical and Geological Study of Ocean Basins: Am. Geophys. Union Trans., v. 19, p. 571-574.

Bullard, E.C., 1962, Richard Montgomery Field: Geol. Soc. London Proc., no. 1602, p. 154-155.

Hess, H.H., 1962, Richard Montgomery Field: Am. Geophys. Union Trans., v. 43, p. 1-3.

Wertenbaker, William, 1974, The floor of the sea - Maurice Ewing and the search to understand the earth: Boston, Little Brown & Co. (See p. 22-23, 27, 32, and 176).

Bullard, E.C., 1975, William Maurice Ewing 1906-1974: Royal Soc. London Biogr. Mem. Fellows, v. 21, p. 269-311.

1931

140 A.C. Seward concluded from his survey of paleobotanical evidence that South America, Africa, Madagascar, India and Australia must have been joined at least as recently as the Triassic and that Africa was connected to South America, and India with Madagascar and Australia as recently as the Triassic.

Seward, A.C., 1931, Plant life through the ages: Cambridge, Cambridge Univ. Press, 601 p.

1932

141 The American Geophysical Union held a symposium in Washington, D.C. on the application of geophysical methods to geological problems of ocean basins.

American Geophysical Union, 1932, Symposium on the application of geophysics to the ocean basins and margins: Amer. Geophys. Union Trans., v. 13, p. 11-49.

142 E.L. DeGolyer suggested that seismic reflection and refraction methods that had been widely used on land would be ideal for investigating the geology of ocean bottoms and margins. He pointed out that both geophones and explosions could be located near the surface, thus taking advantage of water's excellent sound transmission.

DeGolyer, E.L., 1932, The application of seismic methods to submarine geology: Amer. Geophys. Union Trans., v. 13, p. 37-40.

143 L. Dubertet suggested that the Dead Sea fault is a transcurrent fault that has undergone more than 150 km of strike slip.

Dubertet, Louis, 1932, Les formes structurales de la Syrie et de la Palestine; leur origine: Acad. Sci. Comptes Rendus, v. 195, p. 66-68.

144 J. Schmidt, on the basis of data gathered during the 1928-30 Dana Expedition to the Indian Ocean, suggested that there is an Indian Ocean Ridge.

Schmidt, Johannes, 1932, Dana's togt omkring jorden 1928-1930: Copenhagen, 255 p.

145 C. Schuchert concluded that "the marine life of southern North America and southern Europe, and South America and Africa, from the Silurian to the end of Miocene time, backed by the distribution of land plants and animals from the Permian into the Cretaceous, and by the discontinuous localization of Recent life in Africa and South America, is overwhelmingly in favor of the existence of the Gondwana land bridge from Pre-Cambrian until the end of Cretaceous time." A great deal of evidence bearing on the existence of Gondwana land bridges as reported by many workers in various field was discussed.

Schuchert, Charles, 1932, Gondwana land bridges: Geol. Soc. America Bull., v. 43, p. 875-915.

146 B. Willis rejected, largely on geophysical (i.e., isostatic) grounds, the idea of extensive land connection between continents but, in order to explain the great similarities in fossil floras and faunas between now widely separated continents, he developed the concept of narrow "isthmian links." Willis was of the opinion that the isthmian links are preserved on the modern ocean floor in the form of submerged ridges and he described a "Brazil-Guinea Ridge", an "Africa-India Isthmus", and several "East Indian Isthmii" as examples.

Willis, Bailey, 1932, Isthmian links: Geol. Soc. America Bull., v. 43, p. 917-952.

1933

147 N. Nielsen concluded that regional east-west tensional stresses are responsible for the fissure volcanism and other linear structural features of Iceland. He related the tensional tectonics of Iceland to continental drift and contrasted them with the compressional tectonics of the Alps where continental collision between Europe and Gondwanaland was said to have occurred.

Nielsen, N., 1933, Contributions to the physiography of Iceland, with particular reference to the highlands west of Vatnajokull: D. Kgl., Danske Vidensk. Selsk. Skrifter, Naturvidensk. Og. Mathem., Afd., 9 Raekke, IV, 5, p. 183-286 + figures and maps.

148 H.F. Reid suggested that the San Andreas fault is undergoing continuous right lateral slip.

Reid, H.F., 1933, The mechanics of earthquakes: in Nat. Res. Council, Physics of the Earth, v. VI, Seismology, p. 87-103.

1933 - 1934

149 The British John Murray Expedition to the Indian Ocean made soundings from the H.M.S. Mabahiss that showed the Carlsberg Ridge northwest of the Chagos Archipelago has a deep central valley.

Sewell, R.B.S., 1935, Introduction and lists of stations: in The John Murray Expedition 1933-34 Scientific Reports, v. 1, no. 1, British Museum (Natural History), p. 1-42 + chart.

Farquharson, W.I., 1936, Topography with an appendix on magnetic observations: in The John Murray Expedition 1933-34 Scientific Reports, v. 1, no. 2, British Museum (Natural History), p. 43-62 + plates and charts.

1934

150 A.C. Seward concluded that the geographic distribution of fossil plants such as the genus Araucaria is best explained by continental drift.

Seward, A.C., 1934, A phytogeographical problem: fossil plants from the Kergeulen Archipelago: Annals of Botany, v. 48, p. 715-741.

151 G. Wust used hydrologic data to show that the Indian Ocean Ridge is continuous from the central Indian Ocean to the southwest, between Antarctica and Australia.

Wust, Georg, 1934, Anzeichen von Beziehungen swischen Bodenstrom und Relief in der Tiefsee des Indischer Ozeans: Die Naturwissenschaften, v. 22, p. 241-244.

1934 - 1935

152 M. Matuyama carried out gravity surveys in the Japan Trench area and found parallel belts of large negative and positive anomalies that follow the topographic trends. The measurements were made while submerged in the Japanese submarine I-24.

Matuyama, Motonori, 1936, Distribution of gravity over the Nippon Trench and related areas: Imperial Acad. Tokoyo Proc., v. 12, p. 93-95.

153 A. Wade proposed a theory of continental drift that involved two sial masses, one each at the south and north poles. These masses were said to have spread out towards the equator under the influence of gravity and "centrifugal force" and to have formed mountain chains where they encountered each other.

Wade, Arthur, 1934, The distribution of oilfields from the viewpoint of the theory of continental spreading: World Petroleum Conf. Proc., v. 1, p. 73-77.

___ 1935, New theory of continental spreading: Amer. Assoc. Petroleum Geologists Bull., v. 19, p. 1806-1818.

1934 - 1962

154 Various authors devised convection-current theories to explain earth tectonics.

Holmes, Arthur, 1928, Radioactivity and earth movements: Geol. Soc. Glasgow Trans., v. 18, p. 559-606. (See also entry No. 130)

Vening Meinesz, F.A., 1934, Gravity expeditions at sea 1923-1932: Delft, Netherlands Geodetic Comm., v. 2.

Pekeris, C.L., 1935, Thermal convection in the earth's interior: Royal Astron. Soc. Monthly Notices Geophys. Supplement, v. 3, p. 343-367.
Hales, A.L., 1936, Convection currents in the earth: Royal Astron. Soc. Monthly Notices Geophys. Supplement, v. 3, p. 372-379.
Griggs, David, 1939, A theory of mountain building: Amer. Jour. Sci., v. 237, p. 611-650.
Vening Meinesz, F.A., 1947, Major tectonic phenomena and the hypothesis of convection currents in the earth: Geol. Soc. London Quart. Jour., v. 103, p. 191-207.
___ 1952, Convection-currents in the earth and the origin of the continents I: Akad. Weten. Amsterdam Proc., Ser. B, v. 55, p. 527-553.
Heiskanen, W.A., and Vening Meinesz, F.A., 1958, The earth and its gravity field: New York, McGraw-Hill, 470 p. (See Chap. 11, Convection currents in the earth, Origin of continents and oceans, Great geosyncline belts, p. 397-442.
Vening Meinesz, F.A., 1962, Thermal convection in the earth's mantle: in Runcorn, S.K., ed., Continental drift, New York, Academic Press, p. 145-176.

1935

155 B. Choubert published detailed, precise maps of the relationship of the "pre-Atlantic" continents showing, (1) the "fit" of the continental shelves, (2) a "closed" Bay of Biscay, (3) the absence of Iceland, and (4) the continuation between continents of many structural trends such as the Variscan, Appalachian-Caledonian and Hercynian. He pointed out that the Triassic rifting and basalts of the eastern United States may date the onset of separation of North America and Europe-Africa.

Choubert, Boris, 1935, Recherches sur la genese des chaines Paleozoiques et Antecambriennes: Rev. Geographie Phys. et Geologie Dynam., v. 8, fasc. 1, p. 5-50.
Van Houten, F.B., 1975, A precocious Atlantic reconstruction: an historical note: Geology, v. 3, p. 194-195.

156 M. Ewing, A.P. Crary and H.M. Rutherford developed a method of obtaining seismic refraction measurements at sea.

Ewing, M., Crary, A.P., and Rutherford, H.M., 1936, Geophysical investigations in the emerged and submerged Atlantic coastal plain: Part I - Methods and results (Abs.): Geol. Soc. America Proc., p. 75.
___ 1937 Geophysical investigations in the emerged and submerged Atlantic coastal plain: Part I - Methods and results: Geol. Soc. America Bull., v. 48, p. 753-802.

157 The Geological Society of London and the Royal Astronomical Society held a joint meeting to discuss the hypothesis of continental drift.

Anonymous, 1935, A discussion of the hypothesis of continental drift: Geol. Soc. London Quart. Jour., v. 91, p. 6-11.

158 N.H. Heck of the United States Coast and Geodetic Survey announced that the newly discovered Indian Ocean Ridge is seismically active and that the most seismically active part of the Mid-Atlantic Ridge is right on the equator.

Heck, N.H., 1935, A new map of earthquake distribution: Geog. Rev., v. 25, p. 125-130.

159 N. Linden, D. Mushketov, and N. Rajko described the seismicity of the Arctic Ocean basin and showed a belt of epicenters that connects with the Mid-Atlantic Ridge seismic belt.

Mushketov, D., 1935, On seismicity of the Arctic: Acad. Sci. USSR Pub. Seismol. Inst. No. 61, p. 9-15.
Rajko, N., and Linden, N., 1935, On the earthquake of 20 XI 1933 in the Baffin Bay and on the distribution of epicenters in the Arctic: Acad. Sci. USSR Pub. Seismol. Inst. No. 61, p. 1-8.

160 T. Stocks showed how high-frequency acoustic profilers can be used to obtain sub-bottom information on marine sediment thickness, stratification and structure.
Stocks, Theodor, 1935, Erkundungen uber Art und Schichtung des Meresbodens mit Hilfe von Hochfrequenz-Echoloten: Naturwissenschaften, v. 24, p. 383-387.

161 K. Wadati showed that deep- and intermediate-focus earthquakes in the Japanese Islands region tend to fall along a regular surface that dips beneath the islands toward the southwest and then bends around to dip northwest. Attention was directed to the close association of intermediate-depth foci and volcanism.
Wadati, K., 1935, On the activity of deep-focus earthquakes in the Japan Islands and neighbourhoods: Geophys. Mag., v. 8, p. 305-325.

1935 - 1936

162 R. Endo and T. Sugiyama reported that drillholes on the island of Kita-Daito-Zima in the Pacific Ocean south of Japan penetrated 432 m of shallow-water carbonates and bottomed out in late Oligocene sediments.
Endo, Rokuro, 1935, Method of boring of testing the formation in Kita-Daito-Zima: Geol. and Paleo. Inst. Tohaku Imperial Univ. Pub. No. 14 (Transl. by Tadasi Naito).
Sugiyama, Toshio, 1936, On the second boring at Kita-Daito Zima: Geol. and Paleo. Inst. Tohaku Imperial Univ. Pub. No. 25 (Transl.).

1936

163 W.C. Darrah described collections of fossil plants and bituminous coal made by the 2nd R.E. Byrd Expedition to Antarctica within about three degrees of the Pole and suggested continental drift as one possible explanation for the occurrences. In 1941 he recognized a Glossopteris among the specimens and thereby established their Permo-Carboniferous age and relationship to other Gondwana collections. Still later, in 1962, E.P. Plumstead recognized a total of four Glossopterids from photos of the Byrd collections, further confirming their age and affinities.
Byrd, R.E., 1935, Exploring the ice age in Antarctica: Natl. Geog. Mag., v. 68, p. 399-474.
Darrah, W.C., 1936, Antarctic fossil plants: Science, v. 83, p. 390-391.
___ 1941, La paleobotanica Sudamericana: Lilloa, v. 6, p. 624-626.
Plumstead, E.P., 1962, Fossil floras of Antarctica: in Trans-Antarctic Expedition 1955-1958, Sci. Repts. No. 9, Pt. 2, 154 p. + plates.

164 H.C. Stetson reported dredging Cretaceous sedimentary rocks from the walls of submarine canyons cut into Georges Bank, thus demonstrating that the continental slope is an old, truncated feature, not a relatively new, depositional feature.
Stetson, H.C., 1936, Geology and paleontology of the Georges Bank canyons, Part I, Geology: Geol. Soc. America Bull., v. 47, p. 339-366.

165 G. Wust used bottom-water temperature distribution patterns to show conclusively that the Mid-Atlantic Ridge is continuous through most of the Atlantic Ocean.

Wust, G., 1936, Schichtung und Zirkulation des Atlantischen Ozeans, Das Bodenwasser und die Stratosphare: Wiss. Eggebn. Deutsch. Atlantische Exped. Meteor 1925-1927; v. 6, p. 1-188.

1936 - 1955

166 H. Benioff, H.P. Berlage, J. Coulomb, C. Davison, B. Gutenberg, Y. Iwai, A. Leith, C.F. Richter, J.A. Sharp, S.W. Visser and K. Wadati demonstrated convincingly that the world's deep-focus earthquakes occur in zones that dip from oceanic trenches toward an associated continent or volcanic island arc system, thereby confirming Wadati's earlier conclusions.

Leith, Andrew, and Sharpe, J.A., 1936, Deep-focus earthquakes and their geological significance: Jour. Geology, v. 44, p. 877-917.

Visser, S.W., 1936, Some remarks on the deep-focus earthquakes in the International Seismological Summary: Gerlands Beitr. Geophysik, v. 48, p. 254-267.

Berlage, H.P., 1937, A provisional catalogue of deep-focus earthquakes in the Netherlands East Indies 1918-1936: Gerlands, Beitr. Geophysik, v. 50, p. 7-17.

Davison, Charles, 1937, The distribution of deep-focus earthquakes: Geol. Mag., v. 74, p. 316-324.

Gutenberg, Beno, and Richter, C.F., 1938, Depth and geographical distribution of deep-focus earthquakes: Geol. Soc. America Bull., v. 49, p. 249-288.

___ 1939, Depth and geographical distribution of deep-focus earthquakes (2nd paper): Geol. Soc. America Bull., v. 50, p. 1511-1528.

___ 1941, Seismicity of the earth: Geol. Soc. America Spec. Paper 34, 131 p.

Coulomb, Jean, 1945, Seismes profonds et grandes anomalies negatives de la pesanteur peuvent-ils etre attribues a une extension plastique?: Annales Geophysique, v. 1, p. 244-255.

Gutenberg, Beno, and Richter, C.F., 1945, Seismicity of the earth II (Suppl.): Geol. Soc. America Bull., v. 56, p. 603-668.

Benioff, Hugo, 1949, Seismic evidence for the fault origin of oceanic deeps: Geol. Soc. America Bull., v. 60, p. 1837-1866.

Gutenberg, Beno, and Richter, C.F., 1949, Seismicity of the earth and associated phenomena: Princeton, New Jersey, Princeton Univ. Press, 273 p. (2nd ed., 1954, 310 p.).

Benioff, Hugo, 1954, Orogenesis and deep crustal structure - additional evidence from seismology: Geol. Soc. America Bull., v. 65, p. 385-400.

Wadati, K., and Iwai, Y., 1954, The minute investigation of seismicity in Japan (1): Geophys. Mag., v. 25, p. 167-173.

Benioff, Hugo, 1955, Seismic evidence for crustal structure and tectonic activity: in Poldervaart, Arie, ed., Crust of the earth, Geol. Soc. America Spec. Paper 62, p. 61-74.

1937

167 The American Geophysical Union arranged a symposium on geophysical exploration of the ocean bottom at the American Philosophical Society in Philadelphia.

Anonymous, 1938, Symposium on the geophysical exploration of the ocean bottom arranged by the American Geophysical Union: Amer. Philos. Soc. Proc., v. 79, p. 1-144.

168 W.M. Ewing, A.P. Crary, H.M. Rutherford and B.L. Miller reported that explosion-seismic studies made on and off the east coast of the United States tied in with

deep-well records show that the continental margin consists of sedimentary beds chiefly of Mesozoic age with only a relatively thin covering of younger sediments.

Ewing, W.M., Crary, A.P., and H.M. Rutherford, 1937, Geophysical investigations in the emerged and submerged Atlantic coastal plain: Part I Methods and results: Geol. Soc. America Bull., v. 48, p. 753-802.

Miller, B.L., 1937, Geophysical investigations in the emerged and submerged Atlantic coastal plain: Part II Geological significance of geophysical data: Geol. Soc. America Bull., v. 48, p. 803-812.

169 The Geology, Geography, Botany and Zoology sections of the Indian Science Congress held a joint discussion of continental drift in Hyderbad.

Indian Science Congress, 1937, Wegener's theory of continental drift with reference to India and adjacent countries: in Proc. 24th Indian Science Cong., Calcutta, Royal Asiatic Society of Bengal, p. 502-520.

170 J.D.H. Wiseman and R.B.S. Sewell, on the basis of data gathered during the John Murray Expedition to the Indian Ocean (1) confirmed the existence of the Indian Ocean Ridge, (2) discovered that the Ridge between the Gulf of Oman and west of the Chagos Islands has a deep central valley, (3) dredged basalts from the Ridge, (4) called attention to similarities between the Indian Ocean Ridge and the African rift system and showed that they are connected, (5) noted that the Ridge is seismically active, and (6) compared the Ridge to the Mid-Atlantic Ridge.

Wiseman, J.D.H., and Sewell, R.B.S., 1937, The floor of the Arabian Sea: Geol. Mag., v. 74, p. 219-230.

1938

171 W. Iwan described a German program to measure precisely and periodically the width of the Icelandic graben to determine whether it is actively widening.

Iwan, Walter, 1938, Die Bardartalverwerfung in Nordisland: Gesellschaft fur Erdkunde zu Berlin Zeitschrift, 1938, p. 350-362.

172 A.G. McNish and E.A. Johnson reported the results of the first paleomagnetic measurements of oceanic sediments.

McNish, A.G., and Johnson, E.A., 1938, Magnetization of unmetamorphosed varves and marine sediments: Terr. Mag. Atmos. Elec., v. 43, p. 401-407.

1939

173 E.M. Anderson, A.E. Benfield and E.C. Bullard made the first valid determinations of heat flow from the earth's interior. The measurements were made in continental areas and ranged between about .04 and .08 watts per square metre.

Anderson, E.M., 1939-40, The loss of heat by conduction from the earth's crust in Britain: Royal Soc. Edinburg Proc., v. 60, pt. II, p. 192-209.

Benfield, A.E., 1939, Terrestrial heat flow in Great Britain: Royal Soc. London Proc., v. A173, p. 428-450.

Bullard, E.C., 1939, Heat flow in South Africa: Royal Soc. London Proc., v. A173, p. 474-502.

174 F. Bernauer called attention to groups of parallel fissures in the central volcanic region of Iceland and suggested that they indicate an extension of the volcanic zone that is still occurring. The movement was ascribed to a magmatic current flowing beneath the crust, the crust being carried passively along. The result of the stretching was said to allow the sea to penetrate the fissures and come into contact with magma, leading to the formation of palagonitic rocks. The continuation of the central volcanic zone was thought to be the Reykjanes Peninsula and the Mid-Atlantic Ridge which formed in the same way. Bernauer theorized that the Atlantic Ocean might have formed in this manner but felt that it is more likely that the original Atlantic merely expanded somewhat from its original width. It was suggested that further observations in Iceland could determine more precisely the rate of extension of the central volcanic zone.

Bernauer, Ferdinand, 1939, Island und die Frage der Kontinentalverschiebungen: Geol. Rundschau, v. 30, p. 357-358.

175 E.C. Bullard and T.F. Gaskell conceived the idea of placing hydrophones near the surface of the sea, rather than on the sea bottom, thus greatly simplifying the seismic refraction method.

Bullard, E.C., and Gaskell, T.F., 1941, Submarine seismic investigations: Royal Soc. London Proc., v. A177, p. 476-499.

176 A symposium was held in Frankfurt, Germany at which participants were invited to answer the question, "Is it vertical movements of oceanic, or horizontal movements of the continental regions which in the first place determine the paleogeographical development and the present day picture" of the Atlantic Ocean? Papers or discussion on geophysics and geodesy, tectonics, volcanism, earth history and paleogeography, paleontology and biology, geography, marine geology, North America and Europe-Africa, and South America and southern Africa, as well as several general papers, were given. Among the participants were W.H. Bucher, W.A.J.M. van W. van der Gracht, S. von Bubnoff, K. Wegener, B. Willis, H. Stille, W. Staub, H. Cloos, G. Wust, T. Stocks, A.L. Du Toit and H. Kiedel. A majority of those who spoke directly to the question opposed continental drift.

Anonymous, 1939, Atlantis 1939: Geol. Rundschau, v. 30, p. 1-400.

177 W.Q. Kennedy concluded that there has been at least 100 km of strike slip along the Great Glen Fault in Scotland.

Kennedy, W.Q., 1939, The Great Glen Fault: Geol. Soc. London Abstracts of Proc., Sess. 1938-1939, no 1354, p. 42-43.

___ 1946, The Great Glen Fault: Geol. Soc. London Quart. Jour., v. 102, p. 41-76.

178 H.W. Murray discovered a submarine escarpment (the Gorda escarpment) off the coast of northern California. (Note: This is the north-facing part of the Mendocino escarpment.)

Murray, H.W., 1939, Submarine scarp off Mendocino, California: Field Engineers Bull. (U.S. Coast & Geodetic Survey), v. 13, p. 27-33.

179 A.C. Veatch and P.A. Smith showed how a continuous acoustic profiler can obtain sub-bottom reflections and thereby measure sediment thickness and attitude.

Veatch, A.C., and Smith, P.A., 1939, Atlantic submarine valleys of the United States and the Congo submarine valley: Geol. Soc. America Spec. Paper 7, 101 p.

1939 - 1955

180 W.M. Elsasser and E.C. Bullard developed a magnetohydrodynamic theory of the earth's magnetic field that explains the field in terms of the interaction of fluid motions and electrical currents in the earth's core.

Elsasser, W.M., 1939, On the origin of the earth's magnetic field: Physical Review, v. 55, p. 489-498.

Bullard, E.C., 1949, The magnetic field within the earth: Royal Soc. London Proc., v. A197, p. 433-453.
Elsasser, W.M., 1950, The earth's interior and geomagnetism: Rev. Modern Physics, v. 22, p. 1-35.
___ 1955, Hydromagnetism, I: A review: Amer. Jour. Physics, v. 23, p. 590-609.
Bullard, E.C., 1955, The stability of a homopolar dynamo: Cambridge Philos. Soc. Proc., v. 51, p. 744-760.

1940

181 B. Kullenberg and H. Pettersson designed and tested a new piston-type coring device which could obtain longer cores (10-15 m) from deep-sea sediments.

Pettersson, Hans, and Kullenberg, Borje, 1940, A vacuum core sampler for deep-sea sediments: Nature, v. 145, p. 306.
Kullenberg, Borje, 1957, Deep-sea coring: in Pettersson, Hans, ed., Reports of the Swedish deep-sea expedition, v. 4, Bottom investigations No. 2, Goteborg, Elanders, p. 37-96.

182 F. von Huene described occurrences of the small Permian fresh-water reptile Mesosaurus in eastern South America and southwestern Africa and suggested that this case of disjunct endemism can best be explained by continental drift.

Huene, Friedrich von, 1940, Das unterpermische Alter aller Mesosaurier fuhrenden Schichten: Zentralblatt fur Min., Geol. und Palao., Abt. B., Nr. 1, p. 200-210.
___ 1941, Osteologie und systematische Stellung der Mesosaurier: Palaeontographica, v. 92, p. 45-55.

183 J. Tinsley Wilson found virtually no dispersion of Love waves that had crossed ocean basins and concluded that (1) the Atlantic, Pacific and Indian Oceans are similar in structure, and (2) the mantle is much closer to the surface beneath the ocean basins.

Wilson, J. Tinsley, 1940, The Love waves of the South Atlantic earthquake of August 28, 1933: Seismol. Soc. America Bull., v. 30, p. 273-301.

1940 - 1969

184 F.F. de Almeida, J.J. Bigarella, O. Barbosa, P.F. Carvalho, J.C. Crowell, L.A. Frakes, R. Maack, H. Martin, H. Mau, A.C. Rocha-Campos, R. Salumuni, and K. Schalk used studies of striations, pre-glacial valley orientations, roches moutonees, erratic boulder types, till fabrics, and ice-produced folds and imbrications to show conclusively that the direction of ice flow in the South American Parana Basin during the Permo-Carboniferous was to the west and northwest, a direction of flow that strongly supports continental drift. Facies patterns were matched across the Atlantic to Africa and glacial sediment volumes were matched to possible source areas to further support the idea that Africa and South America were adjacent at that time.

Barbosa, Octavio, 1940, Estrias produzidas por gelo Permo-Carbonifero: Rev. Mineracao e Metalurgia, v. 4, p. 272-273.
Carvalho, P.F. de, 1940, Estrias glacias em granodiorito sobposto ao Gonduana de Santa Caterina: Rev. Mineracao e Metalurgia, v. 4, p. 271-272.
Almeida, F.F. de, 1948, A "roche moutonee" de Salto, Estado de Sao Paulo: Bol. Geol. e Metalurgia (Univ. de Sao Paulo) No. 5, p. 112-118.
Maack, Reinhard, 1952, Die entwicklung der Gondwana-Schichten Sudbrasiliens und ihre Beziehungen zur Karruformation Sudafrikas: 19th Internat. Geol. Congr., Symposium sur les series de Gondwana, p. 339-372.

Almeida, F.F. de, Deformacoes causadas pelos geolo na serie de Tubarao em Sao Paulo: Div. Geol. e Min., D.N.P.M. No. 64, p. 1-8.
Maack, Reinhard, 1953, O desenvolvimento das camadas Gondwanicas do sul do Brasil e suas relacoes com as formacoes Karru da Afrika do Sul: Arquiv. de Biol. e Tecnol., v. 7, p. 201-253.
Martin, Henno, 1953, Notes on the Dwyka succession and on some pre-Dwyka valleys in south-west Africa: Geol. Soc. South Africa Trans., v. 56, p. 37-52.
Maack, Reinhard, 1957, Uber vereisungsperioden und vereisungsspuren in Brasilien: Geol. Rundschau, v. 45, p. 547-595.
Martin, Henno, and Schalk, Karl, 1957, Gletscherschliffe an der Wand Eines U-Tales im nordlichen Kaokofeld, Sudwestafrika: Geol. Rundschau, v. 46, p. 571-575.
Maack, Reinhard, 1958, Vorlaufige Ergebnisse einer Forschungsreise durch Sudafrika zum Problem der tangentialen Krustenverschriefungen der Erde: Die Erde, v. 89, p. 284-305.
Mau, Henry, 1958, Contribuicao para a geologia de Ibaiti: Bol. Soc. Bras. Geol., v. 7, p. 13-28.
Maack, Reinhard, 1960, Zur Palaogeographie des Gondwanalandes: 21st Internat. Geol. Congr. Rept., v. 7, p. 35-55.
Martin, Henno, 1961, The hypothesis of continental drift in the light of recent advance of geological knowledge in Brazil and in south-west Africa: Geol. Soc. South Africa Trans., Annexure to v. 64, 47 p.
___ 1964, The directions of flow of the Itarare ice sheets in the Parana Basin, Brazil: Bol. Paranaense Geog., nos. 10-15, p. 25-77.
Bigarella, J.J., and Salumuni, 1967, Some paleogeographic and paleotectonic feature of the Parana Basin: in Bigarella, J.J., and others, eds., Problems in Brazilian Gondwana geology, Curitiba, Roesner, p. 235-301.
Rocha-Campos, A.C., 1967, The Tubarao Group in the Brazilian portion of the Parana Basin: in Bigarella, J.J., and others, eds., Problems in Brazilian Gondwana geology, Curitiba, Roesner, p. 27-102.
Crowell, J.C., and Frakes, L.A., 1968, Late Paleozoic glacial facies and the origin of the south Atlantic Ocean: 23rd Internat. Geol. Congr. Rept., v. 13, Prague, p. 291-302.
Frakes, L.A., and Crowell, J.C., 1968, Late Paleozoic glacial facies and the origin of the south Atlantic basin: Nature, v. 217, p. 837-838.
___ 1969, Late Paleozoic glaciation: I, South America: Geol. Soc. America Bull., v. 80, p. 1007-1042.

1941

185 T. Minakami showed that the Yoridai-Sawa lava flow of 12 July 1940 became magnetized in the direction of the local geomagnetic field on that date.

Minakami, T., 1941, Magnetization of the new lava flows of Miyakesima Island: Tokyo Univ. Earthquake Research Inst. Bull., v. 19, p. 612-618.

186 H.W. Murray reported discovery by vessels of the United States Coast and Geodetic Survey during the years 1925-39 of many submerged seamounts in the Gulf of Alaska, some of which rise more than 3,600 m above the sea floor, and some of which have flat tops which are submerged 3,000 m below sea level.

Murray, H.W., 1941, Submarine mountains in the Gulf of Alaska: Geol. Soc. America Bull., v. 52, p. 333-362.

187 F.P. Shepard and K.O. Emery showed that (1) the north-facing Gorda escarpment changes to a south-facing (Mendocino) escarpment about 100 km west of the California coast, and (2) the south facing scarp extends at least 650 km from the coast.

Shepard, F.P., and Emery, K.O., 1941, Submarine topography off the California coast: Geol. Soc. America Spec. Paper 31, 171 p.

1945

188 B. Gutenberg and C.F. Richter noted that what is now called the East Pacific Rise is seismically active all the way from west of Mexico to at least 152 degrees east longitude and may be continuous into the Indian Ocean. They compared this feature to the Mid-Atlantic Ridge and the Indian Ocean Ridge

Gutenberg, Beno and Richter, C.F., 1945, Seismicity of the Earth: Geol. Soc. America Bull., v. 56, p. 603-668.

189 A.S. Romer compared the late Carboniferous vertebrate fauna of Kounova, Czechoslovakia to that of contemporaneous redbeds of Texas, found the two assemblages to be strikingly similar, and concluded that the two continents must have been connected in such a way as to allow free and rapid faunal interchange.

Romer, A.S., 1945, The late Carboniferous vertebrate fauna of Kounova (Bohemia) compared with that of the Texas redbeds: Amer. Jour. Sci., v. 243, p. 417-442.

190 T.S. Westoll concluded from studies of the freshwater fish of North America and Europe that there was no Atlantic Ocean during the Pennsylvanian.

Westoll, T.S., 1945, The Haplolepidae, a new family of late Carboniferous bony fishes: Am. Mus. Nat. History Bull., v. 83, p. 1-121.

1946

191 B. Gutenberg published the first unclassified report on interpreting seismograph records of atomic explosions, thereby beginning a long and fruitful association that was to vastly improve the quality and quantity of available seismic information and result in millions of dollars being spent on seismic research and instrumentation.

Gutenberg, Beno, 1946, The interpretation of records obtained from the New Mexico atomic bomb test: Seismol. Soc. America Bull., v. 36, p. 327-330.

192 J.B. Hersey, M. Ewing, S. Bergstrom and (independently) W. Weibull obtained the first seismic reflection measurements of sediment thickness from the deep sea.

Weibull, Waloddi, 1947, The thickness of ocean sediments measured by a reflexion method: Meddelanden fran Oceanografiska Institutet I Goteborg 12 (Goteborgs Kungl. Vetenskaps - och Vitterhets-Samhalle Handlingar, Sjatte Foljden. Ser. B, Band 5, No. 4, p. 2-17.

Hersey, J.B., and Ewing, Maurice, 1949, Seismic reflections from beneath the ocean floor: Am. Geophys Union Trans., v. 30, p. 5-14.

Weibull, Waloddi, 1957, Sound explorations: in Pettersson, Hans, Reports of the Swedish deep-sea expedition, v. 4, Bottom investigations no. 1, Goteborg, Elanders, p. 1-31.

193 H.H. Hess announced discovery of 160 submerged (up to 2,000 m below sea level), flat-topped volcanic mountains in the Pacific Ocean. Hess coined the term "guyots" for these submarine mountains.

Hess, H.H., 1946, Drowned ancient islands of the Pacific basin: Amer. Jour. Sci., v. 244, p. 772-791.

194 N.S. Shatski identified a new class of sedimentary basins characterized by small widths, comparatively great lengths (up to 800 km), thick sedimentary sequences and primarily vertical tectonics. The basins were named "aulacogens" and the subsurface

Pachelma and Dnieper-Donets basins of Russia and the Anadarko basin of the United States were cited as examples.

Shatski, N.S., 1946, Osnovnyye cherty stroyeniya i razvitiya vostochno-yevropeyskoy platformy: sravnitel'naya tektonika drevnikh platform: Akad. Nauk SSSR Izv. Ser. Geol., no. 1, p. 5-62.

___ 1946, Bol'shoy donbass i sistema vichita: sravnitel'naya tektonika drevnikh platform: Akad. Nauk SSSR Izv. Ser. Geol., no. 6, p. 57-90.

___ 1947, O strukturnykh svyazyakh platform so skladchatymi geosinklinal'nymi oblastyami: Akad. Nauk. SSSR Izv. Ser. Geol., no. 5, p. 37-56.

(Note: The articles listed above were reprinted in the following book.)

___ 1961, Vergleichende Tektonik alter Tafeln: Berlin, Akad.-Verlag, 220 p.

1946 - 1970

195 A large number of scientists, working in various parts of the world, firmly established the theory of rock magnetism, developed the necessary measuring devices and techniques, demonstrated that determinations of paleomagnetism can be used to determine paleolatitudes and pole directions, and showed that such studies strongly support the hypothesis of continental drift as it was developed on the basis of other evidence. (Following are some important reviews of paleomagnetic studies.)

Irving, E. 1958, Rock magnetism: a new approach to the problems of polar wandering and continental drift: in Carey, S.W., ed., Continental drift - a symposium, Hobart, Geol. Dept., Univ. of Tasmania, p. 24-61.

Cox, Allan, and Doell, R.R., 1960, Review of paleomagnetism: Geol. Soc. America Bull., v. 71, p. 645-768.

Nagata, Takesi, 1961, Rock magnetism: Tokyo, Maruzen Co. Ltd., 350 p.

Runcorn, S.K., 1962, Paleomagnetic evidence for continental drift and its geophysical cause: in Runcorn, S.K., ed., Continental drift, New York, Academic Press, p. 1-40.

Irving, E., 1964, Paleomagnetism and its application to geological and geophysical problems: New York, John Wiley & Sons, Inc., 399 p.

Creer, K.M., 1965, Paleomagnetic data from the Gondwana continents: Royal Soc. London Philos. Trans., v. A258, p. 27-40.

___ 1970, A review of paleomagnetism: Earth Science Reviews, v. 6, p. 369-466.

Tarling, D.H., 1971, Principles and applications of paleomagnetism: London, Chapman and Hall, 164 p.

McElhinny, M.W., 1973, Paleomagnetism and plate tectonics: Cambridge, The University Press, 358 p.

1947

196 The Woods Hole Atlantis Expedition led by M. Ewing dredged basalts, gabbros and serpentines from the Mid-Atlantic Ridge.

Shand, S.J., 1949, Rocks of the Mid-Atlantic Ridge: Jour. Geol., v. 57, p. 89-92.

197 M. Ewing (in the deep Atlantic) and M.N. Hill and P.L. Willmore (on the European continental shelf) successfully tested a new method of seismic refraction shooting at sea where both shot and hydrophones are positioned near the water's surface. Good preliminary results were obtained in both cases and it was pointed out that the new technique would vastly simplify seismic refraction work at sea, particularly in deep water.

Ewing, Maurice, 1948, Exploring the Mid-Atlantic Ridge: Nat. Geog. Magazine, v. 94, p. 275-294.

Hill, M.N., and Willmore, P.L., 1947, Marine seismic prospecting: Nature, v. 159, p. 707-708.

198 F. Press and M. Ewing suggested that the airborne magnetometer, modified for towing behind a ship, would be an extremely useful device for studying oceanic structures of various kinds, including oceanic ridges.

Press, Frank, and Ewing, Maurice, 1952, Magnetic anomalies over oceanic structures: Amer. Geophys. Union Trans., v. 33, p. 349-355.

199 The Swedish Deep-Sea Expedition, using the ship Albatross, began the era of post-World War II oceanographic exploration with a round-the-world cruise. Among other results, the Expedition (1) established the fact that the sea floor "is, in general, distinctly rugged", suggesting "that a resculpturing of the deep-sea floor goes on, producing and maintaining more or less abrupt differences in level", (2) took cores of Tertiary age, (3) used a seismic reflection method to determine that sea-floor sediment is surprisingly thin in the Pacific and Indian Oceans, (4) made the first heat flow measurements in the deep sea and found values near those for continents, (5) made the first continuous echo-sounding profiles across the Indian Ocean Ridge, and (6) demonstrated that some areas of the sea floor are devoid of sediment cover.

Pettersson, H.A., 1949, Exploring the bed of the ocean: Nature, v. 164, p. 468-470.

___ 1953, The Swedish Deep-Sea Expedition 1947-48: Deep-Sea Research, v. 1, p. 17-24.

Koczy, F.F., 1954, A survey on deep-sea features taken during the Swedish Deep-Sea Expedition: Deep-Sea Research, v. 1, p. 176-184.

___ 1956, Echo soundings: in Pettersson, H.A., ed., Reports of the Swedish Deep-Sea Expedition, v. 4, Bottom Investigations No. 3, Goteborg, Elanders, p. 97-158 + profiles.

Kullenberg, Borje, 1957, Deep-sea coring: in Pettersson, H.A., ed., Reports of the Swedish Deep-Sea Expedition, v. 4, Bottom Investigations No. 2, Goteborg, Elanders, p. 35-96.

Pettersson, H.A., ed., 1957-?, Reports of the Swedish deep-sea expedition 1947-48: Goteborg, Elanders boktr.

1948

200 S. Bergstrom, R.S. Edwards, M. Ewing, B.C. Heezen, J.B. Hersey, H. Johnson, E.T. Miller and G. Oakley converted an airborne magnetometer for towing behind a ship and measured the strength of the magnetic field on cruises by the R/V Atlantis across the Mid-Atlantic Ridge. In 1952 a profile was made continuously from Dakar, Senegal to Barbados, British West Indies. The anomaly pattern obtained was one of alternating peaks and valleys which bore no known relationship to topography or structure.

Heezen, B.C., Ewing, Maurice, and Miller, E.T., 1953, Trans-Atlantic profile of total magnetic intensity and topography, Dakar to Barbados: Deep-Sea Research, v. 1, p. 25-33.

Ewing, Maurice, 1949, New discoveries on the Mid-Atlantic Ridge: Natl. Geog. Mag., v. 96, p. 611-640.

See also:

Alldredge, L.R., and Keller, F., Jr., 1949, Preliminary report on magnetic anomalies between Adak, Alaska and Kwajalein, Marshall Islands: Amer. Geophys. Union Trans., v. 30, p. 444-500

201 A Russian expedition to the Arctic, headed by Y.Y. Hakkel, discovered the Lomonosov Ridge.

Webster, C.J., 1954, The Soviet expedition to the central Arctic, 1954: Arctic, v. 7, p. 59-80.

Sachs, V.N., Belov, N.A. and Lapina, N.N., 1955, Present concepts of the geology of the central Arctic: Priroda (1955), no. 7, p. 13-22 (Defense Research Board of Canada Translation T226R by E.R. Hope).

202 E.A. Johnson, T. Murphy and O.W. Torreson used studies of varved clays and Antarctic Ocean red clays to demonstrate that sedimentary clay particles become magnetized in the direction of the earth's field at the time of their deposition and that paleomagnetic pole positions vary around the present north rotational pole back to -15,000 y.

Johnson, E.A., Murphy, T., and Torreson, O.W., 1948, Pre-history of the earth's magnetic field: Terr. Magn. Atmos. Elec., v. 53, p. 349-372.

203 H.S. Ladd, J.I. Tracey and G.G. Lill reported that drilling on Bikini Atoll in the Marshall Islands penetrated up to 783 m of shallow-water carbonates and bottomed out in lower Miocene sediments. M.B. Dobrin, B. Perkins, Jr., and B.L. Snavely later reported that seismic studies indicate there are probably about 2,000 to 4,000 m of sediment above basement beneath the atoll.

Ladd, H.S., Tracey, J.I., and Lill, G.G., 1948, Drilling on Bikini Atoll, Marshall Islands: Science, v. 107, p. 51-55.

Dobrin, M.B., Perkins, Beauregard, Jr., and Snavely, B.L., 1949, Subsurface constitution of Bikini Atoll as indicated by a seismic refraction survey: Geol. Soc. America Bull., v. 60, p. 807-828.

204 R.L. Sherlock concluded that "the fact of Continental Drift has been proved, although the mechanism by which it occurs and the detailed results are still matters open to discussion" and described a variety of evidence (chiefly stratigraphic and paleontologic) supporting the theory.

Sherlock, R.L., 1948, The Permo-Triassic formations: London, Hutchinson's, 367 p.

205 C.A. Whitten reported that triangulation surveys conducted periodically since 1882 show that the San Andreas fault north of about 36.5 degrees north latitude is undergoing right lateral slip at an average rate of about 5 cm/yr. The data show that this movement continued after the 1906 earthquake.

Whitten, C.A., 1948, Horizontal earth movement, vicinity of San Francisco, California: Am. Geophys. Union Trans., v. 29, p. 318-323.

1948 - 1961

206 G. Arrhenius, S. Bergstrom, M. Bramlette, R.S. Edwards, D.B. Ericson, J. Ewing, M. Ewing, T.F. Gaskell, E.D. Goldberg, G.R. Hamilton, B.C. Heezen, J.B. Hersey, M.N. Hill, J. Hirschman, H.R. Johnson, S. Katz, R.L. Kovach, A.S. Laughton, B. Luskin, C.B. Officer, F. Press, R.W. Raitt, R.R. Revelle, A. Roberts, G.G. Shor Jr., G.H. Sutton, J.C. Swallow, M. Talwani, I. Tolstoy, G. Wollin, J.L. Worzel, and P.C. Wuenschel used earthquake surface wave studies, seismic refraction and reflection measurements and sediment samples from the Atlantic, Pacific and Indian Oceans to show that (1) deep-sea sediments are much thinner than expected, averaging only 300 m thick in the Pacific and 500 m in the Atlantic; (2) the sediments are no older than Cretaceous; (3) the sediments are almost entirely undeformed; and (4) sediments are thin to absent over the central parts of the oceanic ridge system.

Ewing, Maurice, 1948, Exploring the Mid-Atlantic Ridge: Natl. Geog. Mag., v. 94, p. 275-294.

Ewing, Maurice, 1949, New discoveries on the Mid-Atlantic Ridge: Natl. Geog. Mag., v. 96, p. 611-640.

Raitt, R.W., 1949, Studies of ocean-bottom structure off southern California with explosive waves (abs.): Geol. Soc. America Bull., v. 60, p. 1915.

Ewing, Maurice, Worzel, J.L., Hersey, J.B., Press, Frank, and Hamilton, G.R., 1950, Seismic refraction measurements in the Atlantic Ocean basin, Part I: Seismol. Soc. America Bull., v. 40, p. 233-242.

Gaskell, T.F., and Swallow, J.C., 1951, Seismic refraction experiments in the north Atlantic: Nature, v. 167, p. 723-724.

Gaskell, T.F., and Swallow, J.C., 1952, Seismic refraction experiments in the Pacific: Nature, v. 170, p. 1010-1012.

Hersey, J.B., Officer, C.B., Johnson, H.R., and Bergstrom, S., 1952, Seismic refraction observations north of the Brownson Deep: Seismol. Soc. America Bull., v. 42, o. 291-306.

Hill, M.N., 1952, Seismic refraction shooting in an area of the eastern Atlantic: Royal Soc. London Philos. Trans., v. A244, p. 561-596.

Officer, C.B., Ewing, Maurice, and Wuenschel, P.C., 1952, Seismic refraction measurements in the Atlantic Ocean, Part IV, Bermuda, Bermuda Rise and Nares Basin: Geol. Soc. America Bull., v. 63, p. 777-808.

Gaskell, T.F., and Swallow, J.C., 1953, Seismic refraction experiments in the Indian Ocean and in the Mediterranean Sea: Nature, v. 172, p. 535-537.

Tolstoy, Ivan, Edwards, R.S., and Ewing, Maurice, 1953, Seismic refraction measurements in the Atlantic Ocean, Part 3: Seismol. Soc. America Bull., v. 43, p. 35-48.

Gaskell, T.F., 1954, Seismic refraction work by H.M.S. Challenger in the deep oceans: Royal Soc. London Proc., v. A222, p. 356-361.

Ewing, Maurice, Sutton, G.H., and Officer, C.B., 1954, Seismic refraction measurements in the Atlantic Ocean, Part VI: Typical deep stations, North America basin: Seismol. Soc. America Bull., v. 44, p. 21-38.

Hill, M.N., and Laughton, A.S., 1954, Seismic observations in the eastern Atlantic, 1952: Royal Soc. London Proc., v. A222, p. 348-356.

Raitt, R. W., 1954, Some results of seismic studies on Capricorn Expedition (abs.): Geol. Soc. America Bull., v. 65, p. 1348.

Ewing, Maurice, and Press, Frank, 1955, Geophysical contrasts between continents and ocean basins: in Poldervaart, Arie, ed., Crust of the earth, Geol. Soc. America Spec. Paper 62, p. 1-6.

Ewing, Maurice, and Press, Frank, 1955, Seismic measurements in ocean basins: Jour. Marine Research, v. 14, p. 417-422.

Revelle, R.R., Bramlette, G., Arrhenius, G., and Goldberg, E.D., 1955, Pelagic sediments of the Pacific, in Poldervaart, Arie, ed., Crust of the earth, Geol. Soc. America Spec. Paper 62, p. 221-235.

Katz, S., and Ewing, Maurice, 1956, Seismic refraction measurements in the Atlantic Ocean, Part VII: Atlantic Ocean basin, west of Bermuda: Geol. Soc. America Bull., v. 67, p. 475-510.

Raitt, R.W., 1956, Seismic refraction studies of the Pacific basin, Part I: Crustal thickness of the central equatorial Pacific: Geol. Soc. America Bull., v. 67, p. 1623-1640.

Hill, M.N., 1957, Recent geophysical exploration of the ocean floor: in Ahrens, L. H. and others, eds., Physics and chemistry of the earth, v. 2, London, Pergamon Press, p. 129-163.

Gaskell, T.F., Hill, M.N., and Swallow, J.C., 1958, Seismic measurements made by H.M.S. Challenger in the Atlantic, Pacific and Indian Oceans and in the

Mediterranean Sea, 1950-1953: Royal Soc. London Philos. Trans., v. A251, p. 23-83.

Ewing, John, and Ewing, Maurice, 1959, Seismic refraction measurements in the Atlantic Ocean basins, in the Mediterranean Sea, on the Mid-Atlantic Ridge, and in the Norwegian Sea: Geol. Soc. America Bull., v. 70, p. 291-317.

Kovach, R.L., 1959, Surface wave dispersion for an Asio-African and a Eurasian path: Jour. Geophys. Research, v. 64, p. 805-813.

Ewing, John, Luskin B., Roberts, A., and Hirschman, J., 1960, Sub-bottom reflection measurements on the continental shelf, Bermuda banks, West Indies arc, and in the west Atlantic basins: Jour. Geophys. Research, v. 65, p. 2849-2859.

Ericson, D.B., Ewing, Maurice, Wollin, Goesta, and Heezen, B.C., 1961, Atlantic deep-sea sediment cores: Geol. Soc. America Bull., v. 72, p. 193-286.

Ewing, Maurice, 1963, Sediments of ocean basins: in Higginbotham, S.W., ed., Man Science, Learning and education, Houston, Texas, Rice Univ. Press, p. 41-59.

Ewing, Maurice, and Ewing, John, 1964, Distribution of oceanic sediments: in Yoshida, Kozo, ed., Studies on oceanography, Univ. Tokyo Press, p. 525-537 (U.S. Edition in 1965 by Univ. of Washington Press, Seattle).

Ewing, Maurice, Ewing, J.I., and Talwani, Manik, 1964, Sediment distribution in the oceans - the Mid-Atlantic Ridge: Geol. Soc. America Bull., v. 75, p. 17-36.

1948 - 1967

207 Scientists on research vessels from Germany (Meteor), Great Britain (Discovery), Sweden (Albatross) and the United States (Atlantis, Atlantis II, Chain, and Oceanographer) discovered pools of hot (up to 56 degrees C), highly saline (up to 318 parts per thousand) water in relatively small but deep basins in the Red Sea, cored highly metalliferous sediments in the basins, determined that the sediments have great potential economic value, related the water and heavy metal deposits to magmatic activity associated with sea-floor spreading, and suggested that certain other sedimentary ore deposits may be related to episodes of spreading.

Degens, E.T., and Ross, D.A., 1969, Hot brines and recent heavy metal deposits in the Red Sea: New York, Springer-Verlag, 600 p.

1949

208 E.C. Bullard suggested to scientists at Scripps Institution of Oceanography that a program of heat-flow measurements be initiated at sea. Acting on this proposal, Bullard, J.D. Isaacs, A.E. Maxwell and J.M. Snodgrass developed a probe for determining heat flows in deep-sea sediments.

Revelle, R.R., and Maxwell, A.E., 1952, Heat flow through the floor of the North Pacific Ocean: Nature, v. 170, p. 199-200.

Bullard, E.C., 1954, The flow of heat through the floor of the Atlantic Ocean: Royal Soc. London Proc., v. A222, p. 408-429.

Von Herzen, R.P., and Langseth, M.G., 1965, Present status of oceanic heat-flow measurements: in Ahrens, L.H., and others, eds., Physics and chemistry of the earth, v. 6, Pergamon Press, Oxford, p. 365-407.

209 K.O. Emery predicted, on the basis of continuation of the epicenter belt across the Arctic basin, that the Mid-Atlantic Ridge also continues along that belt.

Emery, K.O., 1949, Topography and sediments of the Arctic basin: Jour. Geol., v. 57, p. 512-521.

210 M. Ewing, J.L. Worzel, J.B. Hersey, F. Press, and G.R. Hamilton used a seismic refraction method northwest of Bermuda to determine that the Moho is only five km beneath the ocean floor and that continental basement rocks are absent. They hypothesized that this structure is characteristic of ocean basins generally.

Ewing, Maurice, Worzel, J.L., Hersey, J.B., Press, Frank, and Hamilton, G.R., 1949, (untitled): Geol. Soc. America Bull., v. 60, p. 1303.

___ 1950, Seismic refraction measurements in the Atlantic Ocean Basin, Part one: Seismol. Soc. America Bull., v. 40, p. 233-242.

211 J.W. Graham demonstrated that certain types of sedimentary rocks retain the magnetic orientation they acquired upon deposition for at least 200 million years and suggested additional study might: (1) reveal the history of the earth's magnetic field, (2) throw light on questions of continental drift, (3) allow determination of the distortional history of rocks, and (4) be useful in stratigraphic dating.

Graham, J.W., 1949, The stability and significance of magnetism in sedimentary rocks: Jour. Geophys. Research, v. 54, p. 131-167.

212 R.W. Raitt reported that seismic refraction studies in deep water beyond the continental slope off California found sediments only a few hundred meters thick and crustal rocks only five km thick.

Raitt, R.W., 1949, Studies of ocean-bottom structure off southern California with explosive waves (abs.): Geol. Soc. America Bull., v. 60, p. 1915.

213 Society for the Study of Evolution held a symposium in New York on the problem of land bridges across the south Atlantic Ocean.

Mayr, E., ed., 1952, The problem of land connections across the south Atlantic Ocean with special reference to the Mesozoic: Am. Mus. Nat. History Bull., v. 99, Art. 3, p. 79-258.

214 I. Tolstoy and M. Ewing discovered the topographic expression of a fracture zone that cuts across the Mid-Atlantic Ridge and noted that it is seismically active.

Tolstoy, Ivan and Ewing, Maurice, 1949, North Atlantic hydrography and the Mid-Atlantic Ridge: Geol. Soc. America Bull., v. 60, p. 1527-1540.

215 R.E. Wallace suggested that the San Andreas fault may have been moving at a rate of about 4 mm/yr for the last 30 My.

Wallace, R.E., 1949, Structure of a portion of the San Andreas rift in southern California: Geol. Soc. America Bull., v. 60, p. 781-806.

216 H.W. Wellman reported that the Alpine fault in New Zealand has undergone over 300 km of strike slip since the Jurassic.

Benson, W.N., 1950, Meeting of the Geological Division of the Pacific Science Congress in New Zealand, February, 1949: Geol. Soc. America Interim Proc., Feb., 1950, o. 11-13. (A brief mention of Wellman's work given here seems to be the first written description that appeared before 1954.)

Wellman, H.W., 1954, Active transcurrent faulting in New Zealand (abs.): Geol. Soc. America Bull., v. 65, p. 1322.

___ 1955, New Zealand Quaternary tectonics: Geol. Rundschau: vs. 43, p. 248-257.

1949 - 1950

217 J. Goguel and F.A. Vening-Meinesz showed that empirical gravity profiles across the east African rift valley agree much better with the tensional theory than with compressional theory.

Goguel, Jean, 1949, La structure des fosses Africains et la gravimetrie: Annales Geophysique, v. 5, p. 174-175.

Vening Meinesz, F.A., 1950, Les graben Africains, resultat de compression ou de tension dans la croute terrestre?: Inst. Royale Colonial Belge Bull. Seances, v. 21, p. 539-552.

1950

218 "At the Birmingham meeting of the (British) Association (for the Advancement of Science) the Sections of Geology, Zoology, Geography and Botany held a joint meeting on September 1, to discuss the position of the theory of continental drift. Each of six speakers in the symposium considered the possible validity, and implications of the hypothesis from the viewpoint of a particular branch of science." The six speakers were J.H.F. Umbgrove (geology), R.D. Good (botany), H.E. Hinton (zoology), H. Jeffreys (geophysics), S.W. Wooldridge (geomorphology), and J.R.F. Joyce (geology).

Anonymous, 1951, Symposium on the status of the continental drift hypothesis: Advancement of Science (Rept. of the B.A.A.S.), v. 8, p. 67-88.

219 M. Ewing and F. Press showed that (1) dispersion data for surface waves that have traveled oceanic paths can be accounted for by a layered model consisting of sea water, a thin layer of sediments and a relatively thin oceanic crust; (2) the Atlantic and Pacific Oceans have very similar crustal structures; and (3) the conclusions based on surface-wave dispersion data agree closely with explosion seismic refraction measurements made at sea.

Ewing, Maurice, and Press, Frank, 1950, Crustal structure and surface-wave dispersion: Seismol. Soc. America Bull., v. 40, p. 271-280.

___ 1952, Crustal structure and surfacewave dispersion, Part II ,Solomon Islands earthquake of July 29, 1950: Seismol. Soc. America Bull., v. 42, p. 315-325.

Press, Frank, and Ewing, Maurice, 1955, Earthquake surface waves and crustal structures: in Poldervaart, Arie, ed., Crust of the Earth, Geol. Soc. America Spec. Paper 62, p. 51-60.

220 Scripps Institution of Oceanography R/V Horizon and U.S. Navy ship Epce obtained the first extensive recorded bathymetric profiles from the Pacific Ocean during the Mid-Pacific Expedition.

Dietz, R.S., Menard, H.W., and Hamilton, E.L., 1954, Echograms of the Mid-Pacific Expedition: Deep-Sea Research, v. 1, p. 258-272.

221 P.H. Kuenen estimated that the average rate of sedimentation in the oceans beyond the continental shelves is 5 micrometers/yr and that the equivalent of 10 km of uncompacted sediment (3 km of solid sediment) has been deposited during the two billion years of geologic time.

Kuenen, P.H., 1950, Marine Geology: New York, John Wiley & Sons, 568 p.

1951

222 E.L. Hamilton reported finding rounded boulders and an integrated, shallow-water reef fauna of mid-Cretaceous age on the tops of guyots at 1,800 m depths west of the Hawaiian Islands.

Hamilton, E.L., 1951, Sunken islands of the mid-Pacific mountains (Abs.): Geol. Soc. America Bull., v. 62, p. 1502.

___ 1956, Sunken islands of the mid-Pacific Mountains: Geol. Soc. America Mem. 64, 97 p.

___ 1953, Upper Cretaceous, Tertiary and recent planktonic foraminifera from Mid-Pacific flat-topped sea mounts: Jour. Paleontology, v. 27, p. 204-237.

223 J. Hospers reported that (1) Icelandic lava flows and sediments of Miocene to recent age contain a record of periodic 180-degree reversals of Earth's magnetic field (four periods and three reversals were determined); (2) the reversal pattern is consistent within stratigraphic levels as far as correlations can be made; (3) zones of normal and reversed magnetic orientation have comparable thicknesses; (4) field and laboratory studies show no significant differences (other than magnetic orientation) between flows and sediments with normal and reversed magnetic orientation; (5) the period of reversals is 250 to 500 thousand years with reversal occurring within one-fiftieth of that time; and (6) Iceland has undergone no rotation or change in magnetic latitude since Miocene time and polar wandering has not exceeded 10 degrees since the Eocene.

Hospers, J., 1951, Remanent magnetism of rocks and the history of the geomagnetic field: Nature, v. 168, p. 1111-1112.
___ 1953-1954, Reversals of the main geomagnetic field: Konink L. Nederlandse Akad. Wetensch. Proc., Ser. B, v. 56, p. 467-491 and v. 57, p. 112-121.
___ 1955, Rock magnetism and polar wandering: Jour. Geology, vs. 63, p. 59-74.

224 S. Akimoto, K. Kobayashi, T. Nagata, N.D. Opdyke, M. Ozima, A. Roche, S.K. Runcorn, Y. Shimizu and S. Uyeda developed the concept of a magnetic reversal chronology by studying the polarity and relative ages of lava flows in France, the southwestern United States and Japan.

Roche, A., 1951, Sur les inversions de l'aimantation remante des roches volcaniques dans les monts d-Auvergue: Acad. Sci. Comptes Rendus, v. 233, p. 1132-1134.
___ 1956, Sur la date de la derniere inversion du champ magnetique terrestre: Acad. Sci. Comptes Rendus, v. 243, p. 812-814.
Opdyke, N.D., and Runcorn, S.K., 1956, New evidence for reversal of the geomagnetic field near the Pliocene-Pleistocene boundary: Science, v. 123, p. 1126-1127.
Nagata, T., Akimoto, S., Uyeda, S., Shimizu, Y., Ozima, M., and Kobayashi, K., 1957, Paleomagnetic study on a Quaternary volcanic region in Japan: Philos. Mag. Supp. Adv. Physics, v. 6, p. 255-263.

1952

225 P.M.S. Blackett developed an astatic, fluxgate magnetometer to be used for measuring very weak magnetic fields. The device was later found to be useful for measuring the remanent magnetism of weakly magnetized rocks and became the basis for a research program in rock magnetism that produced many important data which strongly supported continental drift and also led to the training of many researchers who started up their own laboratories and produced independent results from many parts of the world.

Blackett, P.M.S., 1952, A negative experiment relating to magnetism and the earth's rotation: Royal Soc. London Philos. Trans., v. 245 (A897), p. 309-370.
Lovell, Bernard, 1975, Patrick Maynard Stuart Blackett, Baron Blackett of Chelsea 1897-1974: Biog. Mem. Fellows Royal Soc. London, v. 21, p. 1-115.
Royal Society of London, 1975, Memorial meeting for Lord Blackett, O. M., C. H., F. R. S., at the Royal Society on 31 October 1974: Royal Socl. London Notes and Records, v. 29, p. 135-162.

226 R.S. Dietz reported that continuous echo soundings made in the south Pacific Ocean by the U.S.S. Henderson in 1946-47 showed the existence of large fault scarps, one of which is 3,200 m high and has an escarpment dipping at about 60 degrees to a foredeep.

Dietz, R.S., 1952, Some Pacific and Antarctic sea-floor features discovered during the U.S. Navy Antarctic expedition, 1946-47: Proc. 7th Pacific Sci. Cong., p. 335-344.

227 A symposium on the Gondwana Series was held in Algiers at the 19th International Geological Congress under the sponsorship of the International Commission on the Correlation of the Karroo (Gondwana) System.

Teichert, Curt, ed., 1952, Symposium sur les Series de Gondwana: XIX Internat. Geol. Cong., 395 p.

228 H.S. Ladd, E.Ingerson, R.C. Townsend, M. Russell and H.K. Stephenson reported that two drill-holes on Eniwetok atoll penetrated 1,405 and 1,266 m respectively of shallow-water carbonates before encountering olivine basalt basement directly beneath Eocene rocks.

Ladd, H.S., 1952, Foundation of Eniwetok Atoll (abs.): Geol. Soc. America Bull., v. 63, p. 1273.

Ladd, H.S., Ingerson, Earl, Townsend, R.C. Russell, Martin, and Stephenson, H.K., 1953, Drilling on Eniwetok Atoll, Marshall Islands: Amer. Assoc. Petroleum Geologists Bull., v. 37, p. 2257-2280.

229 H.W. Menard and R.S. Dietz reported that the previously discovered Mendocino escarpment off the California coast extends for at least 2,000 km westward and has a maximum relief of 3,200 m.

Menard, H. W., and Dietz, R. S., 1952, Mendocino submarine escarpment: Jour. Geology, v. 60, p. 266-278.

Dietz, R.S., 1952, The Pacific Floor: Sci. American, v.186, no. 4, p. 19-23.

230 R.R. Revelle, A.E. Maxwell and D.W. Butler reported that data gathered from the Horizon on Scripps' Mid-Pacific Expedition of 1950 show that heat flow through the floors of the deep oceans is approximately equal to that through the continents.

Revelle, R.R. and Maxwell, A.E., 1952, Heat flow through the floor of the north Pacific Ocean: Nature, v. 170, p. 199-200.

Bullard, E.C., 1952, (Discussion of paper cited above): Nature, v. 170, p. 200.

___ 1954, The flow of heat through the floor of the Atlantic Ocean: Royal Soc. London Proc., v. A222, p. 408-429.

1953

231 D.R. Carr and J.L. Kulp, used a uranium-helium age measurement technique to determine that a basalt boulder dredged from the Mid-Atlantic Ridge is between 15 and 45 million years old.

Carr, D.R. and Kulp, J.L., 1953, Age of a Mid-Atlantic Ridge basalt boulder: Geol. Soc. America Bull., v. 64, p. 253-254.

232 R.A. Fisher developed statistical techniques for handling directional data on a sphere, thereby aiding studies of paleomagnetism.

Fisher, R.A., 1953, Dispersion on a sphere: Royal Soc. London Proc., v. A217, p. 295-306.

233 M.L. Hill and T.W. Dibblee, Jr. cited conclusive evidence that at least 280 km of right lateral slip has taken place on the San Andreas Fault in California since the early Miocene and stated that as much as 560 km of movement may have taken place since the Jurassic.

Hill, M.L. and Dibblee, T.W. Jr., 1953, San Andreas, Garlock and Big Pine Faults, California: Geol. Soc. America Bull., v. 64, p. 443-458.
Hill, M.L., 1954, Tectonics of faulting in southern California: in Jahns, R. H., ed., Geology of southern California, Calif. Div. Mines Bull. 170, p. IV-5 to IV-13.
(See also)
Crowell, J.C., 1962, Displacement along the San Andreas Fault, California: Geol. Soc. America Spec. Paper 71, 61 p.

234 A. Rittman found that the calc-alkaline character of lavas increases regularly away from the Java Trench (toward Borneo) and suggested that study of the geochemistry of volcanics in relation to their structural position would contribute greatly to the understanding of "theoretical geology".

Rittman, Alfred, 1953, Magnetic character and tectonic position of the Indonesian volcanoes: Bull. Volcanologique, v. 14, p. 45-58.

235 The Royal Society of London held a discussion about the floor of the Atlantic Ocean. The participants were B.C. Browne, E.C. Bullard, M. Ewing, T.F. Gaskell, H.H. Hess, M.N. Hill, W.B.R. King, P.H. Kuenen, G.M. Lees, J.P. Rothe, F.P. Shepard, J.D.H. Wiseman, G.P. Wollard and J.L. Worzel.

Hill, M.N., 1953, The floor of the Atlantic Ocean: Nature, v. 171, p. 857-860.
Bullard, E.C. (Intro.), 1954, A discussion on the floor of the Atlantic Ocean: Royal Soc. London Proc., v. A222, p. 287-407.

236 M. Tharp, B.C. Heezen and (independently) M.N. Hill discovered that a deep graben valley occurs along the center of the Mid-Atlantic Ridge. Heezen and Tharp also realized that the rift valley coincides with the earthquake epicenter belt.

Hill, M.N., 1954, The topography of the Mid-Atlantic Ridge: Assoc. Oceanog. Phys. Proces-Verbaux, no. 6, Rome, p. 269.
Ewing, Maurice, and Heezen, B.C., 1956, Mid-Atlantic Ridge seismic belt (abs.): Amer. Geophys. Union Trans., v. 37, p. 343.
Hill, M.N., 1956, Notes on the bathymetric chart of the N.E. Atlantic: Deep-Sea Research, v. 3, p. 229-235.
Freeman, I.H., 1957, Crack in world is found at sea: New York Times, Feb. 1.
Heezen, B.C., 1958, A 45,000 mile crack: in Grolier Society, The story of our time, p. 285-287.
____ 1960, The rift in the ocean floor: Sci. American, v. 203, no. 4, p. 98-110.
Hill, M.N., 1960, A median valley of the Mid-Atlantic Ridge: Deep-Sea Research, v. 6, p. 193-205.

1954

237 J.A. Clegg, M. Almond and P.H.S. Stubbs determined from paleomagnetic studies of British late Paleozoic and early Mesozoic sedimentary rocks that (1) the magnetic field of the time was oriented 30 degrees east of its present direction; (2) the magnetic inclination was considerably less than it is now; and (3) the magnetic field has frequently reversed. A brief description was given of work by J. L. Lloyd in which sandstone was powdered and redeposited. The resulting magnetic field was found to be essentially parallel to the earth's present field.

Clegg, J.A., Almond, Mary, and Stubbs, P.H.S., 1954, The remanent magnetization of some sedimentary rocks in Britain: Philos. Mag., v. 45, p. 583-598.

238 Columbia University Geology Department held "an international gathering (in New York) for the purpose of crystallizing present knowledge of the earth's crust and its past history." The subject of continental drift was hardly mentioned.

Poldervaart, Arie, ed., 1955, Crust of the Earth: Geol. Soc. America Spec. Paper 62, 762 p.

239 K.M. Creer, E. Irving and S.K. Runcorn used paleomagnetic field orientation data from sedimentary rocks and lavas of various ages from Great Britain to construct the first polar wandering curve based on such evidence. They cited evidence that the magnetizations have been stable and also noted that the polarity of the magnetic field has frequently reversed.

Creer, K.M., Irving, E. and Runcorn, S.K., 1954, The direction of the geomagnetic field in remote epochs in Great Britain: Jour. Geomagnetism and Geoelectricity, v. 6, p. 163-168.

240 R.S. Dietz described a 2,500 km long chain of seamounts (some of which are guyots) that has a relief of 5,000 m and ties in with the Hawaiian Islands. He named them the Emperor Seamounts.

Dietz, R.S., 1954, Marine geology of the northwestern Pacific: description of Japanese bathymetric chart 6901: Geol. Soc. America Bull., v. 65, p. 1199-1224.

241 M. Ewing, D. B. Ericson, and B. C. Heezen discovered the Sigsbee Knolls in the deepest part of the Gulf of Mexico.

Ewing, Maurice, Ericson, D.B., and Heezen, B.C., 1958, Sediments and topography of the Gulf of Mexico: in Weeks, L.G., ed., Habitat of oil, Tulsa, Okla., Amer. Assoc. Petr. Geologists, p. 995-1053.

242 B. Luskin, B.C. Heezen, M. Ewing and M. Landisman described a sonic sounding device accurate to one part in 3,000.

Luskin, B., Heezen, B.C., Ewing, M. and Landisman, M., 1954, Precision measurement of ocean depth: Deep-sea Res., v. 1, p. 131-140.

243 M. Packard and R. Varian reported development of a simple, yet sensitive proton precession magnetometer that can measure the absolute value of the earth's magnetic field to about one part in 15,000.

Packard, Martin and Varian, Russell, 1954, Free nuclear induction in the earth's magnetic field (abs.): Phys. Rev., v. 93, p. 941.

244 R.W. Raitt reported that seismic refraction studies carried out during Scripps Capricorn expedition to the tropical Pacific in 1952-53 found no continental crust anywhere in the deep Pacific Ocean and determined that the oceanic crust is typically 5-10 km beneath the sea floor.

Raitt, R.W., 1954, Some results of seismic studies on Capricorn expedition (abs.): Geol. Soc. America Bull., v. 65, p. 1348.

___ 1956, Seismic-refraction studies of the Pacific Ocean basin, part I; crustal thickness of the central equatorial Pacific: Geol. Soc. America Bull., v. 67, p. 1623-1640.

245 J.P. Rothe used epicenter locations to demonstrate that the Indian Ocean Ridge and the Mid-Atlantic Ridge are connected.

Rothe, J.P., 1954, La zone seismique mediane Indo-Atlantique: Royal Soc. London Proc., v. A222, p. 387-399.

1955

246 S.W. Carey suggested that Spain has rotated counterclockwise since the Mesozoic and cited as supporting evidence several geologic "match-ups" between the northern Iberian coast and the western coast of Brittany.

Carey, S.W., 1955, The orocline concept in geotectonics: Royal Soc. Tasmania Proc., v. 89, p. 255-288.

247 S.W. Carey prepared plastic continental overlays whose boundaries were the 200 and 2,000 m isobaths and which were curved to fit on a 76 cm globe. The model demonstrated conclusively that the "fit" of South America and Africa is excellent, with very little overlap or underfit.

Carey, S.W., 1955, Wegener's South America-Africa assembly, fit or misfit?: Geol. Magazine, v. 92, p. 196-200.

248 H.H. Hess recognized that the Romanche "Trench" on the equatorial Mid-Atlantic Ridge is a fracture zone similar to those in the northeastern Pacific.

Hess, H.H., 1955, The oceanic crust: Jour. Marine Research, v.14, p. 423-439.

249 H.W. Menard described the Murray fracture zone (3,000 km long, maximum relief of 2,100 m), Clarion fracture zone (4,800 km, 1,700 m) and the Clipperton fracture zone (5,300 km, 750 m) that trend westward from the North American coast. Two of these fracture zones were reported to have scarps which are about 1,600 m high continuously for over 1,600 km of their length.

Menard, H.W., 1955, Deformation of the northeastern Pacific basin and the west coast of North America: Geol. Soc. America Bull., v. 66, p. 1149-1198.

___ 1955, Fractures in the Pacific floor: Sci. American, v. 193, no. 1, p. 36-41.

250 J. Oliver, M. Ewing and F. Press showed that seismic surface-wave evidence indicates the Arctic Ocean basin is underlain by oceanic crust.

Oliver, J., Ewing, M. and Press, F., 1955, Crustal structure of the Arctic regions from Lg phase: Geol. Soc. America Bull., v. 66, p. 1063-1074.

251 United States Coast and Geodetic Survey vessel Pioneer began cruising a series of closely-spaced east-west lines off the west coast of the United States. By agreement with Scripps Institution of Oceanography, a newly developed magnetometer was towed behind the ship for the purpose of gathering data to be used in making the first detailed magnetic anomaly map of a large oceanic area. The magnetometer, originally designed for airborne detection of submarines, was modified according to plans developed by the U. S. Naval Ordinance Laboratory.

Mason, R.G., 1958, A magnetic survey off the west coast of the United States between latitudes 32 and 36 degrees N and longitudes 121 and 128 degrees W: Geophys. Jour., v. 1, p. 320-329.

Raff, A. D., 1961, The magnetism of the ocean floor: Sci. American, v. 205, no. 4, p. 146-156.

252 R.W. Raitt, R.I. Fisher and R.G. Mason reported that data gathered during the 1952-53 Scripps Capricorn expedition established that the Tonga Trench (1) contains less than 200 m of sediment; (2) is topographically steeper on the west side; and (3) has a stronger magnetic field on the west that may indicate depression of the Curie isothermal surface.

Raitt, R.W., Fisher, R.I., and Mason, R.G., 1955, Tonga Trench: in Poldervaart, Arie, ed., Crust of the earth: Geol. Soc. America Spec. Paper 62, p. 237-254.

1956

253 V.L. Bosazza used a suggested pre-drift fit of Madagascar with Africa to predict the occurrence of certain rare radioactive minerals in Malawi. The predicted occurrences were verified in 1957.

Bosazza, V.L., 1959, Radioactive minerals in southern Nyasaland: Mining Mag. (London), v. 101, p. 4955.

___ 1970, Continental drift and un-continental thinking: Econ. Geol., v. 65, p. 892.

254 E.C. Bullard, A.E. Maxwell and R.R. Revelle determined that heat flows are as much as double on oceanic ridges as the average in ocean basins and on continents.

Bullard, E.C., Maxwell, A.E., and Revelle, R.R., 1956, Heat flow through the deep sea floor: in Landsberg, H.E., ed., Advances in Geophysics, v. 3, New York, Academic Press, p. 153-181.

Bullard, E.C., and Day, A., 1961, The flow of heat through the floor of the Atlantic Ocean: Royal Astron. Soc. Geophys. Jour., v. 4, p. 282-292.

255 Cambridge University Department of Geodesy and Geophysics developed a ship-towed nuclear-spin (proton precession) magnetometer.

Hill, M.N., 1959, A ship-borne nuclear-spin magnetometer: Deep-Sea Research, v. 5, p. 309-311.

256 S.W. Carey convened a symposium on continental drift at the University of Tasmania, Hobart.

Carey, S.W., 1958, Continental drift - a symposium: Geol. Dept., Univ. of Tasmania, Hobart, 375 p.

257 J.A. Clegg, E.R. Deutsch and D.H. Griffiths concluded from studies of the paleomagnetism of the Deccan basalts that India has drifted northward through over 50 degrees of latitude (5,000 km) and rotated 25 degrees counterclockwise since the late Cretaceous-Eocene.

Clegg, J.A., Deutsch, E.R., and Griffiths, D.H., 1956, Rock magnetism in India: Philos. Mag., v. 1., p. 419-431.

Deutsch, E.R., Radakrishnamurty, C., and Sahasrabudhe, P.W., 1959, Paleomagnetism of the Deccan Traps: Annales Geophysique, v. 15, p. 39-59.

258 F.B. Estabrook proposed that a well be drilled from an oceanic island through the moho.

Estabrook, F.B., 1956, Geophysical research shaft: Science, v. 124, p. 686.

259 M. Ewing and B. Heezen inferred that the oceanic ridge system is a continuous, world-encircling feature that can be traced by the pattern of associated earthquake epicenters where sounding information is not available and suggested that the earthquakes are concentrated along the central rift valley.

Ewing, Maurice, and Heezen, Bruce, 1956, Mid-Atlantic Ridge seismic belt (abs.): Am. Geophys. Union Trans., v. 37, p. 343.

___ 1956, Some problems of Antarctic submarine geology: in Crary, A., and others, eds., Antarctica in the International Geophysical Year, Am. Geophys. Union Geophys. Mon. Ser., v. 1, p. 75-81.

Freeman, I.H., 1957, Crack in world is found at sea: New York Times, Feb. 1.

Heezen, B.C., Tharp, Marie, and Ewing, Maurice, 1959, The floors of the oceans, I The North Atlantic: Geol. Soc. America Spec. Paper 65, 122 p.

260 S.H. Haughton and C. Teichert organized a symposium on the Karroo (Gondwana) System for the 20th International Geological Congress in Mexico City.

Anonymous, 1958 (actually 1961 or 62), Comision para la correlacion del Sistema Karroo (Gondwana): XX Internat. Geol. Cong., 245 p.

261 E. Irving showed that paleomagnetic pole location data from various continents suggest the continents have moved with respect to each other. Europe and North America, for example, were said to have been much closer before the early Mesozoic.

Irving, E., 1956, Paleomagnetic and paleoclimatological aspects of polar wandering: Geofisica Pura e Applicada, v. 33, p. 23-41.

Irving, E., 1956, Rock magnetism: a new approach to the problems of polar wandering and continental drift: in Carey, S.W., ed., Continental drift - a symposium, Hobart, Geol. Dept., Univ. of Tasmania, p. 24-61.

262 R.G. Mason, A.D. Raff, V. Vacquier and R.E. Warren discovered a pattern of north-south linear magnetic anomalies in the deep ocean off the west coast of the United States and used offsets in the pattern to determine apparent displacements along the Murray (155 km offset), Pioneer (265 km) and Mendocino (1160 km) fracture zones.

Mason, R.G., 1958, A magnetic survey off the west coast of the United States between latitudes 32 degrees and 36 degrees N, longitudes 121 degrees and 128 degrees W: Geophys. Jour., v. 1, p. 320-329.

Vacquier, Victor, 1959, Measurement of horizontal displacement along faults in the ocean floor: Nature, v. 183, p. 452-453.

Mason, R.G., and Raff, A.D., 1961, Magnetic survey off the west coast of North America, 32 degrees N. latitude to 42 degrees N. latitude: Geol. Soc. America Bull., v. 72, p. 1259-1266.

Raff, A.D., and Mason, R.G., 1961, Magnetic survey off the west coast of North America, 40 degrees N. latitude to 52 degrees N. latitude: Geol. Soc. America Bull., v. 72, p. 1267-1270.

Vacquier, Victor, Raff, A.D., and Warren, R.E., 1961, Horizontal displacements in the floor of the northeastern Pacific Ocean: Geol. Soc. America Bull., v. 72, p. 1251-1258.

Vacquier, Victor, 1962, Magnetic evidence for horizontal displacements in the floor of the Pacific Ocean: in Runcorn, S. K., ed., Continental drift, New York, Academic Press, p. 135-144.

___ 1965, Horizontal displacements in the earth's crust: in Blackett, P.M.S., and others, eds., A symposium on continental drift, Royal Soc. London, p. 77-81.

263 A.E.M. Nairn concluded that paleomagnetic pole positions determined from Jurassic rocks from England, Arizona, Rhodesia and Tasmania are better explained by continental drift than by polar wandering.

Nairn, A.E.M., 1956, Relevance of paleomagnetic studies of Jurassic rocks to continental drift: Nature, v. 178, p. 935-936.

264 John Reynolds developed a mass spectrometer in which the containing mechanism was all glass and could be heated under vacuum to drive off absorbed atmospheric argon, thus allowing precise dating by the K-Ar method of rocks younger than ten million years.

Reynolds, John, 1956, High sensitivity mass-spectrometer for noble gas analysis: Rev. Sci. Instruments, v. 27, p. 928-934.

265 S. K. Runcorn showed that the paleomagnetic North Pole path for North America falls systematically west of the European path and concluded that North America was displaced westward relative to Europe by 24 degrees since the Triassic.

Runcorn, S.K., 1956, Paleomagnetic comparisons between Europe and North America: Geol. Assoc. Canada Proc., v. 8, p. 77-85.

1957

266 J.A. Clegg, E.R. Deutsch, C.W.F. Everitt, and P.H.S. Stubbs measured the magnetic orientation of certain Triassic sandstones from Spain and found that they confirm the postulated counter-clockwise rotation of the Iberian peninsula.

Clegg, J.A., Deutsch, E.R., Everitt, C.W.F., and Stubbs, P.H.S., 1957, Some recent paleomagnetic measurements made at Imperial College, London: Advances in Physics, v. 6, p. 219-231.

267 A team of U. S. geophysicists from Lamont Geological Observatory, headed by M. J. Davidson and working from IGY drifting station Alpha, discovered the Alpha Ridge.

Anonymous, 1957, Submarine mountain range: Science, v. 126, p. 499.

268 M. Ewing, B.C. Heezen and J. R. Hirschman reported that a characteristic positive magnetic anomaly coincides with the rift valley at the crest of the Mid-Atlantic Ridge and suggested that such an anomaly may be characteristic of oceanic ridge crests generally.

Ewing, Maurice, Heezen, B.C., and Hirschman, J.R., 1957, Magnetic anomalies and seismicity in the Mid-Atlantic Ridge and its extensions: Assoc. Seismol. Phys. Interieur de la Terre, Assemblee de Toronto, Resume 110 bis.

Ewing, Maurice, Hershman, Julius, and Heezen, B.C., 1959, Magnetic anomalies of the mid-ocean rift: in Sears, Mary, ed., International Oceanographic Congress preprints, Am. Assoc. Adv. Sci., Washington, D. C., p. 24.

Heezen, B.C., Tharp, Marie, and Ewing, Maurice, 1959, The floors of the oceans, I The north Atlantic: Geol. Soc. America Spec. Paper 65, 122 p.

269 The International Union of Geodesy and Geophysics adopted resolutions that (1) urged "the nations of the world and especially those experienced in deep drilling to study the feasibility and cost of an attempt to drill to the Mohorovicic discontinuity at a place where it approaches the surface", and (2) recommended the development and carrying out of deep ocean drilling.

International Union of Geodesy and Geophysics, 1957, Resolutions Numbers 11 and 12: in Comptes Rendus, XI Assembly Generale, Internat. Union Geodesy and Geophysics, Toronto, 3-14 Sept., p. 181.

270 E. Irving and R. Green described paleomagnetic evidence from Australian Eocene rocks that suggest there has been as much as 2,000 km of relative movement between Europe and Australia since that time and great relative movement between Australia and North America. They found evidence of many reversals of the earth's magnetic field.

Irving, E., and Green, R., 1957, Paleomagnetic evidence from the Createaceous and Cainozoic: Nature, v. 179, p. 1064-1065.

271 W. Munk conceived the idea of drilling through the Mohorovicic discontinuity and sampling the mantle.

Bascom, Willard, 1959, The Mohole: Sci. American, v. 200, no. 4, p. 41-49.

Hess, H.H., 1959, The AMSOC hole to the earth's mantle: Amer. Geophys. Union Trans., v. 40, p. 340-345.

Bascom, Willard, 1961, A hole in the bottom of the sea: Garden City, New York, Doubleday, 352 p.

Hess, H.H., 1966, The Mohole Project, Phase II: in Findlay, D. C., and Smith, C. H., eds., Drilling for scientific purposes, Geol. Surv. of Canada Paper 66-13, p. 146-150.

272 P. St. Amand and H. Benioff suggested, on the basis of strike-slip movements along transcurrent faults, that the entire Pacific basin may be rotating counterclockwise relative to the surrounding continental masses.

Benioff, Hugo, 1957, Circum-Pacific tectonics: in Hodgson, J.H., The mechanics of faulting, with special reference to fault plane work, Publications of the Dominion Observatory, Ottowa, v. 20, p. 395-402.

St. Amand, P., 1957, Geological and geophysical synthesis of the tectonics of portions of British Columbia, the Yukon Territory and Alaska: Geol. Soc. America Bull., v. 68, p. 1343-1370.

1957 - 1958

273 L.J.B. LaCoste and Anton Graf independently developed spring-type gravimeters suitable for sea-surface work. The first successful measurements were made as an outgrowth of International Geophysical Year activities.

Harrison, J.C., 1959, Tests of the LaCoste-Romberg surface-ship gravity meter, I: Jour. Geophys. Research, v. 64, p. 1875-1881.

IGY Bulletin, 1958, First sea-surface gravimeter: Amer. Geophys. Union Trans., v. 39, p. 175-178.

LaCoste, L.J.B., 1959, Surface ship gravity measurements on the Texas A. and M. College ship, the Hidalgo: Geophysics, v. 24, p. 309-322.

Worzel, J.L., 1959, Continuous gravity measurements on a surface ship with the Graf sea gravimeter: Jour. Geophys. Research, v. 64, p. 1299-1315.

1958

274 The Alberta Society of Petroleum Geologists sponsored a symposium in Calgary on polar wandering and continental drift.

Raasch, G.O., ed., 1958, Polar wandering and continental drift, a symposium: Alberta Soc. Petroleum Geologists Jour., Pt. 1, v. 6, p. 139-162; Pt. 2, p. 170-188.

275 An international conference was held in Geneva by scientists and other representatives from the United States, the United Kingdom, France, Canada, the U.S.S.R., Czechoslovakia, Roumania and Poland for the purpose of determining whether existing techniques for detecting atomic explosions and discriminating them from earthquakes were adequate to insure that a proposed atomic-test-ban agreement was observed. An outgrowth of the conference was the realization among geophysicists, the military and even some politicians that seismology was in rather poor condition. As a result of this realization and the importance of seismology to national defense and foreign policy, tens of millions of dollars were made available for seismic research and instrumentation, and rapid progress was made in many branches of the field, some of them crucial to plate tectonics.

United Nations Report on the Conference of Experts to Study the Methods of Detecting Violations of a Possible Agreement on the Suspension of Nuclear Tests, 1958: New York, United Nations.

(See also)

Bolt, B.A., 1976, Nuclear explosions and earthquakes - the parted veil: San Francisco, W. H. Freeman, 309 p.

276 E. Irving and R. Green showed that variations in Australia's geomagnetic latitude (from about 30 degrees N to about 80 degrees S) agree with the paleoclimatic evidence. For example, both lines of evidence were said to show that Australia was at high latitudes during the late Paleozoic. The lack of agreement of Australian paleopole positions with those from other continents was cited as evidence favoring continental drift.

Irving, E., and Green, R., 1958, Polar movement relative to Australia: Royal Astron. Soc. Geophys. Jour., v. 1, p. 64-72.

277 M. Landisman and Y. Sato used observations of Love waves from a 1931 Sinkiang earthquake as a basis for confirming the existence of a low-velocity layer between 100 and 200 km below the earth's surface.

Landisman, M., and Sato, Y., 1958, Shear wave velocities in the upper mantle: Amer. Geophys. Union Trans., v. 39, p. 522.

278 C.D. McClure, H.F. Nelson, W.B. Huckaby and L.R. Padberg described continuous-recording, sub-bottom seismic reflection profilers that make use of an electro-acoustic transducer to generate high-energy pulses of low-frequency sound that penetrate sea-floor sediments and show their stratification and thickness (up to about 1 km).

McClure, C.D., Nelson, H.F., and Huckaby, W.B., 1958, Marine Sonoprobe system, new tool for geologic mapping: Am. Assoc. Petroleum Geologists Bull., v. 42, p. 701-706.

Padberg, L.R., 1958, Method of subsurface exploration by sonic means: U.S. Patent No. 2,866,512, 7 p.

279 H.W. Menard demonstrated that the oceanic ridge system follows the median line of the Atlantic, Antarctic and Indian oceans with remarkable consistency and that even in the Pacific there is considerable agreement between the two.

Menard, H.W., 1958, Development of median elevations in ocean basins: Geol. Soc. America Bull., v. 69, p. 1179-1186.

1958 - 1959

280 United States' nuclear submarines Nautilus and Skate obtained the first continuously recorded echo-sounding profiles across and within the Arctic Ocean basin during their historic under-the-ice voyages. The profiles confirmed earlier predictions on the basis of earthquake epicenter locations that a continuation of the Mid-Atlantic Ridge would be found in the area north of Spitzbergen and showed that the predicted rift valley is well-developed.

Anderson, W.R., 1959, Nautilus-90-north: Cleveland, World Pub., 251 p.

Lalor, W.G., 1959, Submarine through the North Pole: Natl. Geol. Mag., v. 115, p. 1-24.

Dietz, R.S., and Shumway, G.A., 1959, Arctic basin geomorphology: in Sears, Mary, ed., International Oceanographic Congress Preprints, Am. Assoc. Adv. Sci., Washington, D. C., p. 18-19.

Dietz, R.S. and Shumway, George, 1961, Arctic basin geomorphology: Geol. Soc. America Bull., v. 72, p. 1319-1330.

1958 - 1971

281 B.C. Heezen and M. Tharp produced a series of four physiographic diagrams of the world's ocean basins (north Atlantic, south Atlantic, Indian, and western Pacific). The diagrams were interpretive in part and illustrated the various bathymetric features

(oceanic ridges, fracture zones, volcanic chains and trenches) which were then being newly interpreted in terms of sea-floor spreading.

Heezen, B.C., and Tharp, Marie, 1958, Physiographic diagram of the north Atlantic Ocean: Geol. Soc. America, accompanied Spec. Paper 65. (revised 1968.)
___ 1962, Physiographic diagram of the south Atlantic Ocean: Geol. Soc. America.
___ 1964, Physiographic diagram of the Indian Ocean: Geol. Soc. America.
___ 1971, Physiographic diagram of the western Pacific Ocean: Geol. Soc. America.

1959

282 M. Ewing and M. Landisman reported that the oceanic ridge system is a continuous world-encircling feature whose existence had been confirmed by sounding in many remote parts of the world ocean where its occurrence had previously been suggested only by the location of the associated epicenter belt. It was also said that the ridge characteristically is 1,500 km wide, 3 km high, has a deep central rift valley, and usually bisects the oceans in which it is found. Several extensions of the system to nearby continents were also described.

Ewing, Maurice, and Landisman, Mark, 1961, Shape and structure of ocean basins: in Sears, Mary, ed., Oceanography - Invited lectures presented at the International Oceanographic Congress held in New York, 31 August - 12 September 1959, Amer. Assoc. Adv. Sci. Pub. No. 67, p. 3-38.

283 B. C. Heezen hypothesized that ocean basins may form by a continual process of expansion of rift valleys (and the earth), starting with a form like the Gulf of Aden and proceding to something like the Atlantic Ocean.

Heezen, B.C., 1959, Paleomagnetism, continental displacements, and the origin of submarine topography: in Sears, Mary, ed., International Oceanographic Congress preprints, Am. Assoc. Adv. Sci., Washington, D. C., p. 26-28.

284 An International Oceanographic Congress was held in New York. Among the subjects discussed were: shape and structure of the ocean basins and the forces involved; stratigraphy of the deep sea and the marine climatic record; and the origin, distribution, constituents and processes affecting deep-sea sediments. Curiously, the two most striking facts about deep-sea sediments, their unexpected and then only recently discovered thinness and youthfulness, were hardly mentioned in the preprints.

Sears, Mary, 1959, International Oceanographic Congress preprints: Washington, D. C., Amer. Assoc. Adv. Sci., 1022 p.

285 H. Kuno showed that there is a systematic increase in the alkali content of basalts across the Japanese Islands and inferred that the trend is related to the depth of the underlying earthquake zone.

Kuno, Hisashi, 1959, Origin of Cenozoic petrographic provinces of Japan and surrounding areas: Bull. Volcanologique, v. 20, p. 37-76.

286 M.G. Rutten introduced the use of K-Ar dates to reversal chronology studies and found that certain Italian lavas extended the present interval of normal polarity back to almost -.5 My and that there may have been another interval of normal polarity at about -2.4 My.

Rutten, M.G., 1959, Paleomagnetic reconnaissance of mid-Italian volcanoes: Geologie en Mijnbouw, v. 21, p. 373-374.

287 R.P. Von Herzen reported finding a broad band of high heat flows along 6,000 km of the East Pacific Rise and low flows in some trenches.

Von Herzen, R.P., 1959, Heat flow values from the southeastern Pacific: Nature, v. 183, p. 882-883.

288 R.A.M. Wilson inferred that the Troodos Massif of Cyprus formed by the injection of numerous dikes as the area was subjected to tensional stress.

Wilson, R.A.M., 1959, The geology of the Xeros-Troodos area: Geol. Survey Dept. Cyprus Memoir No. 1, 135 p.

1960

289 M. Ewing and B.C. Heezen showed that the Indian Ocean Ridge and the Mid-Atlantic Ridge are topographically connected, that the central rift valley also occurs in the Indian Ocean south of Rodriguez Island, and that the rift follows the epicenter belt as previously predicted by them.

Ewing, Maurice and Heezen, B.C., 1960, Continuity of mid-oceanic ridge and rift valley in the southwestern Indian Ocean confirmed: Science, v. 31, p. 1677-1678.

290 J.I. Ewing and G.B. Tirey with assistance from A. Stockel, D. Loelsch, S. Gerber and J.L. Worzel developed the apparatus and field techniques for doing continuous explosion seismic reflection work at sea and obtaining profiles automatically. The method was shown to be capable of working in deep water and revealing the thickness (up to 5 km) and stratification of unlithified sediments and the topography of the surface on which they were deposited.

Ewing, J.I. and Tirey, G.B., 1961, Seismic profiler: Jour. Geophys. Research, v. 66, p. 2917-2927.

291 H.H. Hess advanced what is now called the sea-floor spreading hypothesis.

Hess, H.H., 1960, Evolution of ocean basins: mimeographed pre-publication manuscript, 39 p.

___ 1962, History of ocean basins: in Engel, A.E.J. and others, eds., Petrological studies - a volume in honor of A.F. Buddington, Geol. Soc. America, p. 599-620. (Reprinted in Cloud, 1970, and Cox, 1973.)

___ 1965, Mid-oceanic ridges and tectonics of the sea-floor: in Whittard, W. F., and Bradshaw, R., eds., Submarine geology and geophysics, London, Butterworths, p. 317-322.

Vine, F.J., and Hess, H.H., 1970, Sea-floor spreading: in Maxwell, A.E., The Sea, v. 4, part II, New York, Wiley-Interscience, p. 587-622.

292 H. W. Menard showed that offsets of depth contours in the northeast Pacific agree with offsets of the magnetic anomaly pattern and identified the Gorda and Juan de Fuca ridges as being extensions of the crest of the East Pacific Rise.

Menard, H.W., 1960, The East Pacific Rise: Science, v. 132, p. 1737-1746.

___ 1961, The East Pacific Rise: Sci. American, v. 205, no. 6, p. 52-61.

293 The Society of Economic Paleontologists and Mineralogists held a symposium on "The Mineralogic and Paleontologic Aspects of Continental Drift and Polar Wandering" at Atlantic City, New Jersey.

Munyan, A.C., ed., 1963, Polar wandering and continental drift: Soc. Econ. Paleontologists and Mineralogists Spec. Pub. No. 10, Tulsa, Okla., 169 p.

___ 1960, Polar wandering and continental drift: Science, v. 132, p. 1264-1266.

1960 - 1961

294 P.M.S. Blackett, J.A. Clegg and P.H.S. Stubbs evaluated five possible causes for the apparent variation in orientation of Earth's magnetic field through time and found that continental drift was the most plausible cause, although the possibility of a time-dependent, non-dipole field could not be excluded. Blackett then compared the paleolatitudes as indicated by rock magnetic results with those determined by paleoclimatic indicators and found general agreement.

Blackett, P.M.S., Clegg, J.A., and Stubbs, P.H.S., 1960, An analysis of rock magnetic data: Royal Soc. London Proc., v. A256, p. 291-322.

Blackett, P.M.S., 1961, Comparison of ancient climates with the ancient latitudes deduced from rock magnetic measurements: Royal Soc. London Proc., v. A263, p. 1-30.

1961

295 R. S. Dietz authored the first published account of and named the hypothesis of sea-floor spreading. He suggested that underthrusting at geosynclines such as at the eastern continental margin of the United States during the Paleozoic could be the cause of mountain building.

Dietz, R.S., 1961, Continent and ocean basin evolution by spreading of the sea-floor: Nature, v. 190, p. 854-857.

___ 1961, The spreading ocean floor: Saturday Evening Post, v. 242, no. 42 (Oct. 21), p. 34-35, 94, 96.

___ 1962, Ocean-basin evolution by sea-floor spreading: in Runcorn, S. K., ed., Continental drift, New York, Academic Press, p. 289-307.

___ 1964, Commotion in the ocean - the growth of continents and ocean basins: in Yoshida, Kozo, ed., Studies on oceanography, Tokyo, Univ. of Tokyo Press, p. 465-478. (U.S. ed. in 1965 by Univ. of Washington Press, Seattle)

___ 1972, Geosynclines, mountains and continent building: Sci. American, v. 226, no. 3, p. 30-38.

296 W. Hamilton concluded from geologic evidence that Baja California was once a part of mainland Mexico, with the Gulf of California having formed by a combination of 480 km of strike slip and up to 160 km of cross strike separation with "subcontinental material having welled up into the rift gap" behind the "moving plate". The San Andreas fault was said to have undergone 560 km of strike slip since the late Cretaceous.

Hamilton, Warren, 1961, Origin of the Gulf of California: Geol. Soc. America Bull., v. 72, p. 1307-1318.

297 B.C. Heezen, M. Tharp and R.D. Gerard used detailed sounding surveys to show that the equatorial Mid-Atlantic Ridge is cut by several fracture zones and that the ridge crest may be offset by as much as 1,000 km.

Heezen, B.C., Tharp, Marie, and Gerard, R.D., 1961, Equatorial Atlantic fracture zones (abs.): Geol. Soc. America Spec. Paper 68, p. 195-196.

Heezen, B.C., and Tharp, Marie, 1961, Physiographic diagram of the South Atlantic Ocean: Boulder, Colorado, Geol. Soc. America.

Heezen, B.C., Gerard, R.D., and Tharp, Marie, 1964, The Vema fracture zone in the equatorial Atlantic: Jour. Geophys. Research, v. 69, p. 733-739.

Heezen, B.C., Bunce, E.T., Hersey, J.B., and Tharp, Marie, 1964, Chain and Romanche fracture zones: Deep-Sea Research, v. 11, p. 11-33.

298 H. Martin compared the stratigraphy of Brazil to that of southwestern Africa and found that every change in stratigraphic knowledge of the Silurian-Cretaceous interval since the time of Du Toit (1937) has increased the similarity between the two areas, to the point that "The stratigraphic and lithologic columns for this period of some 200 my have become almost identical." He also suggested that a definitive test of the drift theory in this area could be made by deep-sea drilling in the south Atlantic.

Martin, Henno, 1961, The hypothesis of continental drift in the light of recent advances of geological knowledge in Brazil and in south-west Africa: Geol. Soc. South Africa Trans., Annexure to v. 64, p. 1-47.

299 Project Mohole conducted test drilling from Global Marine Exploration Company's drilling ship Cuss I in waters up to 3,500 m deep near Guadalupe Island off Baja California. Five holes were drilled and up to 310 m of sediment penetrated. Vesicular basalt was encountered beneath 170 m of sediment in one hole and was said to coincide with the top of "layer 2" as determined by seismic methods.

AMSOC Committee, 1961, Experimental drilling in deep water at La Jolla and Guadalupe sites: Natl. Acad. Sci. - Natl. Research Council Pub. No. 914, Washington, D. C., 183 p.

Anonymous, 1961, Preliminary Mohole Project drilling successful: Geotimes, v. 5, no. 8, p. 10-13.

Horton, E. E., 1961, Preliminary drilling phase of Mohole Project, I Summary of drilling operations: Am. Assoc. Petroleum Geologists Bull., v. 45, p. 1789-1792.

Riedel, W.R., Ladd, H.S., Tracey, J.I. Jr., and Bramlette, M. N., 1961, Preliminary drilling phase of Mohole Project, II Summary of coring operations: Am. Assoc. Petroleum Geologists Bull., v. 45, p. 1793-1798.

300 A.D. Raff suggested that the oceanic linear magnetic anomalies may generally be parallel to the crest of the oceanic ridge system.

Raff, A.D., 1961, The magnetism of the ocean floor: Sci. American, v. 205, no. 4, p. 146-156.

301 The United States Coast and Geodetic Survey, with financial support from the Defense Department, began installation of a World-Wide Standardized Seismic Network (WWSSN).

Murphy, L.M., 1966, World-wide seismic network: in Environmental Science Services Administration, Symposium on Earthquake Prediction, U.S. Govt. Printing Ofc., Washington, D.C., p. 53-56.

1962

302 H. Benioff used epicenter locations and first-motion studies to identify and locate an active, dextral, transcurrent fault off the west coast of North America from about 47 degrees N to about 60 degrees N.

Benioff, Hugo, 1962, Movements on major transcurrent faults: in Runcorn, S.K., ed., Continental Drift, New York, Academic Press, p. 103-134.

303 R.R. Coats deduced from structural, seismologic, mineralogical and chemical data that eugeosynclinal sediments and basaltic volcanics have been carried down a major thrust fault that dips at about 30 degrees northward beneath the Aleutian arc to depths of at least 100 km.

Coats, R.R., 1962, Magma type and crustal structure in the Aleutian Arc: in MacDonald, G.A., and Kuno, Hishashi, eds., Crust of the Pacific Basin, Am. Geophys. Union, Geophys. Mon. Ser., v. 6, p. 92-109.

304 J.P. Eaton reported that earthquake foci associated with Hawaiian volcanism occur as deep as 60 km and suggested that the lavas originate in the mantle.

Eaton, J.P., 1962, Crustal structure and volcanism in Hawaii: in Macdonald, G. A., and Kuno, Hisashi, eds., The crust of the Pacific basin, Am. Geophys. Union Geophys. Mon. 6.

305 R.W. Girdler suggested that a 50 km-wide strip of ocean may be forming in the Red Sea as a result of convective upwelling and intrusion of basic material into a central rift that is continuously opening. As evidence he cited seismic and gravity data and reported that magnetic anomaly patterns parallel the coastlines of both the Red Sea and the Gulf of Aden. The tension crack filled with oceanic crust was said to open to the south, suggesting a relative rotation between Africa and Arabia. High heat flows in both the Red Sea-Gulf of Aden area and in the Gulf of California were cited as additional evidence that both areas are examples of an early stage of continental drift.

Girdler, R.W., 1962, Initiation of continental drift: Nature, v. 194, p. 521-524.

306 R.H. Gunst and E.R. Engdahl described the computer techniques developed at and used by the United States Coast and Geodetic Survey to determine epicenter locations to within about .1 degree and focal depths to within 25 km.

Gunst, R.H. and Engdahl, E.R., 1962, Progress report of USC & GS hypocenter computer program: Seismol. Soc. America, Eastern Section, Earthquake Notes, v. 33, p. 93-96,

307 The International Union of Geodesy and Geophysics sponsored a symposium in Kyoto, Japan on paleomagnetism.

Nagata, Takesi, 1962, Symposium on rock-and paleo-magnetism: Jour. Geomagnetism and Geoelectricity, v. 13, p. 63-165.

308 L.C. King put together an integrated picture of continental drift through time and cited numerous lines of evidence from stratigraphy, paleontology, glacial geology, geomorphology, geophysics and structural geology in support of his ideas.

King, L.C., 1962, The morphology of the earth: New York, Hafner Pub. Co., 699 p. (2nd ed., Edinburgh & London, Oliver & Boyd, 1967, 726 p.)

309 W.E. Long reported the discovery of a 250 m thick late Paleozoic deposit of tillite (including striated boulders and a boulder pavement) resting on a striated and grooved pavement in the Horlick Mountains of Antarctica, suggesting that Antarctica experienced a late Paleozoic ice age like the other Gondwana continents.

Long, W.E., 1962, Sedimentary rocks of the Buckeye Range, Horlick Mountains, Antarctica: Science, v. 136, p. 319-321.

___ 1964, The statigraphy of the Horlick Mountains: in Adie, R.J., ed., Antarctic geology, Amsterdam, North Holland Pub. Co., p. 353-363.

___ 1965, Stratigraphy of the Ohio Range, Antarctica: in Hadley, J. B., ed., Geology and paleontology of the Antarctic, v. 6 of Antarctic Research Series, Washington, D. C., Amer. Geophys Union, p. 71-116.

310 E.P. Plumstead concluded from her studies of fossil plants collected in 1955-1958 by the British Trans-Antarctic Expedition that those fossils strongly support the theory of continental drift and that they cannot be adequately explained by any other hypothesis. She also reviewed the paleobotanical evidence in general for drift and found it to be even stronger than generally supposed.

Plumstead, E.P., 1962, Fossil floras of Antarctica: in Trans-Antarctic Expedition 1955-1958, Sci. Repts. No. 9, Pt. 2, 154 p. + plates.

311 S. K. Runcorn edited a book whose purpose was to assemble in one volume "a discussion of the geophysical evidence relating to horizontal movements in the earth's crust in its widest sense."

Runcorn, S.K., ed., 1962, Continental drift: New York, Academic Press, 338 p.

312 V. Vacquier postulated that the linear magnetic anomalies discovered in the northeast Pacific constitute a record of the formation of rift oceans by some kind of regenerative process and that they should be parallel to the oceanic ridges.

Vacquier, Victor, 1962, Magnetic evidence for horizontal displacements in the floor of the Pacific Ocean: in Runcorn, S. K., ed., Continental drift, New York, Academic Press, p. 135-144.

1962 - 1974

313 V.V. Beloussov attacked various aspects of plate tectonics theory. Among the points raised were the following: (1) displacement of oceanic ridges seems mechanically unlikely; (2) crustal-age distribution in the Pacific cannot be accounted for; (3) consumption of a spreading center by a subduction zone (as, for example, in the Aleutians) is unexplainable; (4) the magnetic anomaly pattern in the Red Sea is oblique to its supposed spreading center; (5) it is unjustifiable to extrapolate modern seismicity patterns into the past; (6) Many details of the theory are untestable; and (7) continental geological history is completely ignored. Beloussov advocated a theory of vertical movements in continental interiors and periodic melting of the upper mantle with accompanying envelopment of continental crust.

Beloussov, V.V., 1962, Basic problems in geotectonics: New York, Mcgraw-Hill, 816 p.

___ 1970, Against the hypothesis of ocean-floor spreading: Tectonophysics, v. 9, p. 498-511.

___ 1974, Sea-floor spreading and geologic reality: Kahle, C.F., ed., Plate tectonics: assessments and reassessments, Tulsa, Oklahoma, Amer. Assoc. Petroleum Geologists Mem. 23, p. 155-166.

___ 1979, Why I do not accept plate tectonics: EOS, v. 60, p. 207-208.

1963

314 A. Cox, R.R. Doell and G.B. Dalrymple proposed the first polarity reversal time scale. The time scale extended back to -3.5 My.

Cox, Allan, Doell, R.R., and Dalrymple, G.B., 1963, Geomagnetic polarity epochs and Pleistocene geochronometry: Nature, v. 198, p. 1049-1051. (Reprinted in Cox, 1973.)

315 R.S. Dietz suggested that ophiolite suites are fragments of the oceanic layer and mantle that have been tectonically emplaced in orogenic belts by the process of sea-floor spreading.

Dietz, R.S., 1963, Alpine serpentines as oceanic rind fragments: Geol. Soc. America Bull., v. 74, p. 947-952.

316 C.L. Drake, J.R. Heirtzler and J. Hirschman discovered a pattern of linear magnetic anomalies in the Atlantic Ocean off the east coast of the United States and found the pattern offset by 160 km along a 1,000 km long fault.

Drake, C.L., Heirtzler, J.R., and Hirschman, J., 1963, Magnetic anomalies off eastern North America: Jour. Geophys. Research, v. 68, p. 5259-5275.

317 I.G. Gass and D. Masson-Smith suggested that the Troodos massif of Cyprus and its ophiolite suite is an exposed mass of oceanic crust that has been uplifted far above its level of formation.

Gass, I.G., and Masson-Smith, D., 1963, The geology and gravity anomalies of the Troodos Massif, Cyprus: Royal Soc. London Philos. Trans., v. A255, p. 417-467.

Gass, I.G., and Smewing, J.D., 1973, Intrusion, extrusion and metamorphism at constructive margins: evidence from the Troodos massif, Cyprus: Nature, v. 242, p. 26-29.

318 J. Gilluly calculated that there is at least five times as much sediment and sedimentary rock underlying the Atlantic Ocean off the United States as there is beneath the Pacific and pointed out that this disparity is exactly what plate tectonics theory predicts since western North America has been over-riding the Pacific plate.

Gilluly, James, 1963, Tectonic evolution of the western United States: Geol. Soc. London Quart. Jour., v. 119, p. 133-174.

___ 1969, Oceanic sediment volumes and continental drift: Science, v. 166, p. 992-994.

319 A. Hallam suggested that the worldwide Cretaceous trangression of the seas was caused by growth of the world's oceanic ridge system and that the subsequent Tertiary regression was brought about by the decay of the convection cells that formed the ridges.

Hallam, A., 1963, Major epeirogenic and eustatic changes since the Cretaceous and their possible relationship to crustal structure: Am. Jour. Sci., v. 261, p. 397-423.

320 W. Hamilton concluded from his study of new field information on Antarctic tectonics that Antarctica's structural pattern is basically that suggested by Du Toit's reassembly of the southern hemisphere continents.

Hamilton, Warren, 1963, Tectonics of Antarctica: in Childs, O. E., and Beebe, B. W., eds., Backbone of the Americas, Amer. Assoc. Petroleum Geologists Mem. 2, p. 4-15.

___ 1967, Tectonics of Antarctica: Tectonophysics, v. 4, p. 555-568.

321 M.N. Hill (editor) assembled a series of articles which described the techniques and results of geological and geophysical exploration of the world's ocean basins up to about 1962. The overall work (The Sea, 3 volumes) was the first comprehensive treatise on nonbiological oceanography in 20 years and v. 3 (The earth beneath the sea, History) described the state of knowledge that existed immediately before crucial breakthroughs were made in these areas of study.

Hill, M.N., ed., 1963, The sea - ideas and observations on progress in the study of the seas, v. 3, The earth beneath the sea, History: New York, Interscience Publishers, 963 p.

322 A symposium on the evolution of mammals on the southern continents of Africa, Australia and South America was held at the 16th International Congress of Zoology in Washington, D. C.. Considerable discussion was devoted to the relationship of continental drift and the character of southern hemisphere mammalian faunas.

Keast, Allen, Erk, F.C., and Glass, Bentley, eds., 1972, Evolution, mammals and southern continents: Albany, New York, State Univ. of New York Press, 543 p.

323 D. H. Matthews showed that the Carlsberg Ridge is offset about 300 km by the Owen Fracture Zone.

Matthews, D.H., 1963, A major fault scarp under the Arbian Sea displacing the Carlsberg Ridge near Socotra: Nature, v. 198, p. 950-952.

___ 1966, The Owen Fracture Zone and the northern end of the Carlsberg Ridge: Royal Soc. London Philos. Trans., v. A259, p. 172-186.

324 R. N. Pflug called attention to great similarities between structural patterns in the eastern Brazilian shield and those in the Angolan shield of west Africa. The similarities were said to persist over several thousand kilometers.

Pflug, R.N., 1963, Prakambrische Strukturen in Afrika und Sudamerika, eine gegenuberstellung: Neuse Jarhbuch fur Geol. und Paleo. Monatshefte, Stuttgart, v. 7, p. 355-358.

325 Global Marine drilling vessel Submarex drilled and cored a sedimentary section in 610 m of water off the coast of Jamaica. The test penetrated 56.4 m of sediment and recovered 20.7 m of core that consisted largely of late Pliocene and early Pleistocene calcareous fossil oozes.

Bolli, H.M., Boudreaux, J.E., Emiliani, Cesare, Hay, W.W., Hurley, R.J., and Jones, J.I., 1968, Biostratigraphy and paleotemperatures of a section cored on the Nicaragua Rise, Caribbean Sea: Geol. Soc. America Bull., v. 79, p. 459-470.

326 L.R. Sykes used improved epicenter location data provided by the establishment of several new seismograph stations in Antarctica and the availability of computers and programs to calculate least squares locations to show that: (1) epicenters associated with oceanic ridges follow the crest; (2) ridge crests in both the south Pacific and the equatorial Atlantic are offset large distances by fracture zones (e.g., by 1,000 km by the Eltanin fracture zone); and (3) epicenters along fracture zones are restricted to the segment of the fracture zone between ridge crests.

Sykes, L.R., 1963, Seismicity of the south Pacific Ocean: Jour. Geophys. Research, v. 68, p. 5999-6006.

327 F.J. Vine and D.H. Matthews, and (independently) L. W. Morley explained the oceanic linear magnetic anomalies by a combination of sea-floor spreading and magnetic field reversal. (Note: Morley's paper on this subject was rejected by two prestigeous scientific journals and was first published in full in 1981.)

Vine, F.J., and Matthews, D.H., 1963, Magnetic anomalies over oceanic ridges: Nature, v. 199, p. 947-949. (Reprinted in Cox, 1973)

Morley,L.W., and Larochelle, A., 1964, Paleomagnetism as a means of dating geological events: in Osborne, F.F., ed., Geochronology in Canada, Royal Soc. Canada Spec. Pub. 8, Univ. Toronto Press, p. 39-51.

Morley, L.W., 1967, (untitled exerpt): in Lear, John, Canada's unappreciated role as scientific innovator, Saturday Review mag., 2 Sept., p. 45-50. (This exerpt was reprinted in Cox, 1973, p. 224-225.)

___ 1981, An explanation of magnetic banding in ocean basins: in Emiliani, Cesare, ed., The oceanic lithosphere, v. 7 of The Sea, New York, John Wiley & Sons, p. 1717-1719.

328 R.P. Von Herzen measured relatively high heat flows in the Gulfs of Aden and California, showing that the Carlsberg Ridge and East Pacific Rise respectively, continue into those features.

Von Herzen, R.P., 1963, Geothermal heat flow in the Gulfs of California and Aden: Science, v. 140, p. 1207-1208.

329 J.T. Wilson suggested that the ages of some oceanic volcanic islands increase directly with distance from their associated oceanic ridge and that certain chains of islands and/or aseismic ridges are "streamlines" of spreading motion and show where two separated continents were previously joined. It was suggested that this arises from the motion of rigid plates over fixed "hot spots".

Wilson, J.T., 1963, Evidence from islands on the spreading of ocean floors: Nature, v. 197, p. 536-538.

___ 1963, Hypothesis of earth's behavior: Nature, v. 198, p. 925-929.

___ 1963, A possible origin of the Hawaiian Islands: Canadian Jour. Physics, v. 41, p. 863-870.

___ 1963, Continental drift: Sci. American, v. 208, no. 4, p. 86-100. (Reprinted in Continents Adrift, 1972, W. H. Freeman & Co., 172 p.)

___ 1965, Evidence from ocean islands suggesting movement in the earth: in Blackett, P.M.S., and others, eds., A symposium on continental drift, The Royal Soc., London, p. 145-167.

1963 - 1969

330 H.L. Alsopp, F.H. Chamalaun, A. Cox, G.H. Curtis, G.B. Dalrymple, R.R. Doell, J.F. Evernden, C.S. Gromme, R.L. Hay, D.M. Hopkins, G.T. James, I. McDougall, D.E. Savage, D.H. Tarling, and H. Wensink developed a magnetic reversal time scale extending back to -4.5 My that included both longer periods (epochs) and shorter ones (events).

Cox, Allan, Doell, R.R., and Dalrymple, G.B., 1963, Geomagnetic polarity epochs - Sierra Nevada II: Science, v. 142, p. 382-385.

Gromme, C.S., and Hay, R.L., 1963, Magnetization of basalt of Bed I, Olduvai Gorge, Tanganyika: Nature, v. 200, p. 560-561.

McDougall, Ian and Tarling, D.H., 1963, Dating of polarity zones in the Hawaiian Islands: Nature, v. 200, p. 54-56.

Cox, Allan, Doell, R.R., and Dalrymple, G.G., 1964, Geomagnetic polarity epochs: Science, v. 143, p. 351-352.

___ 1964, Reversals of the earth's magnetic field: Science, v. 144, p. 1537-1543.

Evernden, J.F., Savage, D.E., Curtis, G.H., and James, G.T., 1964, Potassium-argon dates and the Cenozoic mammalian chronology of North America: Amer. Jour. Sci., v. 262, p. 145-198.

McDougall, Ian, and Tarling, D.H., 1964, Dating geomagnetic polarity zones: Nature, v. 202, p. 171-172.

Cox, Allan, Hopkins, D.M., and Dalrymple, G.B., 1966, Geomagnetic polarity epochs - Pribilof Islands, Alaska: Geol. Soc. America Bull., v. 77, p. 883-910.

Chamalaun, F.H., and McDougall, Ian, 1966, Dating geomagnetic polarity epochs in Reunion: Nature, v. 210, p. 1212-1214.

Doell, R.R., Dalrymple, G.B., and Cox, Allan, 1966, Geomagnetic polarity epochs - Sierra Nevada data, 3: Jour. Geophys. Research, v. 71, p. 531-541.

Doell, R.R., and Dalrymple, G.B., 1966, Geomagnetic polarity epochs a new polarity event and the age of the Brunhes-Matuyama boundary: Science, v. 152, p. 1060-1061.

McDougall, Ian, Alsopp, H.L., and Chamalaun, F.H., 1966, Isotopic dating of the newer volcanics of Victoria, Australia and geomagnetic polarity epochs: Jour. Geophys. Research, v. 71, p. 6107-6118.

McDougall, Ian, and Chamalaun, F.H., 1966, Geomagnetic polarity scale of time: Nature, v. 212, p. 1415-1418.

McDougall, Ian, and Wensink, H., 1966, Paleomagnetism and geochronology of the Pliocene-Pleistocene lavas in Iceland: Earth and Planetary Sci. Letters, v. 1, p. 232-236.

Cox, Allan, 1969, Geomagnetic reversals: Science, v. 163, p. 237-245.

1963 - 1974

331 A large number of scientists from many countries and oceanographic institutions, working from both ships and planes, mapped the detailed magnetic field of the earth's oceans. They found that a linear magnetic anomaly pattern parallel to and generally symmetrical about the oceanic ridge crests is characteristic of most areas. On the basis of the measurements, they deciphered the history of the earth's magnetic field and worked out the detailed history of continental displacements. If any single line of evidence can be said to have decided the question of continental drift, the marine magnetic anomalies would be that evidence. (Some important review papers are listed below.)

Bullard, E.C., and Mason, R.G., 1963, The magnetic field over the oceans: in Hill, M. N., ed., The sea, v. 3, New York, Wiley, p. 175-217.
Ostenso, N.A., 1966, Geomagnetism: in Status report of Special Committee on Geophysical Studies, Am. Geophys. Union Trans., v. 47, p. 303-332.
Bullard, E.C., 1968, Reversals of the earth's magnetic field: Royal Soc. London, Philos. Trans., v. A263, p. 481-524.
Allan, T.D., 1969, A review of marine geomagnetism: Earth Sci. Reviews, v. 5, p. 217-254.
Heirtzler, J.R., 1970, Magnetic anomalies measured at sea: in Maxwell, A. E., ed., The sea, v. 4, pt. 1, New York, Wiley-Interscience, p. 85-128.
Heirtzler, J.R., and Vogt, P.R., 1971, Marine magnetic anomalies and their bearing on polar wandering and continental drift: EOS (Am. Geophys. Union Trans.), v. 52, p. (IUGG) 220-224.
McKenzie, D.P., and Sclater, J.G., 1971, The evolution of the Indian Ocean since the late Cretaceous: Royal Astron. Soc. Geophys. Jour., v. 25, p. 437-528.
Ostenso, N.A., and Wold, R.J., 1971, Aeromagnetic survey of the Arctic Ocean, techniques and interpretations: Marine Geophys. Research, v. 1, p. 47-88.
Larson, R.L., and Chase, C.G., 1972, Late Mesozoic evolution of the western Pacific Ocean: Geol. Soc. America Bull., v. 83, p. 3627-3644.
Larson, R.L., and Pitman, W.C. III, 1972, World-wide correlation of Mesozoic magnetic anomalies and its implications: Geol. Soc. America Bull., v. 83, p. 3645-3662.
Pitman, W.C. III, and Talwani, Manik, 1972, Sea-floor spreading in the north Atlantic: Geol. Soc. America Bull., v. 83, p. 619-646.
Vacquier, Victor, 1972, Geomagnetism in marine geology. New York, Elsevier, 185 p.
Pitman, W.C. III, Larson, R.L., and Herron, E.M., 1974, Map showing age of the ocean basins determined from magnetic anomaly lineations (2 sheets with explanatory text): Geol. Soc. America, Boulder, Colorado.

1964

332 P.M.S. Blackett, E.C. Bullard and S.K. Runcorn conducted "A Symposium on Continental Drift" for the Royal Society, London.

Blackett, P.M.S., Bullard, E.C., and Runcorn, S.K., eds., 1965, A Symposium on Continental drift: Royal Society London, 330 p.

333 G. Bodvarsson and G.P.L. Walker concluded that the oldest (Eocene?) volcanic rocks in Iceland, which are now separated by some 400 km from east to west, were extruded in an active belt only a few tens of km wide and have since been carried apart by crustal drift resulting mostly from crustal extension through the injection of dikes. The younger volcanics are said to have been separated by a smaller amount and current crustal drift is confined to the volcanically active median belt.

Bodvarsson, G. and Walker, G.P.L., 1964, Crustal drift in Iceland: Geophys. Jour., v. 8, p. 285-300.

334 K.M. Creer, suggested that the locations and movement of North America, South America, Africa and Europe as determined from paleomagnetic pole positions agree broadly with the continental drift proposals of Du Toit and others. The North Atlantic was said to have begun opening in the Pennsylvanian and Europe to have closed up the Tethys sea beginning in the Permian. Gondwana began to break up in the Permian.

Creer, K.M., 1964, A reconstruction of the continents for the upper Paleozoic from paleomagnetic data: Nature, v. 203, p. 1115-1120.

335 C.L. Drake and R.W. Girdler used bathymetric, gravity, magnetic and seismic information from the Red Sea to put together a detailed structural model of the area. They concluded that crustal tension has produced a fissure in continental crust (the deep central trough) into which basic igneous material flowed, resulting in the strong positive Bouguer gravity anomaly and linear magnetic anomalies parallel to the central rift. It was pointed out that all the data studied strongly support the idea that the Red Sea is a part of the world's oceanic ridge system.

Drake, C.L., and Girdler, R.W., 1964, A geophysical study of the Red Sea: Royal Astron. Soc. Geophys. Jour., v. 8, p. 473-495.

336 M. Ewing, J.I. Ewing and M. Talwani reported that seismic profiler studies show that the rough topography characteristic of the Mid-Atlantic Ridge is continuous from the crest of the Ridge to near the continental margin. At progressively greater distances from the Ridge the roughness is gradually hidden beneath a cover of sediment.

Ewing, Maurice, Ewing, J.I., and Talwani, Manik, 1964, Sediment distribution in the oceans: the Mid-Atlantic Ridge: Geol. Soc. America Bull., v. 75, p. 17-36.

337 C.G.A. Harrison and B.M. Funnell found reversals of the earth's magnetic field preserved in sediment cores from the central Pacific Ocean.

Harrison, C.G.A. and Funnell, B.M., 1964, Relationship of paleomagnetic reversals and micropaleontology in two late Cenozoic cores from the Pacific Ocean: Nature, v. 204, p. 566.

338 Scientists from Lamont, Miami, Scripps and Woods Hole organized the "Joint Oceanographic Institutions Deep Earth Sampling Program" (JOIDES) for the purpose of promoting and carrying out a program of deep-sea core drilling.

JOIDES, 1967, The deep-sea drilling project: Amer. Geophys. Union Trans., v. 48, p. 817-832.

Van Andel, T.H., 1968, Deep-sea drilling for scientific purposes: a decade of dreams: Science, v. 160, p. 1419-1424.

339 G.J.F. MacDonald concluded from his studies of heat flow and gravity that "continental structure extends to depths on the order of 500 km" and that it is consequently "no longer possible to imagine thin continental blocks sailing over a fluid mantle."

MacDonald, G.J.F., 1964, The deep structure of continents: Science, v. 143, p. 921-929.

340 I. McDougall used the K-Ar method to determine the age of lava flows on the Hawaiian Islands chain and found that eruption migrated systematically from Kauai to Hawaii at 10-15 cm/yr.

McDougall, Ian, 1964, Potassium-argon ages from lavas of the Hawaiian Islands: Geol. Soc. America Bull., v. 75, p. 107-128.

341 H.W. Menard concluded that the crest of the East Pacific Rise disappears as an identifiable topographic feature at the southern end of the Gulf of California and reappears off the Oregon coast north of the Mendocino fracture zone. Evidence cited as supporting this view included seismicity, heat flow and topography.

Menard, H.W., 1964, Marine geology of the Pacific: New York, McGraw-Hill, 271 p.

342 The Royal Society of Canada sponsored a symposium on continental drift at Charlottetown, Canada.

Garland, G.D., 1966, Continental drift: Royal Soc. Canada Spec. Pub. No. 9, 140 p.

343 L.R. Sykes and M. Landisman used new computational methods and data from new seismographic stations and were able to locate epicenters in northeast Africa and adjacent seas to an accuracy of about 10 km. They showed that the epicenters are concentrated along narrow, linear belts that precisely follow the crest of the Carlsberg Ridge into the Gulf of Aden and along faults that offset the ridge crest.

Sykes, L.R., and Landisman, Mark, 1964, The seismicity of east Africa, the Gulf of Aden and the Arabian and Red Seas: Seismol. Soc. America Bull., v. 54, p. 1927-1940.

344 R.P. Von Herzen found high heat flows north of the Mendocino fracture zone, suggesting that a segment of the East Pacific Rise is located there.

Von Herzen, R.P., 1964, Ocean-floor heat-flow measurements west of the United States and Baja California: Marine Geol., v. 1, p. 225-239.

1964 - 1975

345 Several histories of the development of continental drift, sea-floor spreading and plate tectonics, and a bibliography of the subject were published. A recurrent theme in these publications was that the science of the solid earth had undergone a revolution and that a new ruling theory (some called it a "paradigm") had been developed.

Bullard, E.C., 1964, Continental drift: Geol. Soc. London Quart. Jour., v. 120, p. 1-33.

Carozzi, A.V., 1970, New historical data on the origin of the theory of continental drift: Geol. Soc. America Bull., v. 81, p. 283-286.

Rupke, N.A., 1970, Continental drift before 1900: Nature, v. 227, p. 349-350.

Tarling, Don & Maureen, 1971, Continental drift - a study of the earth's moving surface: London, G. Bell, 140 p.

Horsfield, Brenda, and Stone, P.B., 1972, The great ocean business: New York, Coward, McCann & Geoghegan, 360 p.

Kasbeer, Tina, 1972, Bibliography of continental drift and plate tectonics: Geol. Soc. America Spec. Paper 142, 96 p.

Cox, Allan, ed., 1973, Plate tectonics and geomagnetic reversals: San Francisco, W. H. Freeman & Co., 702 p.

Hallam, Anthony, 1973, A revolution in the earth sciences: Oxford, Clarendon Press, 127 p.

Marvin, U.B., 1973, Continental drift - the evolution of a concept: Washington, D. C., Smithsonian Inst. Press, 239 p.

Schlee, Susan, 1973, The edge of an unfamiliar world - a history of oceanography: New York, E. P. Dutton & Co., 398 p.

Sullivan, Walter, 1974, Continents in motion - the new earth debate: New York, McGraw-Hill Book Co., 399 p.

Wertenbaker, William, 1974, The floor of the sea - Maurice Ewing and the search to understand the earth: Boston, Little-Brown and Co., 275 p.

Bullard, E.C., 1975, The emergence of plate tectonics: a personal view: in Donath, F. A., and others, eds., Annual review of earth and planetary sciences, v. 3, Palo Alto, Calif., Ann. Rev. Inc., p. 1-30.
Bullard, E.C., 1975, William Maurice Ewing 1906-1974: Royal Soc. London Biogr. Mem. Fellows, v. 21, p. 269-311.
Kasbeer, Tina, 1975, Bibliography of continental drift and plate tectonics (v. II): Geol. Soc. America Spec. Paper 164, 151 p.

1965

346 E.C. Bullard, J.E. Everett and A.G. Smith used a computer technique to show that the fit of the continents around the Atlantic Ocean, using the 1000 m depth contour as the boundary, is extremely good. The Iberian peninsula was rotated clockwise to improve the fit of Europe with North America.

Bullard, E.C., Everett, J.E. and Smith, A.G., 1965, The fit of the continents around the Atlantic: in Blackett, P.M.S., and others, eds., A symposium on continental drift, The Royal Society, London, p. 41-51.

347 JOIDES group used drilling vessel Caldrill to conduct a series of successful test drillings in the Atlantic Ocean off Florida in water depths up to 1,032 m with bottom penetrations of 120 to 320 m and core recovery averaging 36 percent.

JOIDES, 1965, Ocean drilling on the continental margin: Science, v. 150, p. 709-716.
JOIDES, 1967, Deep-sea drilling project: Amer. Assoc. Petr. Geologists Bull., v. 51, p. 1787-1802.

348 A.M. Coode explained the restriction of fracture zone earthquakes to segments between offset ridge crests by postulating that, because of sea-floor spreading, those segments are the only places where there is relative motion across the fault. It was also suggested that this type of mechanism explains the abrupt termination of these faults at continental margins.

Coode, A.M., 1965, A note on oceanic transcurrent faults: Canadian Jour. Earth Sci., v. 2, p. 400-401.

349 From 18 to 28 August the U.S.N.S. Eltanin, towing a Varian proton precession magnetometer, traversed the Pacific-Antarctic Ridge on its 19th run of a series to obtain a profile almost 3,000 km long from 62 degrees S, 140 degrees W to 48 degrees S, 171 degrees W. On board were geologists/geophysicists George Carpenter, David Massey, Lawrence Oblinger and Thomas Wustenberg.

Heirtzler, J.R., Hayes, D.E., Herron, E.M., and Pitman, W.C. III, 1969, Preliminary report of vol. 20 U.S.N.S. Eltanin cruises 16-21, January, 1965 - January 1966, Part A - Navigation and Part B - Bathymetric & geomagnetic measurements: Office of Naval Research Tech. Rept. 3-CU-3-69.
(Note: The Eltanin 19 profile was published in the following:)
Pitman, W.C., III, and Heirtzler, J.R., 1966, Magnetic anomalies over the Pacific-Antarctic Ridge: Science, v. 154, p. 1164-1171.
Pitman, W.C. III, Herron, E.M., and Heirtzler, J.R., 1968, Magnetic anomalies in the Pacific and sea-floor spreading: Jour. Geophys. Research, v. 73, p. 2069-2085.
Heirtzler, J.R., Dickson, G.O., Herron, E.M., Pitman, W.C. III, and Le Pichon, Xavier, 1968, Marine magnetic anomalies, geomagnetic field reversals, and motions of the ocean floor and continents: Jour. Geophys. Research, v. 73, p. 2119-2136. (Reprinted in Bird and Isacks, 1972, and Cox, 1973)
(Note: The significance of the Eltanin 19 profile is discussed in:)
Wertenbaker, William, 1974, The floor of the sea: Boston, Little, Brown & Co., 275 p. (See Chapter 11 - Eltanin 19)

Cox, Allan, 1973, Plate Tectonics and Geomagnetic Reversals: San Francisco, W. H. Freeman (See p. 226-227)

Frankel, Henry, 1979, Why drift theory was accepted with the confirmation of Harry Hess's concept of sea-floor spreading: in Schneer, C.J., Two hundred years of geology in America, University Press of New England, p. 337-353.

350 R.W. Girdler showed that paleomagnetic pole location data from Spanish rocks support the idea that Spain has rotated 30-40 degrees counter-clockwise with respect to Europe since the Triassic.

Girdler, R.W., 1965, Continental drift and the rotation of Spain: Nature, v. 207, p. 396-398.

351 B.C. Heezen and M. Tharp showed that arcs connecting points on opposite sides of the Atlantic Ocean that are said to have been separated by continental drift are parallel to recently discovered fracture zones that cut and offset the Mid-Atlantic Ridge. They also noted the general paucity of sediments on the world's oceanic ridges.

Heezen, B.C. and Tharp, Marie, 1965, Tectonic fabric of the Atlantic and Indian oceans and continental drift: in Blackett, P. M. S., and others, eds., A symposium on continental drift, Royal Soc. London, p. 90-106.

352 The International Upper Mantle Committee and UNESCO sponsored a symposium at Nairobi, Kenya on the East African Rift System.

Drake, C.L., 1965, East African rift system: Geotimes, v. 10, no. 4, p. 13-14.

Anonymous, 1965, Report of the UMC/UNESCO seminar on the east African rift system: Nairobi, Kenya, University College, 145 p.

353 The International Upper Mantle Committee sponsored a symposium on the world rift system at Ottawa, Ontario, Canada.

Irvine, T.N., ed., 1966, The world rift system: Geol. Survey Canada Paper 66-14, 471 p.

354 A.N. Khramov, V.P. Rodionov and R.A. Komissarova reported that magnetic paleopole location data from east and west of the Uralide belt respectively indicate that the two regions were once separated and that the Urals are the result of the collision between the two plates during the Permo-Triassic.

Khramov, A.N., Rodionov, V.P., and Komissarova, R.A., 1965, Novyye dannyye o paleozovskov istorii zemnogo magnitnogo polya na territorii SSSR, in Ozerskaya, M. L., ed., Nastoyashcheye i proshloye magnitnogo polya zemli, Moscow, Izdat, Nauka, p. 206-213.

355 A.S. Laughton concluded that the Afar Triangle consists mostly of youthful basaltic rocks that formed after the Yemeni corner of the Arabian block moved away to the northeast, leaving the Danakil Alps behind as a continental fragment. Seismic, magnetic and topographic evidence was described to support the hypothesis.

Laughton, A.S., 1966, The Gulf of Aden in relation to the Red Sea and the Afar depression of Ethiopia: in Irvine, T. N., ed., The world rift system, Geol. Survey of Canada Paper 66-14, p. 78-97.

___ 1966, The Gulf of Aden: Royal Soc. London Philos. Trans. v. A259, p. 150-171.

356 Y.R. Nayudu reported that dredged rock samples taken in the vicinity of the Mendocino fracture zone during Scripps expeditions "Fan Fare" in 1959 and "Mendocino" in 1960 consist of basalts of various kinds and palagonite tuff breccias. No indurated sedimentary rocks were found, thus suggesting that layer 2 is composed of extruded basalt, not consolidated sedimentary rock.

Nayudu, Y.R., 1965, Petrology of submarine volcanics and sediments in the vicinity of the Mendocino fracture zone: in Sears, Mary, ed., Progress in oceanography, v. 3, Oxford, Pergamon, p. 207-220.

357 G. Plafker showed by geologic mapping that the 1964 "Good Friday" earthquake in Alaska occurred by northward underthrusting on a fault that dips at a shallow angle beneath Alaska.

Plafker, George, 1965, Tectonic deformation associated with the 1964 Alaska earthquake: Science, v. 148, p. 1675-1687. (Reprinted in Cox, 1973)

___ 1972, Alaskan earthquake of 1964 and Chilean earthquake of 1960 - implications for arc tectonics: Jour. Geophys. Research, v. 77, p. 901-925. (Reprinted in Bird & Isacks, 1972)

___ 1972, Tectonics: in Comm. on the Alaska Earthquake (Nat. Res. Council), The great Alaska earthquake of 1964 (v. 2), Seismology and geodesy, Washington, D. C., Nat. Acad. Sci., p. 112188. (Also included in (v. 1), Geology, as p. 47-122; originally published as U.S. Geol. Survey Prof. Paper 543-I.)

358 M. Talwani, X. Le Pichon, and J.R. Heirtzler suggested that apparent offsets along oceanic fracture zones do not represent movement but that the various ridge crest segments developed at their present positions and that the fracture zones were lines of weakness that existed before the ridges formed.

Talwani, Manik, Le Pichon, Xavier, and Heirtzler, J.R., 1965, East Pacific Rise - the magnetic pattern and the fracture zones: Science, v. 150, p. 1109-1115.

359 F.J. Vine and J.T. Wilson concluded from study of the magnetic anomaly pattern of the Pacific Ocean off the northwestern U.S. that (1) the crestal magnetic anomaly is stronger because it is composed entirely of young, normally magnetized material, whereas the flank areas have been contaminated by younger flows of differing polarities; (2) the anomalies are probably produced by basalt flows in layer 2 which have a thickness of about 2 km; (3) the anomaly pattern is symmetrical about the crest of the Juan de Fuca Ridge; and (4) the anomaly pattern corresponds well with the already-established reversal time scale and a spreading half-rate of 1.5 cm/yr.

Vine, F.J., and Wilson, J.T., 1965, Magnetic anomalies over a young oceanic ridge off Vancouver Island: Science, v. 150, p. 485-489. (Reprinted in Cox, 1973)

360 J.T. Wilson identified the Juan de Fuca ridge as a section of the East Pacific Rise on the basis of bathymetric, seismic and magnetic evidence. The short section of ridge was said to be connected to the Rise by the San Andreas transform fault and to structures further north by the Queen Charlotte Islands transform fault. The magnetic anomaly pattern was found to be symmetrical about the crest of the newly-identified ridge.

Wilson, J.T., 1965, Transform faults, oceanic ridges, and magnetic anomalies southwest of Vancouver Island: Science, v. 150, p. 482-485.

361 J.T. Wilson suggested that (1) the earth's surface consists of several large, rigid plates that are not readily deformable except at their margins and that are separated by mobile belts (active primary mountain belts and island arcs in compression, oceanic ridges in tension, and active transform faults in horizontal shear); (2) the mobile belts are interconnected; and (3) any type of mobile belt may pass into one of the other types. Wilson proposed the term "transform fault" for a class of strike-slip faults that terminate abruptly at both ends but show great offset; for example, the San Andreas fault was said to be a dextral, ridge-ridge transform fault connecting the East Pacific Rise and the Juan de Fuca Ridge.

Wilson, J.T., 1965, A new class of faults and their bearing on continental drift: Nature, v. 207, p. 343-347. (Reprinted in Cox, 1973)

1965 - 1972

362 W.A. Berggren, A. Bertels, W. Broecker, L.H. Burckle, G.O. Dickson, J. Dymond, J. H. Foster, M.D. Fuller, B. Glass, H.G. Goodell, C.G.A Harrison, J.D. Hays, B.C. Heezen, K.W. Henry, T. Kanaya, J.P. Kennett, K. Kitazawa, K. Kobayashi, T. Ku, T. Lin'-kova, W. Lowrie, Y.R. Nayudu, D. Ninkovich, N.D. Opdyke, J.D. Phillips, T. Saito, T. Sakai, J.D. Smith, B.L.K. Somayajulu, P. Vella, D. Wall, and N.D. Watkins, used studies of the paleomagnetism of deep-sea sediment cores to: (1) establish reversal stratigraphy as a major dating and correlation technique in deep-sea sediments younger than about -5.5 My; (2) show that the reversal sequence agrees with that determined on land from lava flows and at sea from the magnetic anomaly patterns; (3) determine with greater accuracy the ages of the various magnetic epochs and events; (4) extend the polarity reversal time scale back to the middle Miocene (-5.5 My); (5) verify the ages determined by using fossils and radiometric dating; (6) show that the magnetic field of the earth averages to an axial dipole over periods of 27,000 years and has been a geocentric axial dipole field for the last 2 million years; (7) establish that the field decreases in intensity to zero before increasing in intensity with opposite polarity; and (8) determine that reversals take place in 20,000 years or less.

Lin'kova, T.I., 1965, Some results of paleomagnetic study of Arctic Ocean floor sediments, (transl. by E.R. Hope from Natoyascheye i Proshloye Magnitnogo Polia Zemli, p. 279281, Nauka, Moscow): Directorate of Sci. Infor. Services, pub. T463R, Canada.

Dickson, G.O., and Foster, J.H., 1966, The magnetic stratigraphy of a deep sea core from the north Pacific Ocean: Earth and Planetary Sci. Letters, v. 1, p. 458-462.

Harrison, C.G.A., 1966, The paleomagnetism of deep sea sediments: Jour. Geophys. Research, v. 71, p. 3033-3043.

Harrison, C.G.A., and Somayajulu, B.L.K., 1966, Behavior of the earth's magnetic field during a reversal: Nature, v. 212, p. 1193-1195.

Ninkovich, Dragoslov, Opdyke, N.D., Heezen, B.C., and Foster, J.H., 1966, Paleomagnetic stratigraphy, rates of deposition and tephrachronology in north Pacific deep-sea sediments: Earth and Planetary Sci. Letters, v. 1, p. 476-492.

Opdyke, N.D., Glass, Billy, Hays, J.D., and Foster, J.H., 1966, Paleomagnetic study of Antarctic deep-sea cores: Science, v. 154, p. 349-357.

Berggren, W.A., Phillips, J.D., Bertels, A., and Wall, D., 1967, Late Pliocene-Pleistocene stratigraphy in deep sea cores from the southcentral north Atlantic: Nature, v. 216, p. 253-255.

Hays, J.D., and Opdyke, N.D., 1967, Antarctic radiolaria, magnetic reversals and climatic change: Science, v. 158, p. 1001-1011.

Watkins, N.D., and Goodell, H.G., 1967, Inconsistencies resulting from a limited study of magnetic polarity variation in submarine sedimentary cores (abs.): Amer. Geophys. Union Trans., v. 48, p. 90.

___ 1967, Confirmation of the reality of the Gilsa geomagnetic polarity event: Earth and Planetary Sci. Letters, v. 2, p. 123-129.

___ 1967, Geomagnetic polarity change and faunal extinction in the southern ocean: Science, v. 156, p. 1083-1087.

Goodell, H.G., and Watkins, N.D., 1968, The paleomagnetic stratigraphy of the southern ocean: 20 degrees west to 160 degrees east longitude: Deep-Sea Research, v. 15, p. 89-112.

Ku, T., Broecker, W., and Opdyke, N.D., 1968, Comparison of sedimentation rates measured by paleomagnetic and ionium methods of age determination: Earth and Planetary Sci. Letters, v. 4, p. 1-16.

Ninkovich, D., 1968, Pleistocene volcanic eruptions in New Zealand recorded in deep-sea sediments: Earth and Planetary Sci. Letters, v. 4, p. 89-102.

Watkins, N.D., 1968, Short-period geomagnetic polarity events in deep-sea sedimentary cores: Earth and Planetary Sci. Letters, v. 4, p. 341-349.

Dymond, J., 1969, Age determinations of deep-sea sediments: a comparison of three methods: Earth and Planetary Sci. Letters, v. 6, p. 914.

Hays, J.D., Saito, T., Opdyke, N.D., and Burckle, L.H., 1969, Pliocene-Pleistocene sediments of the equatorial Pacific: Their paleomagnetic, biostratigraphic, and climatic record: Geol. Soc. America Bull., v. 80, p. 1481-1514.

Opdyke, N.D., 1969, The Jaramillo event as detected in oceanic cores: in Runcorn, S. K., ed., The application of modern physics to Earth and planetary interiors, New York, John Wiley, p. 549-552.

Opdyke, N.D., and Henry, K.W., 1969, A test of the dipole hypothesis: Earth and Planetary Sci. Letters, v. 6, p. 139-151.

Opdyke, N.D., and Phillips, J.D., 1969, Paleomagnetic stratigraphy of sites 1-7 (Leg 1), preliminary report: in M. Ewing, and others, eds., Initial Reports of the Deep-Sea Drilling Project, v. 1, U. S. Govt. Printing Ofc., Washington, D. C., p. 501-519.

Smith, J.D., and Foster, J.H., 1969, Geomagnetic reversal in Brunhes normal polarity epoch: Science, v. 163, p. 565-567.

Foster, J.H., and Opdyke, N.D., 1970, Upper Miocene to recent magnetic stratigraphy in deep-sea sediments: Jour. Geophys. Research, v. 75, p. 4465-4473

Opdyke, N.D., and Foster, J.H., 1970, Paleomagnetism of cores from the north Pacific: in Hays, J. D., ed., Geological investigations of the north Pacific, Geol. Soc. America Mem. 126, p. 83-119.

Kennett, J.P., Watkins, N.D., and Vella, P., 1971, Paleomagnetic chronology of Pliocene-early Pleistocene climates and the Plio-Pleistocene boundary in New Zealand: Science, v. 171, p. 276-279.

Kobayashi, K., Kitazawa, K., Kanaya, T., and Sakai, T., 1971, Magnetic and micropaleontological study of deep-sea sediments from the west-central equatorial Pacific: Deep-Sea Research, v. 18, p. 1045-1062.

Opdyke, N.D., Ninkovich, D., and Lowrie, W., 1971, The paleomagnetism of two deep-sea cores from the eastern Mediterranean (abs.): EOS (Amer. Geophys. Union Trans.), v. 52, p. 186.

Opdyke, N.D., 1972, Paleomagnetism of deep-sea cores: Reviews of Geophys. and Space Phys., v. 10, p. 213-249.

1965 - 1975

363 J. Auzende, Y. Belmonte, L.W. Butler, K.O. Emery, H. Hoskins, R. Leyden, G. Pautot, P.A. Rona, R.P. Von Herzen, J.M. Ayme, B.D. Bornhold, J. Daniel, M. Ewing, G.L. Johnson, W.J. Ludwig, R. Querol, E.D. Schneider, R. Wenger, T.R. Baumgartner, G. Bryan, J. Dupont, P. Hirtz, X. Le Pichon, J.R. Mascle, V. Renard, T.H. van Andel, and the Societe Cherifienne des Petroles established the existence of salt diapirs beneath the Atlantic Ocean off Africa and North and South America in water depths of up to almost 6,000 m and out to almost 1,500 km from present coastlines.

Ayme, J.-M., 1965, The Senegal salt basin: in anonymous, Salt basins around Africa, London, The Institute of Petroleum, p. 83-90.

Belmonte, Y., Hirtz, P., and Wenger, R., 1965, The salt basins of Gabon and the Congo (Brazzaville): in anonymous, Salt basins around Africa, London, The Institute of Petroleum, p. 55-74.

Querol, Ramon, 1966, Regional geology of the Spanish Sahara: in Reyre, D., ed., Sedimentary basins of the African coasts, Paris, Assoc. African Geol. Surveys, p. 27-39.
Societe Cherifienne des Petroles, 1966, Le bassin du sud-ouest Marocain: in Reyre, D., ed., Sedimentary basins of the African coasts, Paris, Assoc. African Geol. Surveys, p. 5-12.
Rona, P.A., 1969, Possible salt domes in the deep Atlantic off northwest Africa: Nature, v. 224, p. 141-143.
Butler, L.W., 1970, Shallow structure of the continental margin, southern Brazil and Uraguay: Geol. Soc. America Bull., v. 81, p. 1079-1096.
Pautot, Guy, Auzende, Jean-Marie, and Le Pichon, Xavier, 1970, Continuous deep sea salt layer along north Atlantic margins related to early phase of rifting: Nature, v. 227, p. 351-354.
Schneider, E.D., and Johnson, G.L., 1970, Deep-ocean diapir occurrences: Amer. Assoc. Petr. Geologists Bull., v. 54, p. 2151-2169.
Baumgartner, T.R., and van Andel, T.H., 1971, Diapirs of the continental margin of Angola, Africa: Geol. Soc. America Bull., v. 82, p. 793-802.
Leyden, Robert, Ludwig, W.J., and Ewing, Maurice, 1971, Structure of continental margin off Punta del Este, Uraguay and Rio de Janerio, Brazil: Amer. Assoc. Petr. Geologists Bull., v. 55, p. 2161-2173.
Emery, K.O., 1972, Eastern Atlantic continental margin: some results of the 1972 cruise of the R. V. Atlantis II: Science, v. 178, p. 298-301.
Leyden, Robert, Bryan, G., and Ewing Maurice, 1972, Geophysical reconnaissance on African shelf: 2. Margin sediments from Gulf of Guinea to Walvis Ridge: Amer. Assoc. Petr. Geologists Bull., v. 56, p. 682-693.
Von Herzen, R.P., Hoskins, Hartley, and van Andel, T.H., 1972, Geophysical studies in the Angola diapir field: Geol. Soc. America Bull., v. 83, p. 1901-1910.
Mascle, J.R., Bornhold, B.D., and Renard, Vincent, 1973, Diapiric structures off Niger delta: Amer. Assoc. Petr. Geologists Bull., v. 57, p. 1672-1678.
Pautot, Guy, Renard, Vincent, Daniel, Jacques, and Dupont, Jacques, 1973, Morphology, limits, origin and age of salt layer along south Atlantic African margin: Amer. Assoc. Petr. Geologists Bull., v. 57, p. 1658-1671.
(See also)
Kinsman, D.J.J., 1975, Salt floors to geosynclines: Nature, v. 255, p. 375-378.

1966

364 J.F. Bonaparte found two new species of therapsid reptiles in Argentina and found them to be extremely similar to specimens found in the Cynognathus zone of South Africa, from which he concluded that South America and Africa were connected in some way during the Triassic.

Bonaparte, J.F., 1966, Sobre nuevos terapsidos Triasicos hallados en el centro de la provincia de Mendoza, Argentina: Acta Geol. Lilloana, v. 8, p. 91-100.
___ 1967, New vertebrate evidence for a southern transatlantic connexion during the lower or middle Triassic: Paleontology, v. 10, p. 554-563.

365 J.R. Cann and F.J. Vine reported that a detailed magnetic, bathymetric and gravity survey carried out by H.M.S. Owen of a small area on the crest of the Carlsberg Ridge revealed the existence of a reversely magnetized seamount, thus lending support to the hypothesis that polarity reversals are recorded in the oceanic crust.

Cann, J.R., and Vine, F.J., 1966, An area on the crest of the Carlsberg Ridge: petrology and magnetic survey: Royal Soc. London Philos. Trans., v. A259, p. 198-217.

366 D.J. Elvers, G. Peter, R.L. Moses, C.C. Matthewson and R.E. Kohler reported the discovery of a unique bend in the magnetic anomaly pattern south of the Aleutian Islands.

Peter, George, 1966, Magnetic anomalies and fracture pattern in the northeast Pacific Ocean: Jour. Geophys. Research, v. 71, p. 5365-5374.

Elvers, D.J., Peter, George, and Moses, R.L., 1967, Analysis of magnetic lineations in the north Pacific (abs.): Am. Geophys. Union Trans., v. 48, p. 89.

Elvers, D.J., Matthewson, C.C., Kohler, R.E., and Moses, R.L., 1967, Systematic ocean surveys by the USCGS Pioneer 1961-1963: Coast and Geodetic Survey Operational Data Rept. C and CSDRI, 19 p.

Peter, George, Erickson, B.H., and Grim, P.J., 1971, Magnetic structure of the Aleutian Trench and northeast Pacific basin: in Maxwell, A. E., ed., The Sea, v. 4, Part II, New York, Interscience Pubs. Inc., p. 191-222.

367 J. Ewing, J.L. Worzel, M. Ewing and C. Windisch noted that a graph of the width of the Atlantic Ocean as determined from sediment distribution vs. the age of the sediments is a straight line and pointed out that this relationship strongly supports the theory of sea-floor spreading. A spreading rate of one centimeter per year was determined and extrapolated backward to suggest that the deep Atlantic Ocean was born about 190 million years ago.

Ewing, John, Worzel, J.L., Ewing, Maurice, and Windisch, Charles, 1966, Ages of horizon A and the oldest Atlantic sediments: Science, v. 154, p. 1125-1132.

368 J.R. Heirtzler, X. Le Pichon and J.G. Baron described the results of the first detailed magnetic survey carried out over the known crest of an oceanic ridge. The aeromagnetic survey of the Reykjanes Ridge southwest of Iceland found a linear magnetic anomaly pattern parallel to and symmetrical about the crest of the ridge.

Heirtzler, J.R., Le Pichon, Xavier, and Baron, J.G., 1966, Magnetic anomalies over the Reykjanes Ridge: Deep-Sea Research, v. 13, p. 427-443.

369 H.W. Menard reported that newly-obtained depth soundings indicate the crest of the East Pacific Rise is offset in at least 17 places, ten of which are along typical linear fracture zones. Individual offsets range up to 500 km and most are seismically active.

Menard, H.W., 1966, Fracture zones and offsets of the East Pacific Rise: Jour. Geophys. Research, v. 71, p. 682-685.

370 Project Mohole was cancelled by action of the U.S. House of Representatives.

Greenberg, D.S., 1966, NSF appropriation - mutiny on the mohole: Science, v. 152, p. 895-896.

___ 1964, Mohole: the project that went awry: Science, v. 143, p. 115-119, 223-227, 334-337.

Solow, Herbert, 1963, How NSF got lost in Mohole: Fortune, v. 67, no. 5 (May, 1963), p. 138-141, 198, 203, 204, 208, 209.

Greenberg, D.S., 1971, Mohole, geopolitical fiasco: in Gass, I.G., Smith, P.J., and Wilson, R.C.L., eds., Understanding the earth, Cambridge, Mass., Mass. Inst. Tech. Press, p. 343-348.

Congressional Record, 89th Congress, 2nd Session, p. 19919-19926.

371 N.D. Opdyke, B. Glass, J.D. Hays and J.H. Foster confirmed the polarity reversal time scale developed on the basis of lava flows by identifying the four most recent polarity epochs (back to about -3.5 My) and several polarity events in marine sediments. They found that the reversal time scale in sediments agrees consistently with independent ages derived from fossils and sedimentation rates.

Opdyke, N.D., Glass, Billy, Hays, J.D., and Foster, J. H., 1966, Paleomagnetic study of Antarctic deep-sea cores: Science, v. 154, p. 349-357.
Opdyke, N.D., and Foster, J.H., 1967, Reversals of the earth's magnetic field: Science Journal, v. 3, no. 9, p. 56-61.

372 W.C. Pitman III and J.R. Heirtzler published the central 1,000 km of the Eltanin 19 profile and showed that magnetic profiles across the Pacific-Antarctic Ridge correlate well with the Reykjanes Ridge profiles and agree with the established reversal time scale back to -3.4 My, yielding a spreading rate of 4.5 cm/yr. They then assumed a constant spreading rate and derived a reversal time scale back to -10 My.
Pitman, W.C., III, and Heirtzler, J.R., 1966, Magnetic anomalies over the Pacific-Antarctic Ridge: Science, v. 154, p. 1164-1171.

373 A symposium on continental drift was held at Rutgers University.
Johnson, Helgi, and Smith, B.L., 1970, The megatectonics of continents and oceans: New Brunswick, N. Jersey, Rutgers Univ.Press, 282 p.

374 R. Stonley used structural, seismic and gravity information to show that the Niger Delta has probably grown beyond the original continental margin since the separation of South America and Africa and that the computed "overlap" in the area is not, therefore, evidence against continental drift.
Stonley, R., 1966, The Niger Delta region in the light of the theory of continental drift: Geol. Mag., v. 103, p. 385-397.

375 L.R. Sykes showed that first-motion data from earthquakes on fracture zones cutting across oceanic ridges confirm J. T. Wilson's concept of transform faulting and that normal faulting under tensional stress characterizes ridge crest seismic activity away from fracture zones.
Sykes, L. R., 1966, Mechanism of earthquakes and nature of faulting on the mid-oceanic ridges (abs.): Geol. Soc. America Program for 1966 Annual Meeting, p. 216-217.
___ 1967, Mechanism of earthquakes and nature of faulting on mid-ocean ridges: Jour. Geophys. Research, v. 72, p. 2131-2153. (Reprinted in Bird and Isacks, 1972 and Cox, 1973.)

376 L.R. Sykes found that locations of epicenters and first motion studies of earthquakes in the Gulf of California confirm that the East Pacific Rise continues up the Gulf but is offset by a series of transform faults whose location and orientation were previously determined by other evidence.
Sykes, L.R., 1968, Seismological evidence for transform faults, sea floor spreading, and continental drift: in Phinney, R. A., ed., History of the earth's crust, Princeton, New Jersey, Princeton Univ. Press, p. 120-150.

377 L.R. Sykes used data from many new seismograph stations and a newly developed computer program for calculating focal locations to show that the circum-Pacific deep-focus earthquake zones are considerably thinner than previously thought (not more than 50-100 km) and continuous in slope to depths of at least 650 km.
Sykes, L.R., 1966, The seismicity and deep structure of island arcs: Jour. Geophys. Research, v. 71, p. 2981-3006.

378 F.J. Vine demonstrated that the Vine-Matthews hypothesis is correct by showing that (1) the magnetic anomaly patterns of the oceans are remarkably linear, continuous for at least 3,000 km, and perhaps as much as 14,000 km, and symmetrical about oceanic ridge crests such as the Reykjanes, Juan de Fuca and Gorda ridges and the southern East Pacific Rise; (2) the anomaly pattern matches the previously-developed reversal time scale back to about -4 My and can be used to extend the scale back to -11.5 My by extrapolation (3) the spreading rates determined are exactly those needed to account for continental drift; and (4) changes in the direction of sea-floor spreading can be determined from the anomaly patterns.

Vine, F.J., 1966, Proof of ocean-floor spreading? (abs.): Geol. Soc. America Program for Ann. Mtg., p. 229-230.

___ 1966, Spreading of the ocean floor - new evidence: Science, v. 154, p. 1405-1415. (Reprinted in Cox, 1973.)

___ 1968, Magnetic anomalies associated with mid-ocean ridges: in Phinney, R. A., ed., The history of the earth's crust, Princeton, N. Jersey, Princeton Univ. Press, p. 73-89.

379 J.T. Wilson cited faunal and other evidence to show that (1) there was an "Atlantic Ocean" with two distinct faunal realms (European and North American) from the late Pre-Cambrian until medial Ordovician time; (2) the ocean must have closed considerably by then since the realms became one; (3) there was no Atlantic from Permian to Jurassic time; and (4) the continents have separated since the Cretaceous but the break did not take place at the older continental junction. The opening and closing of ocean basins has since become known as a "Wilson cycle."

Wilson, J.T., 1966, Did the Atlantic close and then reopen: Nature, v. 211, p. 676-681.

1966 - 1967

380 During November, 1966 and April, 1967, crucial breakthroughs in several lines of research were described at three important conferences: (1) a meeting in New York sponsored by the Columbia University Geology Department and Goddard Institute for Space Studies; (2) the annual meeting of the Geological Society of America, in San Francisco; and (3) the annual meeting of the American Geophysical Union in Washington D. C.. The significant papers reported quantum advances in such fields as magnetic anomaly interpretation, first motion studies of transform fault earthquakes, sediment-core reversal determinations, and radiometric age-province match-ups between South America and Africa. Most important of all was the demonstration that Earth's magnetic reversal history is preserved in (1) lava flows on land; (2) ocean floor sediments; and (3) marine magnetic anomaly patterns, and that there is agreement between the three. This conclusively confirmed the Vine-Matthews hypothesis and allowed detailed interpretation of the anomaly patterns to begin. Notable also was the marked contrast between the New York Conference and one held there 12 years earlier (also sponsored by Columbia). The latest meeting dealt almost exclusively with continental drift, whereas that subject was conspicuously ignored in 1954. (Note: The specific contributions are detailed in separate entries.)

Geological Society of America, 1966, Program 1966 Annual Meetings: Geol. Soc. America, 252 p.

American Geophysical Union, 1967, (Annual meeting schedule and abstracts): Amer. Geophys. Union Trans., v. 48 (March), p. 6-263.

Phinney, R.A., ed., 1968, The history of the earth's crust: Princeton, New Jersey, Princeton Univ. Press, 244 p.

1967

381 J.M. Ade-Hall, P. Dagley, J. Edwards, R.L. Grasty, S.E. Haggerty, T. Sigurgeirsson, P.J. Smith, G.P.L. Walker, N.D. Watkins, and R.L. Wilson made over 2,000 paleomagnetic measurements on several hundred successive lava flows in eastern Iceland, including the oldest accessible flows dating to about -20 My. They found about 60 reversals in this time span and concluded that analysis of the reversal pattern is consistent with the sea-floor spreading hypothesis.

Dagley, P., Wilson, R.L., Walker, G.P.L., Watkins, N.D., Sigurgeirsson, T., Haggerty, S.E., Smith, P.J., Edwards, J., and Ade-Hall, J., 1967, Analysis of the paleomagnetism of 726 successive lavas in eastern Iceland in terms of the spreading oceanic floor hypothesis (abs.): Amer. Geophys. Union Trans., v. 48, p. 89-90.

Dagley, P., Wilson, R.L., Ade-Hall, J.M., Walker, G.P.L., Haggerty, S.E., Sigurgeirsson, T., Watkins, N.D., Smith, P.J., Edwards, J., and Grasty, R.L., 1967, Geomagnetic polarity zones for Icelandic lavas: Nature, v. 216, p. 25-29.

382 P.J. Barrett found part of a lower jaw of a labyrinthodont amphibian at Graphite Peak in Antarctica in late Paleozoic - early Mesozoic rocks. This was the first find of a land-living vertebrate in Antarctica and it suggested that Antarctica had once been connected to other continents and had a very different climate at that time.

Barrett, P.J., Baillie, R.J., and Colbert, E.H., 1968, Triassic amphibian from Antarctica: Science, v. 161, p. 460-462.

Anonymous, 1968, The ancient relative that preceded man to the South Pole: Saturday Review, 6 April, p. 49-52.

383 L.H. Burckle, J. Ewing, R. Leyden, W.R. Riedel and T. Saito reported that cores taken from the East Pacific Rise demonstrate that (1) sediments are either lacking or very thin near the Rise crest; (2) only Pleistocene sediments are found near and on the crest; (3) sediment thickness increases away from the crest; and (4) maximum sediment age increases away from the crest.

Burckle, L. H., Ewing, John, Saito, Tsunema, and Leyden, Robert, 1967, Tertiary sediment from the East Pacific Rise: Science, v. 157, p. 537-540.

Riedel, W.R., 1967, Radiolarian evidence consistent with spreading of the Pacific floor: Science, v. 157, p. 540-542.

384 Columbia University arranged an international conference at Gander, Newfoundland on "Stratigraphy and Structure Bearing on the Origin of the North Atlantic Ocean".

Woodward, H.P., 1967, Geology along the north Atlantic: Gander Conference: Science, v. 158, p. 1368-1371.

Kay, Marshall, 1969, North Atlantic - Geology and continental drift: Am. Assoc. Petroleum Geologists, Mem. 12, 1082 p.

385 W.R. Dickinson and T. Hatherton showed that increasing depth to a Benioff zone is directly related to an increase in the potassium content of the associated andesitic volcanics and suggested that this fact could be used to determine the location and orientation of inactive subduction zones.

Dickinson, W.R., and Hatherton, Trevor, 1967, Andesitic volcanism and seismicity around the Pacific: Science, v. 157, p. 801-803.

Hatherton, Trevor, and Dickinson, W.R., 1969, The relationship between andesitic volcanism and seismicity in Indonesia, the Lesser Antilles, and other island arcs: Jour. Geophys. Research, v. 74, p. 5301-5310. (Reprinted in Cox, 1973.)

386 L. Fisher, G.L. Johnson and B.C. Heezen showed that the central Indian Ocean Ridge is offset by a northeast-southwest trending series of fracture zones.

Fisher, R.L., Johnson, G.L., and Heezen, B.C., 1967, Mascarene Plateau, western Indian Ocean: Geol. Soc. America Bull., v. 78, p. 1247-1266.

387 The Geoscience Society of Iceland held a symposium on "Iceland and Mid-Ocean Ridges" at Reykjavik.

Bjornsson, Sveinbjorn, 1967, Iceland and midocean ridges: Reykjavik, Prentsmidjan Leiftur H. F., 209 p.

388 C.G.A. Harrison and J.D. Mudie showed that sediment thickness patterns on the Pacific Ocean floor north of the Murray fracture zone support the sea-floor spreading hypothesis.

Harrison, C.G A., and Mudie, J.D., 1967, Sediment thickness and sea-floor spreading in the Pacific (abs.): Amer. Geophys. Union Trans., v. 48, p. 133.

389 P.M. Hurley and colleagues showed that the boundary between two-billion-year-old rocks of one province in Africa and 600-million-year-old rocks of another province is found in South America in exactly the position and orientation predicted by continental drift. They also showed that many other structural trends continue between continents in a way that suggests the continents were once adjacent.

Hurley, P.M., De Almeida, F.F.M., Melcher, G.C., Cordani, U.G., Rand, J. R., Kawashita, K., Vandoros, P., Pinson, W.H. and Fairbairn, H.W., 1967, Test of continental drift by comparison of radiometric ages: Science, v. 157, p. 495-500.

Hurley, P.M., 1968, The confirmation of continental drift: Sci. American, v. 218, no. 4, p. 52-64. (Reprinted in Sci. American, 1973, Continents Adrift, W. H. Freeman, San Francisco, p. 56-67.)

390 The International Astronomical Union in cooperation with the International Union of Geodesy and Geophysics organized a symposium in Stresa, Italy on "Continental drift, secular motion of the pole, and rotation of the earth".

Markowitz, William, and Guinot, B., eds., 1968, Continental drift, secular motion of the pole, and rotation of the earth: Dordrecht, Holland, D. Riedel Pub. Co., and New York, Springer-Verlag, 109 p.

391 Under the auspices of the International Union of Geological Sciences, the Gondwana Subcommission of the International Commission on Stratigraphy organized a symposium on Gondwana stratigraphy and paleontology at Mar del Plata, Argentina.

Anonymous, 1970, Reviews prepared for the first symposium on Gondwana stratigraphy: Haarlem, The Netherlands, Internat. Union Geol. Sci. Secretariat, 304 p.

Teichert, Curt, 1967, International symposium on Gondwana stratigraphy and paleontology: Geotimes, v. 12, no. 10, p. 10-11.

Anonymous, 1969, Gondwana stratigraohy: Paris, United Nations Educational Scientific and Cultural Organization, 1173 p.

392 The International Upper Mantle Committee sponsored a symposium on the world rift system at the 14th general assembly of the International Union of Geodesy and Geophysics at Zurich, Switzerland.

Knopoff, Leon, Heezen, B.C., and Macdonald, G.J.F., eds., 1969, The world rift system: Tectonophysics, v. 8, p. 265-577.

393 B.L. Isacks, L.R. Sykes, and J. Oliver showed that the focal mechanisms of earthquakes in the Fiji-Tonga-Kermadec region indicate that the Pacific crust is being thrust beneath the Tonga-Kermadec arcs at an angle of about 45 degrees.

Isacks, B.L., Sykes, L.R., and Oliver, Jack, 1967, Focal mechanisms of deep and shallow earthquakes in the Fiji-Tonga-Kermadec region (abs.): Am. Geophys. Union Trans., v. 48, p. 219.

___ 1969, Focal mechanisms of deep and shallow earthquakes in the Tonga-Kermadec region and the tectonics of island arcs: Geol. Soc. America Bull., v. 80, p. 1443-1469.

394 G.L. Johnson and B.C. Heezen suggested that the Alpha Ridge may have been a spreading center before the locus of spreading shifted to its present location.

Johnson, G.L., and Heezen, B.C., 1967, The Arctic mid-ocean ridge: Nature, v. 215, p. 724-725.

395 D.H. Matthews, C.A. Williams and A.S. Laughton used sounding and magnetic data gathered by RRS Discovery to show that the Indian Ocean Ridge continues into the Gulf of Aden.

Matthews, D.H., Williams, C.A., and Laughton, A.S., 1967, Mid-ocean ridge in the mouth of the Gulf of Aden: Nature, v. 215, p. 1052-1053.

396 H.W. Menard suggested that irregular crestal topography, relatively thick layer 2 and well developed rift valleys are characteristic of oceanic ridges that have low spreading rates.

Menard, H.W., 1967, Sea-floor spreading, topography, and the second layer: Science, v. 157, p. 923-924.

397 J. Oliver and B. Isacks discovered that earthquake waves are transmitted more efficiently and with higher velocities along Benioff zones than along paths normal to those zones, thereby confirming the "downgoing slab" aspect of sea-floor spreading.

Oliver, Jack, and Isacks, B.L., 1967, Some evidence on the structure of the mantle near an island arc (abs.): Amer. Geophys. Union Trans., v. 48, p. 219.

___ 1967, Deep earthquake zones, anomalous structures in the upper mantle, and the lithosphere: Jour. Geophys. Research, v. 72, p. 4259-4275. (Reprinted in Bird and Isacks, 1972.)

398 R.D. Schuilling suggested that tin belts on continents surrounding the Atlantic Ocean form a coherent pattern when the continents are united in their pre-drift positions.

Schuilling, R.D., 1967, Tin belts on the continents around the Atlantic Ocean: Econ. Geol., v. 62, p. 540-550.

399 The United Nations Educational, Scientific and Cultural Organization (UNESCO) and the International Union of Geological Sciences (IUGS) sponsored a symposium at Montevideo, Uruguay on continental drift in the South Atlantic.

Wilson, J. T. (Convenor), 1972, Continental drift emphasizing the history of the south Atlantic area: EOS (Trans. Am. Geophys. Union), v. 53, p. 164-185. (includes conference abstracts).

American Geophysical Union, 1972, Continental drift emphasizing the history of the south Atlantic area: Washington, D. C., Amer. Geophys. Union, (microfilm), 1088 p.

400 P.G. Van Dongen studied the paleomagnetism of Permian and Triassic rocks from the eastern Pyrenees Mountains and concluded that the Iberian peninsula has undergone a post Permian counter-clockwise rotation of about 30 degrees relative to Europe.

Van Dongen, P.G., 1967, The rotation of Spain: paleomagnetic evidence from the eastern Pyrenees: Paleogeog. Paleoclim. Paleoecol., v. 3, p. 417-432.

401 P.R. Vogt and N.A. Ostenso suggested that the subsidence of oceanic ridges, coral atolls, guyots and continental shelves is determined by contraction due to heat loss of ridge material. Data on sediment ages, magnetic anomalies and spreading rates were cited to support the concept of steady state crustal spreading.

Vogt, P.R., and Ostenso, N.A., 1967, Steady state crustal spreading: Nature, v. 215, p. 810-817.

1967 - 1968

402 G.O. Dickson, J.R. Heirtzler, E.M. Herron, X. Le Pichon and W.C. Pitman, III assembled a large amount of magnetic, bathymetric, seismic and other evidence from the earth's ocean basins to (1) show that there is a world-wide pattern of linear magnetic anomalies parallel to and symmetrical about the oceanic ridge crests and extending over most ocean basins to the base of the continental slope; (2) demonstrate that the patterns are everywhere similar, with differences being explainable by differences in spreading rate; (3) determine spreading rates for many ridge segments using the established reversal time scale back to -3.4 My; (4) extend the reversal time scale back to -76 My (into the Cretaceous) by assuming constant spreading rates; (5) show that the sea floor can be dated anywhere the magnetic anomaly pattern can be correlated; (6) suggest that changes in observed spreading rates may be related to continental orogenic events; (7) relate the anomaly patterns to continental movement; and (8) demonstrate that there is good agreement between Euler poles of rotation determined by fracture zone geometry with those determined by spreading rates. They showed that the Eltanin 19 profile across the Pacific-Antarctic Ridge is almost perfectly symmetrical about the Ridge and that it agrees remarkably with a calculated profile based on the inferred reversal history of the earth back to -45 My.

Pitman, W.C. III, Herron, Ellen, and Heirtzler, J.R., 1967, Magnetic anomalies in the south Pacific Ocean and ocean floor spreading (abs.): Amer. Geophys. Union Trans., v. 48, p. 132.

Dickson, G.O., Pitman, W.C., III, and Heirtzler, J.R., 1967, Magnetic anomalies in the south Atlantic and oceanfloor spreading (abs.): Amer. Geophys. Union Trans., v. 48, p. 132.

Le Pichon, Xavier, and Heirtzler, J.R., 1967, Magnetic anomalies in the Indian Ocean and ocean-floor spreading (abs.): Amer. Geophys. Union Trans., v. 48, p. 132-133.

Heirtzler, J.R., Dickson, G.O., Le Pichon, Xavier, and Pitman, W.C. III, 1967, Magnetic anomalies and ocean-floor spreading since late Cretaceous (abs.): Amer. Geophys. Union Trans., v. 48, p. 133.

Pitman, W.C., III, Herron, E.M., and Heirtzler, J.R., 1968, Magnetic anomalies in the Pacific and sea floor spreading: Jour. Geophys. Research, v. 73, p. 2069-2085.

Dickson, G.O., Pitman, W.C., III, and Heirtzler, J.R., 1968, Magnetic anomalies in the south Atlantic and ocean floor spreading: Jour. Geophys. Research, v. 73, p. 2087-2100.

Le Pichon, Xavier, and Heirtzler, J.R., 1968, Magnetic anomalies in the Indian Ocean and sea floor spreading: Jour. Geophys. Research, v. 73, p. 2101-2117.

Heirtzler, J.R., Dickson, G.O., Herron, E.M., Pitman, W.C., III, and Le Pichon, Xavier, 1968, Marine magnetic anomalies, geomagnetic field reversals, and motions of the ocean floor and continents: Jour. Geophys. Research, v. 73, p. 2119-2136. (Reprinted in Bird & Isacks, 1972, and Cox, 1973.)

1967 - 1969

403 X. Le Pichon, D.P. McKenzie, W.J. Morgan, and R.L. Parker developed a detailed theory of rigid-plate tectonics on a sphere and validated the theory by demonstrating that (1) the Euler poles of rotation of rigid lithosphere plates can be located by relating principles of spherical geometry to fracture zones, spreading rates and focal mechanism solutions of earthquakes; (2) the consequences of relative plate motion where three plates meet at a point (triple junction) can be predicted and tested; (3) the opening of the South Pacific, North Pacific, Arctic and Indian Oceans can each be accurately described by a single rotation of a rigid plate about an Euler pole; (4) the rotations agree with magnetic, physiographic, seismic and other data; (5) the history of periodic spreading since the Mesozoic can be directly related to major phases of orogeny around the world; and (6) oceanic magnetic anomaly patterns can be used to document plate motions through time.

McKenzie, D.P., and Parker, R.L., 1967, The north Pacific: an example of tectonics on a sphere: Nature, v. 216, p. 1276-1280. (Reprinted in Cox, 1973.)

Le Pichon, Xavier, 1968, Sea-floor spreading and continental drift: Jour. Geophys. Research, v. 73, p. 3661-3697. (Reprinted in Bird and Isacks, 1972 and Cox, 1973.)

Morgan, W. J., 1968, Rises, trenches, great faults, and crustal blocks: Jour. Geophys. Research, v. 73, p. 1959-1982. (Reprinted in Bird and Isacks, 1972 and Cox, 1973.)

McKenzie, D.P., and Morgan, W.J., 1969, Evolution of triple junctions: Nature, v. 224, p. 125-133. (Reprinted in Cox, 1973.)

1968

404 The American Philosophical Society held a symposium entitled "Gondwanaland revisited: new evidence for continental drift" at its annual general meeting.

Amer. Philos. Soc., 1968, Gondwanaland revisited: new evidence for continental drift: Amer. Philos. Soc. Proc., v. 112, p. 307-353.

405 F. Aumento, R.K. Wanless, R.D. Stevens, R.L. Fleischer, J.R.M. Viertl and P. B. Price found that K-Ar dating and fission-track dating of extrusive rocks from the northern Mid-Atlantic Ridge agree with each other, are consistent with the hypothesis of sea-floor spreading, and yield spreading rates of about 3 cm/yr near the ridge crest.

Aumento, F., Wanless, R.K., and Stevens, R.D., 1968, Potassium-argon ages and spreading rates on the Mid-Atlantic Ridge at 45 degrees north: Science, v. 161, p. 1338-1339.

Fleischer, R.L., Viertl, J.R.M., Price, P.B., and Aumento, F., 1968, Mid-Atlantic Ridge - age and spreading rates: Science, v. 161, p. 1339-1342.

406 J.N. Brune showed how seismic data can be used to determine the velocity of relative motion between plates along major fault zones.

Brune, J.N., 1968, Seismic moment, seismicity, and rate of slip along major fault zones: Jour. Geophys. Research, v. 73, p. 777-784.

407 E.C. Buffington, R.L. Larson, H.W. Menard, D.G. Moore and S. M. Smith showed that the East Pacific Rise continues up the Gulf of California in a series of enechelon segments cut by transform faults.

Larson, R.L., Menard, H.W., and Smith, S.M., 1968, Gulf of California - a result of ocean-floor spreading and transform faulting: Science, v. 161, p. 781-783.

Moore, D.G., and Buffington, E.C., 1968, Transform faulting and growth of the Gulf of California since the late Pliocene: Science, v. 161, p. 1238-1241.

408 I.G. Gass concluded that the Troodos massif of Cyprus is an upthrust remnant of a Mesozoic mid-Tethyan ocean ridge and correlated parts of the massif with layers 2, 3 and 4 of the ocean floor.

Gass, I.G., 1968, Is the Troodos massif of Cyprus a fragment of Mesozoic ocean floor?: Nature, v. 220, p. 39-42.

409 Geotimes published a "debate about the earth" between J.T. Wilson and V.V. Beloussov. The article consisted of (1) a lecture by Wilson (given to the Canadian Institute of Mining and Metallurgy) which suggested that a revolution in the geological sciences was taking place; (2) a letter by Beloussov to Wilson chiding him for being too eager to rewrite textbooks and reorganize education in conformity to new and unproven ideas; and (3) a reply by Wilson stressing the importance of conclusive new evidence gathered in recent years.

Wilson, J.T. and Beloussov, V.V., 1968, Debate about the earth: Geotimes, v. 13, no. 10, p. 10-22.

410 R.W. Girdler reported that the magnetic anomaly pattern in the Red Sea at about 16 degrees north latitude correlates with the reversal time scale back to -3 My and gives a spreading rate of about 1 cm/yr. The older anomalies disappear to the north until only the central anomaly remains.

Girdler, R. W., 1968, Drifting and rifting of Africa: Nature, v. 217, p. 1102-1106.

411 B. Isacks, J. Oliver and L.R. Sykes surveyed the entire field of seismology to evaluate its data in terms of the "new global tectonics" theory and concluded that the theory accounts for the data extremely well and that no substantial body of data tends to refute the theory.

Isacks, Bryan, Oliver, Jack and Sykes, L.R., 1968, Seismology and the new global tectonics: Jour. Geophys. Res., v. 73, p. 5855-5899. (Reprinted in Bird and Isacks, 1972 and Cox, 1973.)

412 D.H. Matthews and C.A. Williams reported discovery of a series of magnetic anomaly patterns in the Bay of Biscay that are best accounted for by sea-floor spreading in the Bay that caused Spain to rotate counterclockwise.

Matthews, D.H., and Williams, C.A., 1968, Linear magnetic anomalies in the Bay of Biscay; a qualitative interpretation: Earth and Planetary Sci. Letters, v. 4, p. 315-320.

413 H.W. Menard and T.M. Atwater used data from the northeast Pacific to show how changes in the direction of sea-floor spreading are recorded by fracture zones and magnetic anomaly patterns.

Menard, H.W., and Atwater, T.M., 1968, Changes in direction of sea-floor spreading: Nature, v. 219, p. 463-467. (Reprinted in Cox, 1973.)

414 W. C. Pitman III, D.E. Hayes and A.D. Raff suggested a possible plate tectonic origin for the unusual magnetic anomaly pattern previously found south of the Aleutian Islands. The postulated mechanism involved a triple junction that was later consumed in the Aleutian subduction zone.

Pitman, W.C. III, and Hayes, D.E., 1968, Sea-floor spreading in the Gulf of Alaska: Jour. Geophys. Research, v. 73, p. 6571-6580.

Raff, A.D., 1968, Sea-floor spreading: another rift: Jour. Geophys. Research, v. 73, p. 3699-3705.

Hayes, D.E., and Pitman, W.C. III, 1970, Magnetic lineations in the North Pacific: in Hayes, J.D., ed., Geological investigations of the North Pacific, Geol. Soc. America Mem. 126, p. 291-314.

415 F. Press used surface-wave data as a basis for determining that the lithosphere is more dense than the asthenosphere beneath it, a situation that is gravitationally unstable.

Press, Frank, 1968, Earth models obtained by Monte Carlo inverson: Jour. Geophys. Research, v. 73, p. 5223-5234.

___ 1969, The suboceanic mantle: Science, v. 165, p. 174-176.

Forsyth, D.W ., and Press, Frank, 1971, Geophysical tests of petrological models of the spreading lithosphere: Jour. Geophys. Research, v. 76, p. 7963-7979.

1968 - 1974

416 T.M. Atwater, J.M. Bird, M. Bocaletti, J. Bonnin, P.A. Brown, K.C.A. Burke, C.F. Burrett, J.F. Dewey, W.R. Dickinson, P. Elter, W.G. Ernst, J.F. Fitton, G.H. Gale, M. Garson, R.A. Gayer, G. Guazzone, W. Hamilton, W.B. Harland, R.D. Hatcher, Jr., B. Horsfield , K.J. Hsu, D.J. Hughes, W.S.F. Kidd, H.P. Laubscher, W.S. McKerrow, A.H. Mitchell, E.M. Moores, W.C. Pitman III, J. Plant, H.G. Reading, R. Riding, D. Roberts, J. Rodgers, W.B.F. Ryan, P.E. Schenck, S.O. Schlanger, A.G. Smith, R.K. Stevens, M. Takin, H. Williams and A.M. Ziegler developed a variety of detailed models that attempted to integrate plate tectonics and geosynclinal theory with the known and inferred history of various fold-mountain belts. General theories applicable to all orogens were also proposed.

Rodgers, John, 1968, The eastern edge of the North American continent during the Cambrian and early Ordovician: in Zen, E. A., and others, Studies of Appalachian geology, northern and maritime, New York, Inter-science, p. 141-149.

Dewey, J.F., 1969, Evolution of the Appalachian/Caledonian orogen: Nature, v. 222, p. 124-129.

___ 1969, Continental margins: a model for conversion of Atlantic type to Andean type: Earth & Planetary Sci. Letters, v. 6, p. 189-197.

Hamilton, Warren, 1969, Mesozoic California and the underflow of Pacific mantle: Geol. Soc. America Bull., v. 80, p. 2409-2430.

Atwater, T.M., 1970, Implications of plate tectonics for the Cenozoic tectonic evolution of western North America: Geol. Soc. America Bull., v. 81, p. 3513-3536.

Bird, J.M., and Dewey, J.F., 1970, Lithosphere plate-continental margin tectonics and the evolution of the Appalachian orogen: Geol. Soc. America Bull., v. 81, p. 1031-1060.

Dewey, J.F., and Bird, J.M., 1970, Plate tectonics and geosynclines Tectonophysics, v. 10, p. 625-638.

___ 1970, Mountain belts and the new global tectonics: Jour. Geophys. Research, v. 75, p. 2625-2647. (Reprinted in Bird & Isacks, 1972, and Cox, 1973.)

Dewey, J.F., and Horsfield, Brenda, 1970, Plate tectonics, orogeny and continental growth: Nature, v. 225, p. 521-525.

Dickinson, W.R., 1970, Plate tectonic models of geosynclines: Earth & Planetary Sci. Letters, v. 10, p. 165-174.

Ernst, W.G., 1970, Tectonic contact between the Franciscan melange and the Great Valley sequence-crustal expression of a late Mesozoic Benioff zone: Jour. Geophys. Research, v. 75, p. 886-901.

Fitton, J.F., and Hughes, D.J., 1970, Volcanism and plate tectonics in the British Ordovician: Earth & Planetary Sci. Letters, v. 8, p. 223-228.

Hamilton, Warren, 1970, The Uralides and the motion of the Russian and Siberian platforms: Geol. Soc. America Bull., v. 81, p. 2553-2576

Moores, E.M., 1970, Ultramafics and orogeny, with models of the US Cordillera and the Tethys: Nature, v. 228, p. 837-842.

Stevens, R.K., 1970, Cambro-Ordovician flysch sedimentation and tectonics in west Newfoundland and their possible bearing on a proto-Atlantic Ocean: Geol. Soc. Canada Spec. Paper no. 7, p. 165-177.

Ziegler, A.M., 1970, Geosynclinal development of the British Isles during the Silurian period: Jour. Geol., v. 78, p. 445-479.

Bird, J.M., Dewey, J.F., and Kidd, W.S.F., 1971, Proto-Atlantic oceanic crust and mantle: Appalachian/Caledonian ophiolites: Nature Phys. Sci., v. 231, p. 28-31.

Bocaletti, M., Elter, P., and Guazzone, G., 1971, Plate tectonic models for the development of the western Alps and northern Appennines: Nature, Phys. Sci., v. 234, p. 108-111.

Dewey, J.F., 1971, A model for the lower Paleozoic evolution of the southern margin of the early Caledonides of Scotland and Ireland: Scottish Jour. Geol., v. 7, p. 219-240.

Dewey, J.F., and Bird, J.M., 1971, Origin and emplacement of the ophiolite suite: Appalachian ophiolites in Newfoundland: Jour. Geophys. Research, v. 76, p. 3179-3206.

Dickinson, W.R., 1971, Plate tectonic models for orogeny at continental margins: Nature, v. 232, p. 41-42.

Hsu, K.J., 1971, Franciscan melanges as a model for eugeosynclinal sedimentation and underthrusting tectonics: Jour. Geophys. Research, v. 76, p. 1162-1170.

Hsu, K.J., and Schlanger, S.O., 1971, Ultrahelvetic flysch sedimentation and deformation related to plate tectonics: Geol. Soc. America Bull., v. 82, p. 1207-1218.

Laubscher, H.P., 1971, The largescale kinematics of the western Alps and the northern Appennines and its palinspastic implications: Am. Jour. Sci., v. 271, p. 193-226.

McKerrow, W.S., and Ziegler, A.M., 1971, The lower Silurian paleogeography of New Brunswick and adjacent areas: Jour. Geol., v. 79, p. 635-646.

Mitchell, A.H., and Reading, H.G., 1971, Evolution of island arcs: Jour. Geol., v. 79, p. 253-284.

Schenck, P.E., 1971, Southeastern Atlantic Canada, northwestern Africa, and continental drift: Canadian Jour. Earth Sci., v. 8, p. 1218-1251.

Smith, A.G., 1971, Alpine deformation and the oceanic areas of the Tethys, Mediterranean and Atlantic: Geol. Soc. America Bull., v. 82, p. 2039-2070.

Williams, Harold, 1971, Mafic-ultramafic complexes in western Newfoundland Appalachians and the evidence for their transportation: a review and interim report: Geol. Assoc. Canada Proc., v. 24, p. 9-25.

Harland, W.B., and Gayer, R.A., 1972, The Arctic Caledonides and earlier oceans: Geol. Magazine, v. 109, p. 289-314.

Hatcher, R.D., Jr., 1972, Developmental model for the southern Appalachians: Geol. Soc. America Bull., v. 83, p. 2735-2760.

McKerrow, W.S., and Ziegler, A.M., 1972, Paleozoic oceans: Nature Phys. Sci., v. 240, p. 92-94.

Takin, Manoochehr, 1972, Iranian geology and continental drift in the Middle East: Nature, v. 235, p. 147-150.

Brown, P.A., 1973, Possible cryptic suture in south-west Newfoundland: Nature Phys. Sci., v. 245, p. 9-10.

Burrett, C.F., 1973, Ordovician biogeography and continental drift: Paleogeog. Paleoclim. Paleoecol., v. 13, p. 161-201.

Dewey, J.F., and Burke, K.C.A., 1973, Tibetan, Variscan, and Precambrian basement reactivation: products of continental collision: Jour. Geol., v. 81, p. 683-692.

Dewey, J.F., Pitman, W.C. III, Ryan, W.B.F., and Bonnin, Jean, 1973, Plate tectonics and the evolution of the Alpine system: Geol. Soc. America Bull., v. 84, p. 3137-3180.

Garson, M.S., and Plant, Jane, 1973, Alpine type ultramafic rocks and episodic mountain building in the Scottish Highlands: Nature Phys. Sci., v. 242, p. 34-38.

Gale, G.H., and Roberts, David, 1974, Trace element geochemistry of Norwegian lower Paleozoic basic volcanics and its tectonic implications: Earth & Planetary Sci. Letters, v. 22, p. 380-390. Hatcher, R. D., Jr., 1974, North American Paleozoic fold belts and deformational histories: a plate tectonics anomaly?: Am. Jour. Sci., v. 274, p. 135-147.

Riding, Robert, 1974, Model of the Hercynian foldbelt: Earth & Planetary Sci. Letters, v. 24, p. 125-135.

1968 - 1975

417 The Glomar Challenger carried out a deep-sea drilling program in the Atlantic, Pacific and Indian Oceans, the Caribbean, Mediterranean, Black and Red Seas and the Gulf of Mexico, in which 573 holes were drilled at 392 sites. Among the achievements of the Deep-Sea Drilling Project (DSDP, previously JOIDES) were the following: (1) the sea-floor spreading theory was confirmed in detail in the Atlantic, Pacific and Indian Oceans; (2) the magnetic anomaly time scale was confirmed by dating (with fossils) the sediment immediately above the oceanic crust and by radiometrically dating the oceanic crust itself; (3) it was determined that the oceanic crust is a complex mixture of basalt flows and sediment and that the magnetic anomalies are the product of the many different magnetic fields of the flows; (4) the various seismic reflectors were dated and seismic velocities and densities of sediments were determined, placing seismic interpretation on a much firmer basis.

Davies, T.A., 1976, Five hundred and seventy three holes in the bottom of the sea - some results from seven years of deep-sea drilling: Jour. Geol. Education, v. 24, p. 142-155.

American Geological Institute, 1975, Deep Sea Drilling Project Legs 1-25: AGI Reprint Series, No. 1, Washington, D. C., Amer. Geol. Inst., 94 p.

___ 1975, Deep Sea Drilling Project Legs 26-44: AGI Reprint Series, No. 2, Washington, D. C., Amer. Geol. Inst., 82 p.

California. University. Scripps Inst. of Oceanography, La Jolla, 1969- , Initial Reports of the Deep Sea Drilling Project, v.'s 1-44, Washington, D. C., U. S. Govt. Printing Ofc.

1969

418 G.O. Allard and V.J. Hurst found that a complex succession of Precambrian geosynclinal rocks continues its trend from Africa to South America in precisely the location predicted by continental drift.

Allard, G.O. and Hurst, V.J., 1969, Brazil-Gabon geologic link supports continental drift: Science, v. 163, p. 528-532.

419 M. Barazangi and J. Dorman prepared new seismicity maps of the earth based on data obtained during 1961-1967 by the United States Coast and Geodetic Survey-supervised World-Wide Standard Seismograph Network and processed by high-speed computer. The new system was said to be capable of locating epicenters within 1 degree and determining depth to within 25 km.

Barazangi, M. and Dorman, J., 1969, World seismicity maps compiled from ESSA, Coast and Geodetic Survey, epicenter data, 1961-1967: Seismol. Soc. America Bull., v. 59, p. 369-380.

National Earthquake Information Center, 1970, Seismicity of the earth: Map NEIC 300, Rockville, Maryland, NEIC.

420 R.H. Dott, Jr. suggested that studies of the age of tectonic activity on land could shed light on changes in sea-floor spreading directions and centers and predicted from widespread evidence of renewed circum-Pacific tectonic activity at about -25 My that a pronounced change in sea-floor spreading would be found to have occurred at that time.

Dott, R.H., Jr., 1969, Circum-Pacific late Cenozoic structural rejuvenation: implications for sea floor spreading: Science, v. 166, p 874-876.

421 D.H. Elliot, E.H. Colbert, W.J. Breed, J.A. Jensen and J.S. Powell discovered the first assemblage of fossil tetrapods of significant age on the Antarctic continent, including labyrinthodont amphibians and thecodont and therapsid reptiles. Of particular interest was the genus Lystrosaurus which is a Triassic index fossil in South Africa and is also found in China and India.

Sullivan, Walter, 1969, Fossil called proof of continental link: New York Times, Dec. 6, p. 1, 24.

Elliot, D.H., Colbert, E.H., Breed, W.J., Jensen, J.A., and Powell, J.S., 1970, Triassic tetrapods from Antarctica - evidence for continental drift: Science, v. 169, p. 1197-1200.

Lear, John, 1970, The bones on Coalsack Bluff - a story of drifting continents: Saturday Review, 7 Feb., p. 46-51.

Colbert, E.H., 1970, The fossil tetrapods of Coalsack Bluff: Antarctic Jour. U. S.: v. 5, p. 57-61.

422 W. M. Elsasser suggested that the rigid lithosphere overlying a plastic asthenosphere could transmit horizontal stresses over great distances.

Elsasser, W. M., 1969, Convection and stress propagation in the upper mantle: in Runcorn, S. K., ed., The application of modern physics to the earth and planetary interiors, New York, Interscience, p. 223-246.

423 D. F. Falls, D.E. Hayes, E.M. Herron, W.J. Morgan and P.R. Vogt demonstrated that oceanic ridges need not be stationary by showing that two active ridges (the Chile Rise and the Pacific-Antarctic Ridge) parallel each other with no intervening subduction zone.

Herron, E.M., and Hayes, D.E., 1969, A geophysical study of the Chile ridge: Earth and Planetary Sci. Letters, v. 6, p. 77-83.

Morgan, W. J., Vogt, P. R., and Falls, D. F., 1969, Magnetic anomalies and sea floor spreading on the Chile rise: Nature, v. 222, p. 137-142.

424 The Geological Society of America conducted a Penrose Conference at Asilomar, Pacific Grove, California to discuss "the meaning of the new global tectonics for magnetism, sedimentation, and metamorphism in orogenic belts."

Dickinson, W.R., 1970, Global tectonics: Science, v. 168, p. 1250-1259.

425 The Glomar Challenger established by drilling that the Sigsbee Knolls in the deepest part of the Gulf of Mexico (3,572 m) are salt domes composed of Mesozoic salt like the other Gulf of Mexico province domes.

Burk, C.A., Ewing, Maurice, Worzel, J.L., Beall, A.O., Berggren, W.A., Bukry, D., Fischer, A. G., and Pessagno, E. A., Jr., 1969, Deep-sea drilling into the Challenger Knoll, central Gulf of Mexico: Am. Assoc. Petr. Geologists Bull., v. 53, p. 1338-1347.

Ewing, Maurice, Berggren, W.A., Fischer, A.G., Worzel, J. L., Bukry, D., Pessagno, E.A., Jr., Beall, A.O., and Burk, C.A., 1969, Site 2: in Initial Reports of the Deep-Sea Drilling Project, v. 1, Washington, D.C., U.S. National Sci. Foundation, p. 84-111.

426 W. Hamilton put together a detailed interpretation of the Mesozoic geology of California based on the idea that 2,000 linear km of Pacific Ocean plate have been underthrust beneath that area and that the overlying sediment is represented in the eugeosynclinal terranes of California. Specifically, the Franciscan Formation was said to consist of "deep-ocean Late Jurassic to Late Cretaceous sedimentary, volcanic, crustal, and mantle material."

Hamilton, Warren, 1969, Mesozoic California and the underflow of Pacific mantle: Geol. Soc. America Bull., v. 80, p. 2409-2429.

427 P. Molnar and L.R. Sykes used seismicity and focal mechanism studies to delineate the plate tectonics of the Caribbean and middle America regions. It was shown that a Cocos plate is underthrusting Mexico and Guatemala in a north-easterly direction and that a Caribbean plate is moving relatively eastward with respect to the Americas plate. The motion along the Cayman Trough was determined by first motions to be left-lateral strike-slip.

Molnar, Peter and Sykes, L.R., 1969, Tectonics of the Caribbean and Middle America regions from focal mechanism and seismicity: Geol. Soc. America Bull., v. 80, p. 1639-1684.

428 J. Schilling found that rare-earth patterns of tholeiitic basalts from the central trough of the Red Sea resemble very closely patterns found in basalts from oceanic ridges.

Schilling, Jean-Guy, 1969, Red Sea floor origin: rareearth evidence: Science, v. 165, p. 1357-1360.

429 W.P. Sproll and R.S. Dietz used a computerized (smallest average misfit) technique to find that southern Australia fits rather well against the Wilkes Land sector of east Antarctica.

Sproll, W.P. and Dietz, R.S., 1969, Morphological continental drift of Australia and Antarctica: Nature, v. 222, p. 345-348.

430 L.R. Sykes, B.L. Isacks and J. Oliver pointed out that the earthquake zones dipping beneath circum-Pacific island arcs have decreased in apparent thickness as precision of focal locations has improved and concluded that even their recent thickness determination of about 20 km for parts of the Tonga zone was still a measure of focal location uncertainty rather than of actual thickness.

Sykes, L.R., Isacks, B.L., and Oliver, Jack, 1969, Spatial distribution of deep and shallow earthquakes of small magnitude in the Fiji-Tonga region: Seismol. Soc. America Bull., v. 59, p. 1093-1113.

431 R. Van der Voo concluded from his study of Permian rocks from the Spanish Meseta that the Iberian peninsula has rotated counterclockwise about 35 degrees since the Permian.

Van der Voo, R., 1969, Paleomagnetic evidence for the rotation of the Iberian peninsula: Tectonophysics, v. 7, p. 5-56.

1970

432 French geologists S. Beuf, B. Biju-Duval, O. de Charpal, P. Rognon, O. Gariel, and A. Bennacef, working under the auspices of the Institut Francais du Petrole of Paris, discovered extensive evidence of an Ordovician ice age in the central Sahara and arranged for an international group of sedimentologists to examine the evidence in the field and later discuss their findings at Algiers.

Fairbridge, R.W., 1970, South Pole reaches the Sahara: Science, v. 168, p. 878-881.

___ 1970, An ice age in the Sahara: Geotimes, v. 15, no. 6, p.

___ 1971, Saharan Ordovician ice age: Science, v. 172, p. 86.

Beuf, serge, Biju-Duval, Bernard, de Charpal, Oliver, Rognon, Pierre, Gariel, Olivier, and Bennacef, Abdelkrim, 1971, Les Gres du Paleozoique inferieur au Sahara: Paris, Inst. Francais Petrole, 464 p.

433 J. W. Collison, J. W. Kitching, J. Ruben and T. Rich discovered abundant articulated skeletons of reptiles and amphibians near McGregor Glacier, Antarctica. Among the fossils were the land-dwelling, early Triassic Thrinaxodon liorhinus, Lystrosaurus murrayi and Procolopon trigoniceps, all of which had previously been found in South Africa.

Colbert, E.H., 1971, Triassic tetrapods fromMcGregor Glacier: Antarctic Jour. U.S., v. 6, p. 188-189.

Kitching, J.W., 1971, Paleontological investigations in the McGregor Glacier area: Antarctic Jour. U.S., v. 6, p. 118-119.

Kitching, J.W., Collinson, J.W., Elliott, D.H., and Colbert, E.H., 1972, Lystrosaurus zone (Triassic) from Antarctica: Science, v. 175, p. 524-526.

434 The Commission for Marine Geology of the International Union of Geological Sciences and the International Upper Mantle Committee held a symposium on "Global tectonics and sea-floor spreading" in Tokyo as part of "The Ocean World - Joint Oceanographic Assembly."

Wilson, J.T., ed., 1971, Global tectonics and sea-floor spreading: Tectonophysics, v. 12, p. 171-270.

435 A.R. Crawford suggested that plate tectonics provides a new basis on which to search for economic mineral deposits and cited examples from India-Ceylon and Australia to support his contention.

Crawford, A.R., 1970, Continental drift and un-continental thinking: Econ. Geol., v. 65, p. 11-16.

436 R. S. Dietz and J. C. Holden reconstructed a cartographically precise Pangaea for the Permian using absolute geographic coordinates and constructed four similar maps showing its breakup: at -180 My (end of the Triassic), at -135 My (late Jurassic), at -65 My (end of the Cretaceous), and the present.

Dietz, R.S., and Holden, J.C., 1970, Reconstruction of Pangaea - break-up and dispersion of continents, Permian to present: Jour. Geophys. Research, v. 75, p. 4939-4956.

437 R.S. Dietz and W.P. Sproll used a computerized (smallest average misfit) technique to demonstrate that southeastern Africa and the Atlantic Ocean margin of Antarctica "fit" together extremely well at the 1,800 m isobath.

Dietz, R.S. and Sproll, W.P., 1970, Fit between Africa and Antarctica - a continental drift reconstruction: Science, v. 167, p. 1612-1614.

438 J.H. Foster and N.D. Opdyke extended the magnetic reversal time scale found in sediments back to the upper Miocene (further back than -5 My) by using two long (26 and 28 m) cores from the central Pacific.

Foster, H.H., and Opdyke, N.D., 1970, Upper Miocene to Recent magnetic stratigraphy in deepsea sediments: Jour. Geophys. Research, v. 75, p. 4465-4473.

439 A.E. Maxwell and colleagues announced that data gathered during deep-sea drilling from the Glomar Challenger in the South Atlantic Ocean demonstrate that (1) the age of sediments immediately above basalt basement increases directly with distance from the crest of the Mid-Atlantic Ridge and the derived spreading half-rate is two cm per year; (2) stratigraphic patterns are symmetrical about the ridge axis; (3) both age and spreading rate agree with those previously determined from the magnetic anomalies and those necessary to account for continental drift; (4) the south Atlantic sea floor has been spreading at an essentially constant rate since -67 My; and (5) the rate suggests that the south Atlantic Ocean began to form during the early Cretaceous.

Maxwell, A.E., Von Herzen, R.P., Hsu, K.J., Andrews, J.E., Saito, T., Percival, S. F. Jr., Milow, E. D., and Boyce, R. E., 1970, Deep sea drilling in the South Atlantic: Science, v. 168, p. 1047-1059. (Reprinted in Cox, 1973)

___ 1970, Summary and conclusions: Maxwell, A. E. and others, Initial Reports of the deep sea drilling project, v. III, Washington, U.S. Govt. Printing Ofc., p. 441-471.

440 The Gondwana Subcommission of the International Commission on Stratigraphy and the International Union of Geological Sciences (IUGS) held a symposium in Cape Town and Johannesburg, South Africa on Gondwana stratigraphy and paleontology.

International Symposium on Gondwana Stratigraphy and Paleontology, 2nd, Univ. of Cape Town and Univ. of Witwaterstrand, 1970, Proceeding sand papers of the 2nd Gondwana Symposium, South Africa, July to August, 1970, Pretoria, Council for Scientific and Industrial Research, distributed by the Geological Survey of South Africa, 689 p.

Teichert, Curt, 1970, Gondwana Symposium: Geotimes, v. 15, no. 9, p. 26, 29, 30.

441 A. Hallam and A.G. Smith used the 1,000 m depth contour and a computer technique to find the best "fit" among South America, Africa, Antarctica, Australia, Madagascar and India. The fit was found to be geometrically and geologically good and was said to confirm many of Du Toit's (1937) ideas.

Hallam, A. and Smith, A.G., 1970, The fit of the southern continents: Nature, v. 225, p. 139-144.

442 G.L. Johnson, N.A. Ostenso, P.R. Vogt and R.J. Wold used bathymetic, seismic, magnetic and gravimetric data from the Arctic Ocean and North Atlantic Ocean as a basis for concluding that (1) the Alpha Cordillera was an active spreading center between -60 and -40 My; (2) the Alpha Cordillera is cut by transform faults; (3) at about -40 My the Atlantic spreading center bifurcated in the north Atlantic and began spreading from the Nansen Ridge; and (4) the new spreading center intersected the Siberian continental margin and carried a thin slice of it north to form the Lomonosov Ridge.

Vogt, P.R. and Ostenso, N.A., 1970, Magnetic and gravity profiles across the Alpha Cordillera and their relation to Arctic sea-floor spreading: Jour. Geophys. Research, v. 75, p. 4925-4937.

Vogt, P.R., Ostenso, N.A., and Johnson, G.L., 1970, Magnetic and bathymetric data bearing on sea-floor spreading north of Iceland: Jour. Geophys. Research, v. 75, p. 903-920.

Ostenso, N. A. and Wold, R. J., 1971, Aeromagnetic survey of the Arctic Ocean - techniques and interpretations: Marine Geophys. Research, v. 1, p. 47-88.

Ostenso, N.A. and Wold, R.J., 1973, Aeromagnetic evidence for origin of Arctic Ocean basin: in Pitcher, M. G., Arctic Geology, Am. Assoc. Petroleum Geologists Mem. 19, p. 506-516.

443 P.W. Lipman, H.J. Prostka and R.L. Christiansen used potassium oxide/silica variation trends in Cenozoic igneous rocks from the western United States to show that during early and middle Cenozoic time there were probably two subduction zones dipping to the east at about 20 degrees beneath that region. One was said to have reached its maximum depth in eastern Idaho, central Utah and Arizona, the other beneath central Montana, western Wyoming, central Colorado and eastern New Mexico. During the late Cenozoic a pronounced change in the character of volcanic rocks was said to have been caused by overriding of the East Pacific Rise and inception of right-lateral transform faulting.

Lipman, P.W., Prostka, W.J., and Christiansen, R.L., 1970, Cenozoic volcanism and tectonism in the western United States and adjacent parts of the spreading ocean floor, Part 1, early and middle Tertiary (abs.): Geol. Soc. America Abs. with Prog., v. 2, p. 112-113.

Christiansen, R.L., and Lipman, P.W., 1970, Cenozoic volcanism in the western United States and adjacent parts of the spreading ocean floor, Part 2, late Cenozoic (abs.): Geol. Soc. America Abs. with Prog., v. 2, p. 81-82.

Lipman, P.W., Prostka, W.J., and Christiansen, R.L., 1971, Evolving subduction zones in the western United States, as interpreted from igneous rocks: Science, v. 174, p. 821-825.

___ 1972, Cenozoic volcanism and plate-tectonic evolution of the western United States, I, Early and middle Cenozoic: Royal Soc. London Philos. Trans., v. A271, p. 217-248.

Christiansen, R.L., and Lipman, P.W., 1972, Cenozoic volcanism and plate-tectonic evolution of the western United States, II, Late Cenozoic: Royal Soc. London Philos. Trans., v. A271, p. 249-284.

443a B.P. Luyendyk determined that the dip of subducing lithospheric plates is inversely related to the rate of plate convergence and suggested specifically that the dip equals an angle whose sine is the sinking rate (4-6 cm/y) divided by the convergence rate.

Luyendyk, B.P., 1970, Dips of downgoing lithospheric plates beneath island arcs: Geol. Soc. America Bull., v. 81, p. 3411-3416.

444 M.W. McElhinny and G.R. Luck reported that paleomagnetic data from lower Paleozoic rocks from Gondwana continents enable a unique reconstruction of Gondwana to be made and stated that the reassembly agrees with the computer fit of the continents and the matching of geological age provinces.

McElhinny, M.W., and Luck, G.W., 1970, Paleomagnetism and Gondwanaland: Science, v. 168, p. 830-832.

445 D. P. McKenzie used focal locations, first motion determinations and other evidence to work out the plate tectonics of the Mediterranean area. It was demonstrated that the major plate boundary in the area (the Eurasian-African) is a subduction zone west of Italy, splits further east to encircle two smaller plates (the Aegean and Turkish) and is a zone of crustal thickening beginning at the Caucausus.

McKenzie, D.P., 1970, Plate tectonics of the Mediterranean Region: Nature, v. 226, p. 239-243.

446 J.G. Sclater and J. Francheteau demonstrated that average heat flow in the north Pacific decreases regularly with age of oceanic crust.

Sclater, J.G., and Francheteau, Jean, 1970, The implications of terrestrial heat flow observations on current tectonic and geochemical models of the crust and upper mantle of the earth: Royal Astron. Soc. Geophys. Jour., v. 20, p. 509-542. (Reprinted in Cox, 1973.)

447 J.W. Valentine and E.M. Moores theorized that variations in the diversity of marine life through time can be explained by eustatic changes in sea level and patterns of continental fragmentation and reassembly. Increased diversity was said to be due to isolation of the shallow sea surrounding separated continents and to related patterns of climate and ocean upwelling.

Valentine, J.W., and Moores, E.M., 1970, Plate tectonic regulation of faunal diversity and sea level: a model: Nature, v. 228, p. 657-659.

Valentine, J.W., 1971, Plate tectonics and shallow marine diversity and endemism, an actualistic model: Systematic Zoology, v. 20, p. 253-264.

Valentine, J.W., and Moores, E.M., 1972, Global tectonics and the fossil record: Jour. Geology, v. 80, p. 167-184.

Valentine, J.W., 1973, Plates and provinciality, a theoretical history of environmental discontinuities: in Hughes, N.F., ed., Organisms and continents through time, Spec. Papers in Paleo. no. 12, Paleontological Assoc., London, p. 79-92.

448 J.K. Weissel and D.E. Hayes reported the first clear magnetic anomaly record of asymmetrical spreading, from the Indian-Antarctic Ridge south of Australia where 230 km more crust was generated on the north side of the spreading center than on the south during the period -38 My to -10 My.

Weissel, J.K., and Hayes, D.E., 1970, Contrasting zones of sea floor spreading south of Australia (abs.): EOS (Am. Geophys. Union Trans.), v. 52, p. 237.

___ 1971, Asymmetric sea floor spreading south of Australia: Nature, v. 231, p. 518-522. (Reprinted in Cox, 1973.)

449 D.A. White, D.H. Roeder, T.H. Nelson and J.C. Crowell proposed the use of the term "subduction" in its modern, plate-tectonic sense for "an abruptly descending or formerly descended elongate body of lithosphere, together with any existing envelope of inter-plate deformation."

White, D.A., Roeder, D.H., Nelson, T.H., and Crowell, J.C., 1970, Subduction: Geol. Soc. America Bull., v. 81, p. 3431-3432.

1970 -1972

450 Y.P. Aggarwal, M. Barazangi, D.A. Falvey, B. Isacks, D.E. Karig and G. H. Packham (1) demonstrated that the oceanic basins immediately behind at least some island arcs have undergone episodic spreading; (2) conceived the idea that the spreading is the result of the vertical movement of thermal diapirs that originate in subduction zones at depth; and (3) showed that seismic wave attenuation and velocity data support this model.

Karig, D.E., 1970, Ridges and basins of the Tonga-Kermadec island arc system: Jour. Geophys. Research, v. 75, p. 239-254.

Barazangi, Mauwia, and Isacks, Bryan, 1971, Lateral variations of seismic wave attenuation in the upper mantle above the inclined earthquake zone of the Tonga island arc: Deep anomaly in the upper mantle: Jour. Geophys. Research, v. 76, p. 8493-8516.

Karig, D.E., 1971, Structural history of the Mariana island arc system: Geol. Soc. America Bull., v. 82, p. 323-344.

___ 1971, Origin and development of marginal basins in the western Pacific: Jour. Geophys. Research, v. 76, p. 2542-2561.

Packham, G.H., and Falvey, D.A., 1971, An hypothesis for the formation of marginal seas in the western Pacific: Tectonophysics, v. 11, p. 79-109.

Aggarwal, Y.P., Barazangi, Mauwia, and Isacks, Bryan, 1972, P and S traveltimes in the Tonga-Fiji region: a zone of low velocity in the uppermost mantle behind the Tonga island arc: Jour. Geophys. Research, v. 77, p. 6427-6434.

451 P. Barker, Z. BenAvraham, C. Bowin, W.B. Bryan, C.G. Chase, J.W. Hawkins, B.P. Luyendyk, K.C. MacDonald, J. Mammerickx, J.G. Sclater, J. Segawa and S. Uyeda reported the discovery of behind-arc spreading centers in the Scotia and Philippine Seas and in the Lau and Woodlark basins.

Barker, P., 1970, Plate tectonics of the Scotia Sea region: Nature, v. 228, p. 1293-1296.

Ben-Avraham, Zvi, Bowin, C., and Segawa, J., 1972, An extinct spreading center in the Philippine Sea: Nature, v. 240, p. 453-455.

Sclater, J.G., Hawkins, J.W., Mammerickx, Jacqueline, and Chase, C.G., 1972, Crustal extension between the Tonga and Lau ridges: Petrologic and geophysical evidence: Geol. Soc. American Bull., v. 83, p. 505-518.

Uyeda, Seiya, and BenAvraham, Zvi, 1972, Origin and development of the Philippine Sea: Nature Phys. Sci., v. 240, p. 176-178.

Luyendyk, B.P., MacDonald, K.C., and Bryan, W.B., 1973, Rifting history of the Woodlark basin in the southwest Pacific: Geol. Soc. America Bull., v. 84, p. 1125-1134.

452 J. Francheteau, H. Kanamori, F. Press, J. G. Sclater and R.I. Walcott determined the thickness of the lithosphere to be 70-80 km in oceanic areas and about 150 km in continental areas.

Kanamori, H., and Press, Frank, 1970, How thick is the lithosphere?: Nature, v. 226, p. 330-331.

Sclater, J.G., and Francheteau, Jean, 1970, The implications of terrestrial heat flow observations on current tectonic and geochemica models of the crust and upper mantle of the earth: Royal Astron. Soc. Geophys. Jour., v. 20, p. 509-542.

Walcott, R.I., 1970, Flexural rigidity, thickness and viscosity of the lithosphere: Jour. Geophys. Research, v. 75, p. 3941-3954.

Press, Frank, 1972, The earth's interior as inferred from a family of models: in Robertson, E. C., ed., The nature of the solid earth, New York, McGraw-Hill, 677 p.

1970 - 1974

453 D.I. Axelrod, A.A. and H.A. Meyerhoff, R.S. Briggs, Jr., C. Teichert and P. S. Wesson authored a series of lengthy and detailed attacks on the validity of plate tectonics theory.

Alexrod, D.I., 1963, Fossil floras suggest stable not drifting continents: Jour. Geophys. Res., v. 68, p. 3257-3263.

Meyerhoff, A.A., 1970, Continental drift: implications of paleomagnetic studies, meteorology, physical oceanography, and climatology: Jour. Geology, v. 78, p. 151.

___ 1970, Continental drift II: high latitude evaporite deposits and geologic history of Arctic and north Atlantic Oceans: Jour. Geology, v. 78, p. 406-444.

Meyerhoff, A.A., and Teichert, Curt, 1971, Continental drift III: late Paleozoic glacial centers, and Devonian-Eocene coal distribution: Jour. Geology, v. 79, p. 285-421.

Meyerhoff, A.A., and Meyerhoff, H.A., 1972, Continental drift IV: the Caribbean "plate": Jour. Geology, v. 80, p. 34-60.

___ 1972, "The new global tectonics": major inconsistencies: Amer. Assoc. Petr. Geologists Bull., v. 56, p. 269-336.

___ 1972, "The new global tectonics": age of linear magnetic anomalies of ocean basins: Amer. Assoc. Petr. Geologists Bull., v. 56, p. 337-357.

Meyerhoff, A.A., Meyerhoff, H. A., and Briggs, R. S. Jr., 1972, Continental drift V: proposed hypothesis of earth tectonics: Jour. Geology, v. 80, p. 663-692.

Wesson, P.S., 1972, Objections to continental drift and plate tectonics: Jour. Geophys. Research, v. 80, p. 185-197.

Meyerhoff, A.A., 1973, Origin of Arctic and north Atlantic Oceans: in Pitcher, M.G. ed., Arctic geology, Amer. Assoc. Petr. Geologists Mem. 19, p. 562-582.

Meyerhoff, A.A., and Meyerhoff, H.A., 1974, Tests of plate tectonics: in Kahle, C. F., ed., Plate tectonics: assessments and reassessments, Amer. Assoc. Petr. Geologists Mem. 23, p. 43-145.

1971

454 R.N. Anderson, M.L. Bell, C.G.A. Harrison, P.A. Rona and J.G. Sclater showed how plate tectonic theory predicts the relationship between oceanic ridge and flank elevation, spreading rate and angular distance from poles of rotation, confirmed the relationship with data from various ridge segments, and demonstrated how topography can be useful in deciphering plate history.

Rona, P.A., 1971, Depth distribution in ocean basins and plate tectonics: Nature, v. 231, p. 179-180.

Sclater, J.G., Anderson, R.N., and Bell, M.L., 1971, Elevation of ridges and evolution of the central eastern Pacific: Jour. Geophys Research, v. 76, p. 7888-7915.

Sclater, J.G., and Harrison, C.G.A., 1971, Elevation of mid-ocean ridges and the evolution of the southwest Indian Ridge: Nature, v. 230, p. 175-77.

455 R.W. Decker, P. Einarsson and P.A. Mohr detected six to seven cm of spreading across the eastern rift zone of Iceland during the period 1967-70. The spreading may have taken place during the 1970 eruption of Hekla volcano.

Decker, R.W., Einarsson, Pall, and Mohr, P.A., 1971, Rifting in Iceland: new geodetic data: Science, v. 173, p. 530-533.

456 R.L. Fisher, D.P. McKenzie and J.G. Sclater used magnetic anomaly patterns, heat flow data, topography and earthquake epicenter locations to work out the history of the Indian Ocean basin since the late Cretaceous. They found that (1) India moved away from Antarctica at a rate of 18 cm/yr for 20 My in the early Tertiary; (2) Australia separated from Antarctica during the same time; (3) the present episode of spreading began at about -36 My; (4) the pattern of oceanic depths agrees with the cooling plate model and with the history determined from magnetic anomalies; and (5) the history determined from oceanographic data agrees with that determined from paleomagnetic pole position studies.

Fisher, R. L., Sclater, J. G., and McKenzie, D. P., 1971, Evolution of the central Indian Ridge, western Indian Ocean: Geol. Soc. America Bull., v. 82, p. 553-562.

McKenzie, D.P., and Sclater, J.G., 1971, The evolution of the Indian Ocean since the late Cretaceous: Royal Astron. Soc. Geophys. Jour., v. 25, p. 437-528.

____ 1973, The evolution of the Indian Ocean: Sci. American, v. 228, no. 5, p. 62-72. (Reprinted in Sci. American, 1976, Continents adrift and continents aground, San Francisco, W.H. Freeman, p. 138-148.)

457 The Geological Society, Palaeontological Association and Systematics Association held a symposium in London on "Organisms and continents through time - Methods of assessing relationships between past and present biologic distributions and the positions of continents."

Hughes, N.F., ed., 1973, Organisms and continents through time: Palaeo. Assoc., London, 334 p.

458 The Geological Society of America held a "Symposium on plate tectonics in geologic history" at its annual meeting in Washington, D. C.

Anonymous, 1971, Geol. Soc. America Abstracts with Programs, v. 3, p. 449-450.
Dickinson, W.R., 1972, Symposium: plate tectonics: Amer. Jour. Science, v. 272, p. 549-668.

459 The Geophysical Laboratory of the Carnegie Institution of Washington sponsored a symposium in Washington on "the constraints imposed by petrology on the current dynamic model of the earth."

Yoder, H.S., Jr., 1971, Petrologic implications of plate tectonics: Science, v. 173, p. 464-466.

460 R.A. Gibb and R.I. Walcott suggested that the boundary between the Churchill and Superior structural provinces marks the place where two continents collided during the period from -2,390 to -1,640 My. It was suggested that the collision resulted in the Hudsonian orogeny and that the suture is characterized by Aphebian geosynclinal rocks and by basic volcanics and ultramafics and their metamorphosed equivalents.

Gibb, R.A., and Walcott, R.I., 1971, A Precambrian suture in the Canadian shield: Earth and Planetary Sci. Letters, v. 10, p. 417-422.

461 J.F. Gilluly suggested that subduction zone magmas are derived from the melting of consumed tholeiitic basalt and overlying sediment rather than from the mantle and cited the relative scarcity of sediments off the west coast of the Americas, the tremendous volume of andesitic volcanics in circum-Pacific belts, and bulk chemical compositions as evidence supporting this view.

Gilluly, J.F., 1971, Plate tectonics and magmatic evolution: Geol. Soc. America Bull., v. 82, p. 2383-2396. (Reprinted in Cox, 1973.)

462 P.W. Guild suggested that the relationship of known ore deposits to plate tectonics should be studied and used as a basis for locating additional deposits. He also provided a table showing the proposed relationships of some ore types to plate tectonic features.

Guild, P.W. 1971, Metallogeny: a key to exploration: Mining Engineering, Jan. 1971, p. 69-72.

463 The International Upper Mantle Committee conducted a symposium on "Rifting in Africa" at the 15th General Assembly of the International Union of Geodesy and Geophysics in Moscow.

Girdler, R.W., ed., 1972, East African Rifts: Tectonophysics, v. 15, p. 1-79.

464 E.M. Moores and F.J. Vine suggested that the diabase of the Troodos massif of Cyprus consists of a dike swarm in which 100 km of extension is indicated over 100 km of exposure and that the dikes record formation by sea-floor spreading. They also described what may be a fossil transform fault.

Moores, E.M., and Vine, F.J., 1971, The Troodos massif, Cyprus and other ophiolites as oceanic crust: evaluation and implications: Royal Soc. London Philos. Trans., v. A268, p. 443-466.
Vine, F.J., and Moores, E.M., 1972, A model for the gross structure, petrology, and magnetic properties of oceanic crust: in Shagam, R., and others, eds., Studies in earth and space sciences, Geol. Soc. America Mem. 132, p. 195-205.

465 W.J. Morgan hypothesized that "hot spots" are manifestations of convection in the lower mantle and that they are the driving force for plate movement. About 20 hot spots were said to be active today.

Morgan, W.J., 1971, Convection plumes in the lower mantle: Nature, v. 230, p. 42-43. (Reprinted in Cox, 1973.)

466 C.H. Scholz, M. Barazangi and M.L. Sbar showed that the high heat flows, thin crust, low seismic velocities and high attenuation of the upper mantle, normal faulting, seismicity pattern and late Cenozoic basaltic volcanism of the Great Basin of the western United States all support the concept that the area is an "ensialic interarc basin" that developed when the middle Cenozoic subduction zone to the west was terminated and crustal extension due to the rising of a large mantle diapir occurred.

Scholz, C.H., Barazangi, Mauwia, and Sbar, Marc, 1971, Late Cenozoic evolution of the Great Basin, western United States, as an ensialic interarc basin: Geol. Soc. America Bull., v. 82, p. 2979-2990.

1972

467 The Australian Petroleum Exploration Association held a symposium on continental drift in Sydney.

Anonymous, 1972, Continental drift symposium, Sydney, 1972: Australian Petroleum Explor. Assoc. Jour., v. 12, pt. 2.

468 C.G. Chase developed and used a least squares method for finding the relative motions of eight major plates. The method insured internal consistency of results and allowed the calculation of global rates of creation and subduction of lithosphere, both of which were found to equal three square km/yr.

Chase, C.G., 1972, The N plate problem of plate tectonics: Royal Astron. Soc. Geophys. Jour., v. 29, p. 117-122.

469 W.A. Duffield reported that Mauna Ulu crater on Kilauea Volcano (Hawaii) provided a miniature scale model of the earth's plate tectonics on the surface of a 90 x 160 m active lava column. Specific plate features mimicked on the cooling upper surface of the lava included: ridges, subduction zones, triple junctions, transform faults, "fossil" ridges, symmetrical, assymetrical and oblique spreading, conservation of area, and folding at subduction zones.

Duffield, W.A., 1972, Kilauea Volcano provides a model for plate tectonics: Geotimes, v. 17, no. 4, p. 19-21.

___ 1972, A naturally occurring model of global plate tectonics: Jour. Geophys. Research, v. 77, p. 2543-2555.

470 JOIDES scientists reported that boreholes drilled by the Glomar Challenger along the crest of the Ninetyeast Ridge encountered shallow-water sediments (and some non-marine sediments) below calcareous ooze and concluded that the litho-stratigraphic units transgress time toward the south where they are younger and that various lines of evidence suggest the Ridge sank rapidly and moved northward with the Indian plate during the interval Maastrichtian-Middle Paleocene.

Anonymous, 1972, Deep-Sea Drilling Project-Leg 22: Geotimes, v. 17, no. 6, p. 15-17.

___ 1973, Across the southern Indian Ocean abord Glomar Challenger: Geotimes, v. 18, no. 3, p. 16-19.

Sclater, J.G., von der Borch, Chris, Veevers, J.J., Hekinian, Roger, Thompson, R.W., Pimm, A.C., McGowran, Brian, Gartner, Stefan, Jr., and Johnson, D. A., 1974, Regional synthesis of the deep sea drilling results from Leg 22 in the eastern Indian Ocean: in Initial Reports of the Deep Sea Drilling Project, v. 22, Washington, D. C., U.S. National Science Foundation, p. 815-831.

(See also)

Luyendyk, B.P., and Davies, T.A., 1974, Results of DSDP Leg 26 and the geologic history of the Indian Ocean: in Initial Reports of the Deep Sea Drilling Project, v. 26, Washington, D. C., U. S. National Science Foundation, p. 909-943.

471 The International Commission for Geodynamics sponsored a symposium on "Mechanisms of plate tectonics" at the 24th International Geological Congress at Montreal, Canada.

Irving, E., ed., 1973, Mechanisms of plate tectonics: Tectonophysics, v. 19, p. 85-194.

472 The International Union of Geological Science Commission on Experimental Petrology at High Pressures and Temperatures and the Inter-Union Commission on Geodynamics sponsored a symposium on "Experimental petrology and global tectonics at Montreal.

Wyllie, P.J., ed., 1973, Experimental petrology and global tectonics: Tectonophysics, v. 17, p. 187-297,.

473 E.D. Jackson, E.A. Silver and G.B. Dalrymple concluded from their study of the Hawaiian-Emperor chain that (1) the eruptions that produced the chain were not influenced by the structure of the crust they were built upon; (2) the eruptions occurred above a melting spot about 300 km in diameter that moved relative to the overlying plate; and (3) the change in plate motion that produced the bend in the chains occurred at about -25 My, a time of renewed tectonism in the western Pacific island arcs and of changing plate boundaries in the northern and eastern Pacific.

Jackson, E.D., Silver, E.A., and Dalrymple, G.B., 1972, Hawaiian-Emperor chain and its relation to Cenozoic circum-Pacific tectonics: Geol. Soc. America Bull., v. 83, p. 601-618.

474 R.L. Larson and W.C. Pitman III made new correlations of magnetic lineations between the North Atlantic and the Pacific, and extended the geomagnetic reversal time scale back to the late Jurassic (-162 My). Using this new time scale it was then determined that (1) drift in the south Atlantic began between -110 and -85 My; (2) the seaward part of the magnetic quiet zones off the eastern United States and northwestern Africa resulted from sea-floor spreading during the late Jurassic period of normal polarity; (3) spreading in the Pacific during the late Mesozoic occurred from five spreading centers and about half of the crust developed since then has been consumed; and (4) there was a rapid pulse of spreading from about -110 to -85 My in both the Atlantic and Pacific which was related to episodes of plutonism around the Pacific rim.

Larson, R.L., and Pitman, W.C. III, 1972, World-wide correlation of Mesozoic magnetic anomalies and its implications: Geol. Soc. America Bull., v. 83, p. 3645-3662.

475 W.J. Morgan showed that the trends of four widely separated volcanic island chains in the Pacific parallel each other, three of them (the Hawaiian-Emperor, the Tuamoto-Line, and the Austral-Gilbert-Marshall) through a sharp bend in trend, demonstrating that the trends trace motion of the rigid Pacific plate over fixed mantle "hot spots".

Morgan, W.J., 1972, Deep mantle convection plumes and plate motions: Am. Assoc. Petroleum Geologists Bull., v. 56, p. 203-213.

___ 1972, Plate motions and deep mantle convection: in Shagam, R., and others, eds., Studies in earth and space sciences, Geol. Soc. America Mem. 132, p. 7-22.

476 The North Atlantic Treaty Organization (NATO) sponsored an Advanced Study Institute at the University of Newcastle Upon Tyne on "Implications of continental drift to the earth sciences."

Tarling, D.H., and Runcorn, S.K., eds., 1973, Implications of continental drift to the earth sciences: London, Academic Press, 2 vols., 1184 p.

477 D. W. Oldenburg and J. N. Brune conducted experiments with cooling molten wax and, using purely tensional forces, were able to establish that the typical oceanic ridge pattern of straight-line segments offset by transform faults is the preferred mode of separation and that, since the hottest and weakest part of the plates remains at the original ridge crest, separation continues there.

Oldenburg, D.W., and Brune, J.N., 1972, Ridge transform fault spreading pattern in freezing wax: Science, v. 178, p. 301-304.

478 W. C. Pitman III and M. Talwani worked out the spreading history of the North Atlantic on the basis of the magnetic anomaly pattern. They found that the recent Atlantic Ocean rifted into existence at about -200 My and that active spreading began at about -180 My and has proceded at variable rates (2 to 4 cm/yr) since then. Europe and North Africa were found to have separated from North America at different rates, their relative motion being taken up at the Azores-Gibraltar Ridge. All post-Cretaceous spreading in the Arctic Ocean was found to have taken place since -63 My from the Mid-Atlantic Ridge extension.

Pitman, W.C. III and Talwani, Manik, 1972, Sea-floor spreading in the North Atlantic: Geol. Soc. America Bull., v. 83, p. 619-646.

479 R.H. Sillitoe interpreted the occurrence of metallogenetic provinces in western America in terms of plate tectonics.

Sillitoe, R.H., 1972, Relation of metal provinces in western America to subduction of oceanic lithosphere: Geol. Soc. America Bull., v. 83, p. 813-818.

480 P. R. Vogt suggested that there has been global synchronism of mantle plume discharge and that such discharges may control not only plate movements (and, therefore, sea level, orogeny and volcanism) but also global temperature, magnetic reversal frequency, sedimentation rates and faunal and floral extinctions.

Vogt, P.R., 1972, Evidence for global synchronism in mantle plume convection, and possible significance for geology: Nature, v. 240, p. 338-342.

481 C.G. Chase, E.M. Herron, R.L. Larson and S.M. Smith used magnetic anomalies, deep-sea drilling information, topography, heat flow data and seismic reflection profiles to reinterpret the detailed history of the Pacific Ocean basin. Among other results, it was concluded that (1) much of the East Pacific Rise has only been active since about -20 My; (2) an older spreading center within what is now the Nazca plate was active until about -9 My; (3) in the late Mesozoic the Pacific included four plates separated by five spreading centers and two triple points; (4) much of the Pacific crust now lies 35 to 40 degrees north of where it formed; and (5) the Mesozoic reversal sequence occurred from -160 to -110 My between the Jurassic and Cretaceous "quiet" periods.

Herron, E.M., 1972, Sea floor spreading and the Cenozoic history of the east central Pacific: Geol. Soc. America Bull., v. 83, p. 1671-1692.

Larson, R.L., and Chase, C.G., 1972, Late Mesozoic evolution of the western Pacific: Geol. Soc. America Bull., v. 83, p. 3627-364

Larson, R.L., Smith, S.M., and Chase, C.G., 1972, Magnetic lineations of early Cretaceous age in the western equatorial Pacific Ocean: Earth and Planetary Sci. Letters, v. 15, p. 315-319.

Larson, R.L., 1974, An updated time scale of magnetic reversals for the late Mesozoic (abs.): EOS (Am. Geophys. Union Trans.), v. 55, p. 236.

1972 - 1975

482 A group of French and American scientists, including R.D. Ballard, G. Bellaiche, W. B. Bryan, J.L. Cheminee, P. Choukroune, J. Francheteau, J. R. Heirtzler, R. Hekinian, G.

Keller, X. Le Pichon, J. G. Moore, D. Needham, P. Tapponnier, and T. H. van Andel, conceived Project FAMOUS (French-American Undersea Study) whose purpose was "to conduct an integrated and detailed geological, geophysical, and geochemical study of a portion of a plate boundary (specifically the Mid-Atlantic Ridge graben at 36 degrees 50 minutes north latitude) where new oceanic crust is generated and where some processes leading to the formation of the oceanic lithosphere may appear more clearly." The Project was to include the "use, for the first time, of precisely navigated, manned research submersibles in the latter phases of the work at sea and thus enable methods of classical field geology to be applied to the study of the deep ocean floor." Included among the early results obtained were observations suggesting that (1) tension cracks parallel to the major ridge axes are very abundant and active; (2) tectonism is continuous across the entire valley floor and seems to override the episodic volcanism that is confined to a 1 km wide strip down the center of the valley; (3) vertical motions of valley walls of up to 300 m have taken place; (4) the vertical temperature gradient below the valley floor is so high that a magma chamber may exist less than 2 km below the sea floor; (5) hydrothermal deposition of minerals has occurred; (6) the youngest volcanic features in the area studied are hundreds to thousands of years old; and (7) the expected sense of transform fault motion can be confirmed on a very small scale by examination of sediment deformation structures. The submersibles used were the American Alvin and the French Archimede and Cyana. Canadian and British scientists and vessels were also involved in the study.

Bellaiche, G., Cheminee, J.L., Francheteau, J., Hekinian, R., Le Pichon, Xavier, Needham, H. D., and Ballard, R. D., 1974, Inner floor of the rift valley: first submersible study: Nature, v. 250, p. 558-560.

Booda, L.L., 1974, Diving 10,000-ft. to witness the birth of new ocean floor: Sea Technology, v. 15, no. 10, p. 16-19.

Heirtzler, J.R., and Le Pichon, Xavier, 1974, FAMOUS: a plate tectonics study of the genesis of the lithosphere: Geology, v. 2, p. 273-274.

Moore, J.G., Fleming, H.S., and Phillips, J.D., 1974, Preliminary model for extrusion and rifting at the axis of the Mid-Atlantic Ridge, 36 degrees 48 minutes north: Geology, v. 2, p. 437-440.

ARCYANA, 1975, Transform fault and rift valley from bathyscaph and diving saucer: Science, v. 190, p. 108-116.

Ballard, R. D., 1975, Project Famous II - dive into the great rift: Natl. Geog. Mag., v. 147, p. 604-615.

Ballard, R.D., Bryan, W.B., Heirtzler, J.R., Keller, G. Moore, J.G. and van Andel, T. H., 1975, Manned submersible observation in the FAMOUS area: Mid-Atlantic Ridge: Science, v. 190, p. 103-107.

Hammond, A.L., 1975, Project FAMOUS: exploring the Mid-Atlantic Ridge: Science, v. 187, p. 823-825.

Heirtzler, J.R., 1975, Project FAMOUS - man's first voyages down to the Mid-Atlantic Ridge where Earth turns inside out: Natl. Geog. Mag., v. 147, p. 586-603.

Heirtzler, J.R., and Bryan, W.B., 1975, The floor of the Mid-Atlantic rift: Sci. American, v. 233, no. 2, p. 78-90. (Reprinted in Sci. American, 1976, Continents adrift and continents aground, San Francisco, W. H. Freeman, p. 158-170.)

Geological Society of America, 1977, Special issues on Project FAMOUS, Parts I and II: Geol. Soc. America Bull., v. 88, No.'s 4 & 5 (Apr. & May), p. 481-677.

1973

483 R.N. Anderson, D.P. McKenzie, and J.G. Sclater showed that (1) depth over the crests of oceanic ridges is related to the free-air gravity anomalies, shallow depth being associated with positive anomalies and vice versa; and (2) there is no correlation between spreading rate and gravity, or spreading rate and crestal depth.

Anderson, R.N., McKenzie, D.P., and Sclater, J.G., 1973, Gravity, bathymetry and convection in the earth: Earth and Planetary Sci. Letters, v. 18, p. 391-407.

484 T. Atwater, K. Burke, K.S.W. Kidd, M.W. McElhinny, P. Molnar and J. T. Wilson described evidence that hot spots are not fixed in position relative to each other or to the earth's mantle.

Burke, Kevin, Kidd, K.S.W., and Wilson, J.T., 1973, Relative and latitudinal motion of Atlantic hot spots: Nature, v. 245, p. 133-137.

McElhinny, M.W., 1973, Mantle plumes, palaeomagnetism and polar wandering: Nature, v. 241, p. 523-524.

Molnar, Peter, and Atwater, Tanya, 1973, Relative motions of hot spots in the mantle: Nature, v. 246, p. 288-291.

485 M. Barazangi, B.L. Isacks, J. Oliver, J. Dubois, and G. Pascal concluded from their study of seismic wave propagation in the New Hebrides, Tonga-Fiji and New Zealand areas that the prominent gap in foci observed between 300 and 600 km depth is due to the separation of a deeper slab of lithosphere from a shallower one.

Barazangi, Mauwia, Isacks, B.L., Oliver, Jack, Dubois, Jacques, and Pascal, Georges, 1973, Descent of lithosphere beneath New Hebrides, Tonga-Fiji and New Zealand: evidence for detached slabs: Nature, v. 242, p. 98-101.

486 J.D. Hays and W.C. Pitman III used calculated volumes of the oceanic ridge system at different spreading rates to show that the world-wide Cretaceous transgression of the oceans (beginning at about -110 My) was caused by expansion of the oceanic ridges during a time of rapid spreading and that the subsequent regression of the seas (beginning at about -85 My) resulted from a decrease in the spreading rate.

Hays, J.D., and Pitman, W.C. III, 1973, Lithospheric plate motion, sea level changes and climatic and ecological consequences: Nature, v. 246, p. 18-22.

487 B.C. Heezen, I.D. MacGregor, H.P. Foreman, G. Forristal, H. Hekel, R. Hesse, R.H. Hoskins, E.J.W. Jones, A. Kaneps, V.A. Krasheninnikov, H. Okada, and M.H. Ruef developed a broad scale and relatively simple model of oceanic sedimentation that included the following controlling factors: relation of spreading center(s) to the equator, rate of spreading and subsidence of oceanic crust, carbonate and silica compensation depth, sedimentation rates, and proximity to volcanic zones and sources of turbidity currents. The model's predictions were compared to the known north Pacific stratigraphic record as determined by deep-sea drilling, coring and seismic methods and found to agree closely. An important aspect of the model that is confirmed by drilling data is that most pelagic sediment units are broadly diachronous.

Heezen, B.C., MacGregor, I.D., Forman, H.P., Forristal, G., Hekel, H., Hesse, R., Hoskins, R.H., Jones, E.J.W., Kaneps, A., Krasheninnikov, V.A., Okada, H., and Rueff, M.H., 1973, Diachronous deposits: a kinematic interpretation of the post Jurassic sedimentary sequence on the Pacific plate: Nature, v. 241, p. 25-32.

Heezen, B.C., and Macgregor, I.D., 1973, The evolution of the Pacific: Sci. American, v. 229, no. 5, p. 102-112. (Reprinted in Sci. American, 1976, Continents adrift and continents aground, San Francisco, W. H. Freeman, p. 149-157.

488 T.L. Holcombe, P.R. Vogt, J.E. Matthews and R.R. Murchison discovered a 110-km-long spreading center in the Cayman trough south of Cuba.

Holcombe, T.L., Vogt, P.R., Matthews, J.E., and Murchison, R.R., 1973, Evidence for seafloor spreading in the Cayman Trough: Earth and Planetary Sci. Letters, v. 20, p. 357-371.
Heezen, B.C., Perfit, M.R., Dreyfuss, M., and Catalano, R., 1973, The Cayman Ridge: Geol. Soc. America Abs. With Programs, v. 5, p. 663.
Ballard, R.D., 1976, Window on Earth's interior: Natl. Geog. Magazine, v. 150, p. 228-249.
Perfit, M.R., 1977, Petrology and geochemistry of mafic rocks from the Cayman Trench: evidence for spreading: Geology, v. 5, p. 105-110.

489 R.L. Larson and J.W. Ladd concluded from their study of magnetic anomalies off southern Africa that the south Atlantic Ocean began to open up during the early Cretaceous (Valanginian) at about -125 to -130 My.
Larson, R.L., and Ladd, J.W., 1973, Evidence for the opening of the South Atlantic in the early Cretaceous: Nature, v. 246, p. 209-212.

490 R.L. Parker and D.W. Oldenburg suggested that thermal subsidence of oceanic crust is controlled by a continuous thickening of the lithosphere that occurs as the crust moves away from an oceanic ridge.
Parker, R.L., and Oldenburg, D.W., 1973, Thermal model of ocean ridges: Nature Phys. Sci., v. 242, p. 137-139.

491 A.G. Smith, J.C. Briden and G.E. Drewry used topographic, tectonic, paleomagnetic and magnetic anomaly data as a basis for preparing sets of three world maps for each of eight times during the geologic past (Eocene, Cretaceous, Jurassic, Triassic, Permian, Mississippian, early Devonian and early Ordovician/Cambrian).
Smith, A.G., Briden, J.C., and Drewry, G.E., 1973, Phanerozoic world maps: in Hughes, N.F., ed., Organisms and continents through time, London, The Palaeontological Assoc., p. 1-42.

492 The Third Gondwana Symposium was held at Canberra, Australia under the auspices of the International Union of Geological Sciences Sub-Commission for Gondwana Stratigraphy and Paleontology.
Campbell, K.S.W., ed., 1975, Gondwana Geology: Canberra, Australian Nat. Univ. Press, 705 p.

1974

493 The American Association of Petroleum Geologists published a collection of articles which offered "some criticisms of the paradigm of plate tectonics." Included were articles by some of the strongest opponents of plate tectonics, including A.A. and H.A. Meyerhoff, V.V. Beloussov and H. Jeffreys.
Kahle, C.F., ed., 1974, Plate tectonics - assessments and reassessments: Tulsa, Oklahoma, Amer. Assoc. Petr. Geologists Mem. 23, 514 p.

494 F. Aumento, P.H. Reynolds and B.M. Gunn concluded from their studies of core samples from an 802 m hole drilled on the island of Bermuda that the sea-mount formed near the crest of the Mid-Atlantic Ridge by the extrusion of tholeiitic basalt lavas at about -90 My. The seamount was dormant for about 60 My and sank well below sea level before it was reactivated at about -33 My when numerous lamprophyric sheets were intruded into the old lava pile, causing a reemergence of the mount above sea level.
Aumento, Fabrizio, Reynolds, P. H., and Gunn, B. M., 1974, The Bermuda seamount: a reactivated section of an older ocean crust (abs.): EOS (Amer. Geophys. Union Trans.), v. 55, p. 455.

Aumento, Fabrizio, and Sullivan, K. D., 1974, Deep drill investigations of the oceanic crust in the north Atlantic: in Kristjansson, L., ed., Geodynamics of Iceland and the North Atlantic area, Dordrecht (Holland), D. Reidel Pub. Co., p. 83-103.

495 A. Carr and P.J. Coleman suggested that the 2,000 km migration of the green turtle (Chelonia mydas) from its usual habitat on the coast of Brazil to its breeding and nesting grounds on Ascension Island is best explained by the process of sea-floor spreading,during the last 80 million years which gradually separated the South American coastline from volcanic islands produced at the Mid-Atlantic Ridge. The hypothesis holds that the necessary behavioral patterns and long-distance swimming ability gradually evolved as the island nesting sites progressively separated from the ordinary habitat.

Carr, Archie and Coleman, P.J., 1974, Seafloor spreading theory and the odyssey of the green turtle: Nature, v. 249, p. 118-130.

496 A symposium on metallogeny and plate tectonics was held at the annual meeting of the Geological and mineralogical Association of Canada at St Johns.

Strong, D.F., 1976, Metallogeny and plate tectonics: Geol. Assoc. Canada Spec. Paper No. 14, 660 p.

497 Deep sea drilling encountered granitic basement rocks beneath the Falkland Plateau, demonstrating that the Plateau is a part of the South American continent and further improving the fit of South America and Africa.

Deep Sea Drilling Project scientists, 1974, Southwestern Atlantic - Leg 36: Geotimes, v. 19, no. 11, p. 16-18.

Barker, P.F., Dalziel, I.W.D., and others, 1976, Initial Reports of the Deep Sea Drilling Project, v. 36: Washington, U.S. Govt. Printing Office, 1080 p.

498 P. Hoffman, J.F. Dewey and K. Burke suggested that aulacogens originate as the abandoned segments of triple-ridge junctions where mantle plumes rise beneath relatively stationary continents and that they commonly form at reentrant angles of geosynclines or continental margins. Examples cited were the two billion year old Athapuscow basin east of Great Slave Lake, Canada, the late Precambrian to Paleozoic Anadarko Basin of the United States and the relatively modern Benue trough and Ethiopian rift of Africa.

Hoffman, Paul, Dewey, J.F., and Burke, Kevin, 1974, Aulacogens and their genetic relation to geosynclines, with a Proterozoic example from Great Slave Lake, Canada: in Dott, R.H., Jr., and Shaver, R.H., Modern and ancient geosynclinal sedimentation, Tulsa, Okla., Soc. Econ. Paleo. and Mineralogists Spec. Pub. No. 19, p. 38-55.

499 J.B. Minster, T.H. Jordan, P. Molnar and E. Haines developed and used an elaborate numerical method of determining the simultaneous relative motions of 11 major plates. Among other results, it was determined that there is some relative motion between North and South America and that the trends of 20 linear island chains and aseismic ridges show no relative motions with respect to each other since -10 My.

Minster, J.B., Jordan, T.H., Molnar, Peter, and Haines, Eldon, 1974, Numerical modelling of instantaneous plate tectonics: Royal Astron Soc. Geophys. Jour., v. 36, p. 541-576.

500 G.K. Muecke, J.M. Ade-Hall, F. Aumento, A. Macdonald, P.H. Reynolds, R.D. Hyndman, J. Quintino, N. Opdyke and W. Lowrie reported that drilling on the flank of a dormant volcano at 72 m above sea level on Sao Miguel Island in the Azores, 400 km from the crest of the Mid-Atlantic Ridge, penetrated 786 m of subaerial volcanics, then a 107 m transition section and then 88 m of submarine volcanics (chiefly pillow lavas). Based on the normal magnetization of all the rocks encountered, it was concluded that the subsidence indicated took place in less than 690,000 years.

Muecke, G.K., Ade-Hall, J.M., Aumento, F., MacDonald, A., Reynolds, P.H., Hyndman, R.D., Quintino, J., Opdyke, N.D., and Lowrie, W., 1974, Deep drilling in an active geothermal area in the Azores: Nature, v. 252, p. 281-285.

501 NATO sponsored an Advanced Study Institute in Reykjavik on the tectonics of Iceland in relation to the mantle plume hypothesis.

Kristjansson, Leo, ed., 1974, Geodynamics of Iceland and the North Atlantic area: Proceedings (NATO Adv. Study Inst.), Univ. of Iceland, Reykjavik, NATO Adv. Study Inst. Ser. C, Math. Phys. Sci., v. 11, 323 p.

Sylvester, A.G., and Kristjansson, Leo, 1974, What is Iceland?: Geotimes, v. 19, no. 12, p. 26-28.

502 The San Joaquin Geological Society sponsored a short course in Bakersfield, California on "geologic global tectonics with applications for California geology and petroleum exploration."

Dickinson, W.R., 1974, Geological interpretations from global tectonics with applications for California geology and petroleum exploration: Bakersfield, California, San Joaquin Geol. Soc., 70 p.

503 D.R. Seely, P.R. Vail and G.G. Walton used magnetic, earthquake, deep-penetration seismic reflection, DSDP holes, dredge haul, piston core and other data to show that the oceanic lithosphere plate continues beneath trench inner slopes and that thrusts and folds are the primary structures on the landward slopes of trenches. Continental accretion was said to occur as the slope builds outward and older trench slopes are eventually incorporated into plutonic belts.

Seely, D.R., Vail, P.R., and Walton, G.G., 1974, Trench slope model: in Burk, C.A., and Drake, C.L., eds., The geology of continental margins, New York, Springer, p. 249-260.

504 P.R. Vogt and O.E. Avery proposed a detailed spreading history of portions of the north Atlantic Ocean basin based on their study of magnetic anomaly patterns and other data. Among other results, they concluded that (1) Greenland separated from Rockall bank at about -60 My; (2) the initial rifting of the Norwegian Sea at about -55 My is marked by the widespread Brito-Arctic igneous activity; (3) a triple-ridge junction connected the Ran Ridge (which separated Greenland from the North American plate) with the Mid-Atlantic Ridge between about -60 My and -30 My; (4) an older north Atlantic basin, subsequently deeply buried with sediment, extended into the Labrador basin south of Rockall bank; (5) a realignment of spreading centers and transform faults in the area occurred during the period -60 to -50 My; and (6) the spreading history of the area has "been controlled by variable discharge from a mantle convection plume now under Iceland."

Vogt, P.R., and Avery, O.E., 1974, Detailed magnetic surveys in the northeast Atlantic and Labrador Sea: Jour. Geophys. Research, v. 79, p. 363-389.

1975

505 J.F. Harper calculated the individual lithosphere plate velocities that would result if the only forces acting were "suduction zone pull" generated as the dense leading edge of a plate sinks into the less dense asthenosphere and by "mid-oceanic (ridge) push" caused by the plate's tendency to slide down the slope of the ridge. The only resisting force was assumed to be Newtonian viscosity in the asthenosphere. It was concluded that "It is possible to account for the major part of the motion of the eight principal plates . . ."

Harper, J.F., 1975, On the driving forces of plate tectonics: Geophys. Jour. Royal Astron Soc., v. 40, p. 465-474.

506 A symposium on "Ancient Plate Margins" was held at the 16th General Assembly of the International Union of Geodesy and Geophysics at Grenoble, France.

Briden, J.C., 1976, Ancient plate margins: Tectonophysics, v. 33, p. i-iv and 215-385.

507 S.C. Solomon, N.H. Sleep and R.M. Richardson determined current absolute plate velocites and showed that there is general correlation between those velocities and (1) the proportion of plate occupied by continent; (2) the proportion of plate boundary undergoing subduction; and (3) the closeness of the plate to the equator. The implication that these factors play some determinative role in plate movement was negated, however, when the authors also showed that none of these correlations held at -55 My. It was also shown that not all hot spots are fixed with respect to each other.

Solomon, S.C., Sleep, N.H., and Richardson, R.M., 1975, On the forces driving plate tectonics: inferences from absolute plate velocities and intraplate stress: Geophys. Jour. Royal Astron. Soc., v. 42, p. 769-801.

1976

508 The United States National Aeronautics and Space Administration launched a laser-ranging satellite (LAGEOS) for the purpose of making direct measurements of plate motion. The system was designed to attain a ranging accuracy of two centimeters and was expected to remain in orbit for ten million years.

Spencer, R.L., 1977, LAGEOS - a geodynamics tool in the making: Jour. Geol. Education, v. 25, p.38-42.

509 G. Plafker, M.G. Bonilla and S.B. Bonis showed that the 1976 Guatamala earthquake was the result of left lateral strike slip along the Motagua fault which had previously been identified as a transform fault forming the boundary between the Caribbean and North American plates. The maximum amount of slip measured at any one locality was 1.42 m and surface movement was evident along 230 km of the fault.

Plafker, George, Bonilla, M.G., and Bonis, S.B., 1976, Geological effects: in Expinosa, A. F. ed., The Guatemalan earthquake of February 4, 1976, a preliminary report, U. S. Geol. Survey Prof. Paper 1002, Washington, D. C., U. S. Govt. Printing Ofc, p. 31-51.

INDEX

Numbers refer to Chronology entry numbers, **not** to page numbers.

AUTHOR CITATION INDEX

Note: The authors cited in Part VI, Chronology, are not included in this index. The reader should refer to the index on page 324.

SUBJECT INDEX

Note: The Chronology has its own subject and author index beginning on page 324.

About the Editor

JAMES H. SHEA is a professor of geology at the University of Wisconsin–Parkside in Kenosha, where he has done his teaching and research since 1967. He received the B.S. and M.S. degrees in geology from the University of Wisconsin–Madison in 1958 and 1960, respectively, and the Ph.D. in geology from the University of Illinois–Urbana in 1964. Dr. Shea began his geological employment as an exploration geologist at Texaco in New Orleans before undertaking his Ph.D. studies. He then worked for two years at the Earth Science Curriculum Project in Boulder, Colorado, developing a new, laboratory-oriented earth science course for secondary schools. In 1966 he joined the faculty at the University of Tennessee, and in 1967 he moved to UW–Parkside, which was just preparing to open as a new four-year campus. Dr. Shea has published in the fields of curriculum development, sedimentology (grain-size analysis), geological education, and the history and philosophy of geology. Since 1975 he has been editor of the *Journal of Geological Education*. His current research interests lie in the history and development of geological theories.

Benchmark Papers in Geology

Series Editor: Rhodes W. Fairbridge, Columbia University

ENVIRONMENTAL GEOMORPHOLOGY AND LANDSCAPE CONSERVATION, Volume I: Prior to 1900 / *Donald R. Coates*
RIVER MORPHOLOGY / *Stanley A. Schumm*
SPITS AND BARS / *Maurice L. Schwartz*
TEKTITES / *Virgil E. Barnes and Mildred A. Barnes*
GEOCHRONOLOGY: Radiometric Dating of Rocks and Minerals / *C. T. Harper*
SLOPE MORPHOLOGY / *Stanley A. Schumm and M. Paul Mosely*
MARINE EVAPORITES: Origin, Diagenesis, and Geochemistry / *Douglas W. Kirkland and Robert Evans*
ENVIRONMENTAL GEOMORPHOLOGY AND LANDSCAPE CONSERVATION, Volume III: Non-Urban Areas / *Donald R. Coates*
BARRIER ISLANDS / *Maurice L. Schwartz*
GLACIAL ISOSTASY / *John T. Andrews*
GEOCHEMISTRY OF GERMANIUM / *Jon N. Weber*
ENVIRONMENTAL GEOMORPHOLOGY AND LANDSCAPE CONSERVATION, Volume II: Urban Areas / *Donald R. Coates*
PHILOSOPHY OF GEOHISTORY: 1785–1970 / *Claude C. Albritton, Jr.*
GEOCHEMISTRY AND THE ORIGIN OF LIFE / *Keith A. Kvenvolden*
SEDIMENTARY ROCKS: Concepts and History / *Albert V. Carozzi*
GEOCHEMISTRY OF WATER / *Yasushi Kitano*
METAMORPHISM AND PLATE TECTONIC REGIMES / *W. G. Ernst*
GEOCHEMISTRY OF IRON / *Henry Lepp*
SUBDUCTION ZONE METAMORPHISM / *W. G. Ernst*
PLAYAS AND DRIED LAKES: Occurrence and Development / *James T. Neal*
GLACIAL DEPOSITS / *Richard P. Goldthwait*
PLANATION SURFACES: Peneplains, Pediplains, and Etchplains / *George F. Adams*
GEOCHEMISTRY OF BORON / *C. T. Walker*
SUBMARINE CANYONS AND DEEP-SEA FANS: Modern and Ancient / *J. H. McD. Whitaker*
ENVIRONMENTAL GEOLOGY / *Frederick Betz, Jr.*
LOESS: Lithology and Genesis / *Ian J. Smalley*
PERIGLACIAL PROCESSES / *Cuchlaine A. M. King*
LANDFORMS AND GEOMORPHOLOGY: Concepts and History / *Cuchlaine A. M. King*
METALLOGENY AND GLOBAL TECTONICS / *Wilfred Walker*
HOLOCENE TIDAL SEDIMENTATION / *George deVries Klein*

Benchmark Papers in Geology Series

PALEOBIOGEOGRAPHY / *Charles A. Ross*
MECHANICS OF THRUST FAULTS AND DÉCOLLEMENT / *Barry Voight*
WEST INDIES ISLAND ARCS / *Peter H. Mattson*
CRYSTAL FORM AND STRUCTURE / *Cecil J. Schneer*
OCEANOGRAPHY: Concepts and History / *Margaret B. Deacon*
METEORITE CRATERS / *G. J. H. McCall*
STATISTICAL ANALYSIS IN GEOLOGY / *John M. Cubitt and Stephen Henley*
AIR PHOTOGRAPHY AND COASTAL PROBLEMS / *Mohamed T. El-Ashry*
BEACH PROCESSES AND COASTAL HYDRODYNAMICS / *John S. Fisher and Robert Dolan*
DIAGENESIS OF DEEP-SEA BIOGENIC SEDIMENTS / *Gerrit J. van der Lingen*
DRAINAGE BASIN MORPHOLOGY / *Stanley A. Schumm*
COASTAL SEDIMENTATION / *Donald J. P. Swift and Harold D. Palmer*
ANCIENT CONTINENTAL DEPOSITS / *Franklyn B. Van Houten*
MINERAL DEPOSITS, CONTINENTAL DRIFT AND PLATE TECTONICS / *J. B. Wright*
SEA WATER: Cycles of the Major Elements / *James I. Drever*
PALYNOLOGY, PART I: Spores and Pollen / *Marjorie D. Muir and William A. S. Sarjeant*
PALYNOLOGY, PART II: Dinoflagellates, Acritarchs, and Other Microfossils / *Marjorie D. Muir and William A. S. Sarjeant*
GEOLOGY OF THE PLANET MARS / *Vivien Gornitz*
GEOCHEMISTRY OF BISMUTH / *Ernest E. Angino and David T. Long*
ASTROBLEMES—CRYPTOEXPLOSION STRUCTURES / *G. J. H. McCall*
NORTH AMERICAN GEOLOGY: Early Writings / *Robert Hazen*
GEOCHEMISTRY OF ORGANIC MOLECULES / *Keith A. Kvenvolden*
TETHYS: The Ancestral Mediterranean / *Peter Sonnenfeld*
MAGNETIC STRATIGRAPHY OF SEDIMENTS / *James P. Kennett*
CATASTROPHIC FLOODING: The Origin of the Channeled Scabland / *Victor R. Baker*
SEAFLOOR SPREADING CENTERS: Hydrothermal Systems / *Peter A. Rona and Robert P. Lowell*
MEGACYCLES: Long-Term Episodicity in Earth and Planetary History / *G. E. Williams*
OVERWASH PROCESSES / *Stephen P. Leatherman*
KARST GEOMORPHOLOGY / *M. M. Sweeting*
RIFT VALLEYS: Afro-Arabian / *A. M. Quennell*
MODERN CONCEPTS OF OCEANOGRAPHY / *G. E. R. Deacon and Margaret B. Deacon*
OROGENY / *John G. Dennis*
EROSION AND SEDIMENT YIELD / *J. B. Laronne and M. P. Mosley*
GEOSYNCLINES: Concept and Place Within Plate Tectonics / *F. L. Schwab*
DOLOMITIZATION / *Donald H. Zenger and S. J. Mazzullo*
OPHIOLITIC AND RELATED MELANGES / *G. J. H. McCall*
ECONOMIC EVALUATION OF MINERAL PROPERTY / *Sam L. VanLandingham*